THIS
BITTER-
SWEET
SOIL

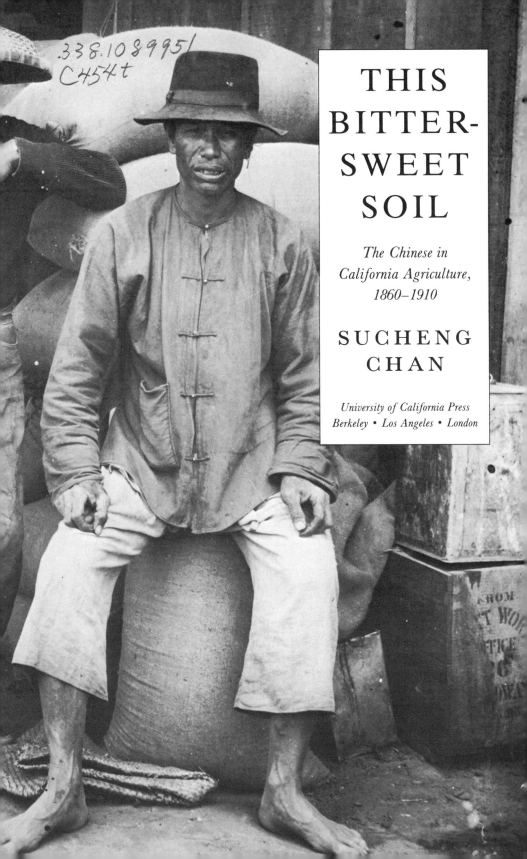

THIS BITTER-SWEET SOIL

*The Chinese in
California Agriculture,
1860–1910*

SUCHENG CHAN

University of California Press
Berkeley • Los Angeles • London

University of California Press
Berkeley and Los Angeles, California

University of California Press, Ltd.
London, England

© 1986 by
The Regents of the University of California

Typeset by Asco Trade Typesetting Ltd., Hong Kong
Printed in the United States of America

Library of Congress Cataloging in Publication Data

Chan, Sucheng.
  This bittersweet soil.

  Bibliography: p.
  Includes index.
  1. Farmers—California—History. 2. Chinese—
California—History. 3. Agriculture—Economic aspects—
California—History. I. Title.
HD8039.F32U62 1986   338.1'089951'0764   85-1084
ISBN 0-520-05376-1

1 2 3 4 5 6 7 8 9

*To the memory of Asian immigrants
who made California so green*

Contents

Illustrations

Figure

Tables

Preface

WHEN I was a child growing up in Shanghai, my favorite treat was Del Monte canned peaches. I was told that cling peaches grew in a faraway place called California. "How wonderful," I thought, "if someday I could go to California and eat peaches to my heart's content!"

Twenty years later, I did settle in California, and I have eaten my fill of fresh and canned peaches. Not until six years after I came to this fabled land of sunshine, however, did I have any inkling that men who might have been my ancestors had planted some of the state's earliest commercial peach orchards. I first learned that Chinese, Japanese, and Filipino immigrants had been important in California's agricultural labor force when I read Carey McWilliams's classic work, *Factories in the Field*, in preparation for teaching Asian American history—a field in which I had received no formal training. As I familiarized myself with the existing literature, I discovered that the five early Asian immigrant groups—Chinese, Japanese, Koreans, Filipinos, and Asian Indians—had each been involved in the development of California agriculture; therefore, scholars interested in a comparative analysis of the Asian American historical experience can gain a great deal of insight by studying how these groups fared in agricultural work. At the same time, I realized that agriculture, an important element in the state's economic growth, has profoundly affected its social development as well because the hierarchical relations among the different groups involved in agricultural production have helped to set the overall pattern of California's race relations. Consequently, a historical fact I had found mildly curious became a topic I wanted to write a book

xv

about because the subject matter touches so many aspects of
both Asian American and California history. Though my original intent was to write a book that gives equal coverage to all five
groups, it soon became apparent that such a task is impossible
because the quality as well as the quantity of available sources
differs greatly for the groups in question. I decided, therefore, to
treat only the Chinese and Japanese in depth.

This book is the first part of a two-volume work on how the
physical imperatives of agricultural production in California
molded a set of social relations that in turn affected the manner
in which that production has been organized. Although each
volume is written to stand alone, together the two will form a
continuous treatise covering an eighty-year period. The first
volume describes the many ways Chinese immigrants participated in the development of agriculture in California; the second
will investigate why the anti-Japanese movement in California
focused on the Japanese farming population.

For more than three years, I searched for and collected some
four hundred items on Asian immigrants in California agriculture, over 95 percent of which are about the Japanese, with only
a dozen or so items about the Chinese. Interestingly, studies of
the Chinese in California agriculture—none a full-length work—
have been done almost exclusively by scholars interested in the
nature of the farm labor supply in California and not by historians of the Chinese in America. Moreover, the historiography
on the topic has been dominated for four decades by the writings
of Varden Fuller, Carey McWilliams, and Lloyd Fisher.[1]
Because these authors wrote about Chinese farm laborers, other
scholars, who have relied on their pioneering works, have
assumed that the Chinese earned a living in agriculture only as
seasonal harvest hands. Though more copious, the literature
on the Japanese is problematic because almost all of it was
partisan—writings either in support of or attacking Japanese immigrants and their tendency to lease and purchase agricultural
land.

In reviewing the existing literature I realized that little that
was new could be said unless different or better primary sources
could be found, for most of the available writings simply have
repeated each other's arguments. A search was made for works
written in Chinese and Japanese, but only one Chinese source—

a superficially written retrospective newspaper article on Chinese farming in California—was unearthed. The Japanese, on the other hand, themselves published many books on their achievements and travails in California agriculture.

Materials seemingly unrelated to the topic under study were also examined, with the hope that some nugget of information might be found buried somewhere. Through the latter search, a set of documents which had never been used before in studies of California's agricultural history was discovered. Without this fortuitous discovery, the present study could not have been done, so it is worth recounting how I came upon the documents.

One day, in looking through Owen Coy, *A Guide to the County Archives of California*, published in 1919, I noticed that among the documents listed were records of leases. I wondered if such records might contain leases of agricultural land, and if so, whether it was possible that these records—inventoried by Coy in 1915–17—might still be available some sixty years later. On a trip to Sacramento to use the State Library, I visited the Sacramento County Recorder's office. The clerk at the front office said she had never heard of such documents, but, she said, perhaps the man who had been microfilming some of their old records might know of them. Indeed, he did. Eighteen volumes of leases signed in Sacramento County covering a seventy-year period had been preserved. He could not find the index, but he allowed me to look through the books themselves. My excitement mounted as I found one Chinese name after another among the lessees. When I came across a document with a signature in Chinese characters, I felt like an adventurer who had found a long-lost treasure. It is hard to describe the purity of my wonder at that moment: it was as though I had made a clairvoyant connection with some almost-ancestors.

Over the next three years, during two of which I was on leave from teaching, I traveled up and down the state and searched through the official archives of forty-one counties. I burrowed through more than 3,000 volumes (about 500 linear feet of documents) spanning an eighty-year period and took notes on some 30,000 documents, approximately one-third of which were about mining and not agriculture. I read these thousands of mining documents because they contained scattered information about the earliest Chinese truck gardens in the state.

I found usable information in seven different types of documents: "Leases," "Chattel Mortgages," "Real Mortgages," "Deeds," "Mining Claims," "Contracts and Agreements," and "Miscellaneous Records." Most of the time, the office staff in the County Recorders' offices did not know of these old books' existence; only by being insistent and persistent was I able to cajole the busy office workers to search their attics and basements to locate the books I wanted. I failed, however, to persuade the lady in the Recorder's office in Ventura County to take time to search for the books I needed; she did not allow me to look through the storage area myself, either. For that reason, the present study does not include Ventura County—an unfortunate gap. There is also no information on San Benito County because the relevant records could not be found despite a careful search through every corner of the premises where the old books are stored. Table 37 lists the records available in each county.

During this period, I breathed an incredible amount of dust in countless basements and strained my eyes reading the faded ink, but it was an extraordinary sleuthing adventure I would not have missed for anything. Even as my eyes teared from fatigue, I kept reading, for what had begun as a conscientious effort to gather data became a genuine labor of love. The narrative in this book has been pieced together like a jigsaw puzzle from these voluminous, nonnarrative sources, all written in dry, legal language. Each document was read, and the usable information was put into a common format so that a quantitative analysis could be made. So far as possible, the findings were plotted on maps, and a narrative was then wrung painstakingly out of the chronological and spatial patterns so revealed.

Since the story told here is reconstructed almost entirely from these county archival documents and from the unpublished manuscript schedules of the censuses of population and agriculture, I have written an essay to tell about the usefulness as well as problems and pitfalls of using the manuscript schedules of the census and to describe the county archival records and what they contain. Although the manuscript population census has long been used by scholars in research, to my knowledge only three persons have used California county archival records.[2] Scholars who have written about the history of land and agriculture in the prairie states have tapped the wealth of information in county records much more extensively than historians of California.[3]

The scope of this study has been dictated partly by the nature of the evidence available. The present volume begins in 1860 because that is the earliest year with systematic data on Chinese who worked in agriculture: the 1860 manuscript population census provides the first quantifiable information on Chinese truck gardeners. The study on the Chinese ends in 1910 because by that time the organizational pattern of California agribusiness had already crystallized, and whatever role Chinese immigrants might have played in the evolution of that pattern had long been institutionalized. Moreover, with the exception of the Sacramento–San Joaquin Delta, Chinese farmers and farm laborers had almost disappeared from the scene. The volume on the Japanese will begin in 1890, when they started working in California's fields, and will end in 1942, when they had to abandon their farms as over 110,000 persons of Japanese ancestry were placed into so-called relocation camps several months after the United States declared war on Japan following the bombing of Pearl Harbor in December 1941.

The agricultural pursuits of both the Chinese and the Japanese will be placed within the larger contexts of their overall economic history in the United States as well as of their migrations overseas. The Chinese had been going abroad to work, trade, and settle for centuries, so those who came to the United States were but a fragment of a far larger phenomenon. The Japanese, on the other hand, with the exception of those who settled Hokkaido and the ones who went to Korea and Manchuria, did not have the same history of international emigration as the Chinese had. Japanese who went to Hawaii and California were therefore pioneer overseas emigrants. Though the Japanese have frequently been depicted as filling a labor vacuum left by the exclusion of the Chinese and were therefore seen by white Americans as the latter's successors, there were in fact notable differences in the experiences of the two groups.

During the 1980–81 academic year, while a visiting scholar at the Asian American Studies Center at UCLA, I had to return to my home campus at Berkeley once or twice a month to chair committee meetings. Traveling by car gave me an opportunity to see the Central Valley and the coastal valleys at all hours of the day and night and in every season of the year. During those long drives, my mind's eye tried to see what this land might have looked like a century ago. At dawn, when a gentle haze hovering over the earth allowed my imagination to float over the landscape in a manner that transcended time, it was sometimes possible to visualize small but sinewy yellow-skinned men (and an occasional woman) stooping in the fields, coaxing the land to pour forth its abundance. Traversing California's fecund valleys, I felt compelled to tell a tale of courage and endurance, and to claim a bit of California history for the Asian immigrants who made this golden state so green.

Some stylistic and substantive choices I have made should be noted. First, I shall refrain from calling the Chinese "sojourners." Though many of the Chinese who came to the United States in the nineteenth century were men and women who did not come with the intention of settling here, they will be called "immigrants" from time to time. To insist that all Chinese who came to America were sojourners—as some scholars have done—is to exclude them categorically from American immigration history. Though Gunther Barth, Stanford Lyman, and others have argued that the Chinese were sojourners with the exception of the small handful who assimilated to Anglo-American culture,[4] in my view neither the term "immigrant" nor the term "sojourner" is completely correct. It is impossible to say if the Chinese were solely one or the other, because to do so requires an accurate knowledge of their motives. Since the Chinese themselves left few records of their perceptions and experiences, it is difficult to determine precisely what percentage who came intended to stay. As with other migrants to America, some who might have wished to settle in the United States could not do so, whereas others, who initially did not intend to stay, lived here until they died. Motives cannot be imputed to an entire group as though it were homogeneous, especially when that group's history has been based largely on sources left by non-Asians who were often hostile or, at best, patronizing toward members of that group.[5]

Second, wherever appropriate, I shall refer to individuals by their names, even though these were often haphazardly or erroneously transliterated in the archival records. Most of the

Chinese who came to America in the nineteenth century were
Cantonese. To this day, Cantonese like to use the diminutive
form "Ah" with their given (first) names to address each other.
For reasons that can only be guessed at, the Chinese in
nineteenth-century America gave these diminutive names but
not their family or full names to census takers and county clerks.
This practice poses an obstacle to research because it is impossible in most cases to trace the lives and careers of particular individuals. The Japanese used their first initials and their family
names, but the latter were not transliterated systematically
either. Despite the inconsistent transliteration and the diminutive forms recorded, names will be given as they appeared in the
documents. A text using names with little biographical information attached may not be much of an improvement over a narrative which refers to the protagonists as "the Chinese," "the
Japanese," or "the Koreans," but this choice has symbolic significance because Asian American history has all too often been
written as a faceless and nameless history. The very act of nam-
ing underscores the fact that Asian Americans have not been just
hordes of cheap labor to be feared; rather, like other human
beings, we are individuals who hope and despair, laugh and cry,
as we attempt to cope with the conditions into which all of us
have been born.

Finally, a few words should be said about what this book does
not attempt to do. Not being a general history of the Chinese in
America, this study will not cover the structure of organizations
in the community, class and generational relations, family life
and the changing role of women, the preservation of traditional
Chinese culture, the development of a uniquely Chinese American literature and visual and performing arts, religious life,
ideology and political activities, reactions to the anti-Chinese
movement, perception of and relationship to different political
groups in China, the process of Americanization, the relationship of Chinese to other Asian immigrants, European immigrants,
or other nonwhite minorities, the Chinese American underworld,
Chinese cuisine, or the tourist trade.[6] The book is not even a
general economic history of the Chinese in America, though agriculture as a means of livelihood will be compared with other
available economic opportunities.

What I have tried to produce is something at once narrower

and broader than most available studies. Narrower, because focusing only on the agricultural sector in California, I shall describe in detail how and why the Chinese took up cultivation of the soil, what they accomplished, the obstacles they faced, their relationship to other groups engaged in agricultural production, and how agriculture as a livelihood influenced the development of Chinese rural communities in California. At the same time, the study is broader because I shall highlight the fact that Chinese emigration was a global phenomenon, and Chinese immigration to California was but a small part of the Chinese diaspora. In California as elsewhere, though Chinese immigrants have often been victims of other people's actions, they and their descendants have nevertheless been active makers of their own history.

Acknowledgments

THIS STUDY would not have been possible without the cooperation of the workers in the forty-one County Recorder's offices in which I did research. I want to thank them as well as the staff members of the Bancroft Library, the Documents Department of Doe Library, the Giannini Agricultural Economics Library, and the Water Resources Center Archives of the University of California, Berkeley; the California State Library; the California State Archives; the Huntington Library; the California Historical Society; the Sacramento Museum and History Room; the Sacramento River Delta Historical Society; and the San Joaquin County Historical Museum for their patience and help.

Whatever human interest this study contains came from the information provided by the elderly persons I interviewed, all of whom were in their late seventies or early eighties when I talked to them. I am truly grateful to all the persons named in the list of interviewees for their keen memories.

I gathered almost all the data in the county offices myself, but a virtual army of undergraduate students assisted me in copying and tallying data from the manuscript schedules of the censuses of population and agriculture. Some worked for only a week or two, others worked for several years. I thank Farzad Bibayan, Tonya Creek, Peter Kang, Jeong Woo Lee, Belinda Leung, Sylvia Leung, Barbara Miller, Monchai Pungaew, Jaime Rodriguez, Lisa Severns, Joe Salazar, Josefina Torio, Thu Thuy Truong, Seema Untewale, and Anna Wong for having put in the most hours. Though the work was extraordinarily tedious, each of them did it carefully.

xxiii

Draft maps were prepared by myself, Dan Holmes, Lisa Larabee, and Noritake Yagasaki. We searched tirelessly and patiently through countless unindexed maps in order to locate as many Chinese and Japanese leased farms as possible. I thank them and Adrienne Morgan who drew the finished versions of the maps.

Many colleagues read this work in its various stages. I thank Gunther Barth, Edna Bonacich, Roger Daniels, Paul W. Gates, Charles Geisler, Philip C. Huang, Norris C. Hundley, Jr., Mark K. Juergensmeyer, James H. Kettner, Karen B. Leonard, Daniel S. Lev, Robert L. Middlekauff, Gary B. Nash, Spencer C. Olin, Jr., James J. Parsons, Jack M. Potter, Alexander Saxton, and Frederic E. Wakeman, Jr., for their insightful—and in several cases, copious—comments that led to improvements in the manuscript. I owe a special debt to Roger Daniels and Bob Middlekauff, who sensed that even though I had been trained as an economist and political scientist, I was really a closet historian. They held my hand while I struggled to come out of the closet. They and Norris Hundley helped to purge my writing of social science jargon: I had to rewrite the manuscript four times before they forgave me my past.

The final version of the manuscript was edited by Gladys Castor, who, together with Phyllis Killen and Barbara Daly Metcalf of the University of California Press, guided the entire publication process. I am also grateful to Helen Hong and Liz Megino, who typed many tables for an earlier version of this work, and to Jenni Currie, who helped to proofread the galleys and prepare the index. The willingness of Carolyn Z. Wong to do some "emergency" computations at one point is appreciated, too.

The research for this study was very expensive because the work was so labor-intensive. I thank the Committee on Research, the Asian American Studies Program, the Institute of Governmental Studies, and the Chancellor's Office at the University of California, Berkeley, the Asian American Studies Center and the Institute of American Cultures at the University of California, Los Angeles, and the Chancellor's Office at the University of California, Santa Cruz, for financial support. Even so, I myself had to bear more than half the cost of this study.

Finally, it took almost a decade to research and write this book because the work had to be done in the "interstices" of time—in between the countless struggles of the Department of Ethnic Studies at the University of California, Berkeley. In those years, a number of friends kept my spirits buoyed up in turbulent seas: Mark, who did things for me he would not have done for himself; Ling-chi, my one true comrade; Dan, who suggested the curious idea that pain, too, can be enjoyed (for, as he asked, "What else can be done with it?"); Rod, with whom I shared a peculiar friendship; and Brandenburg, Rajah, and Rab, who gave warm, furry love.

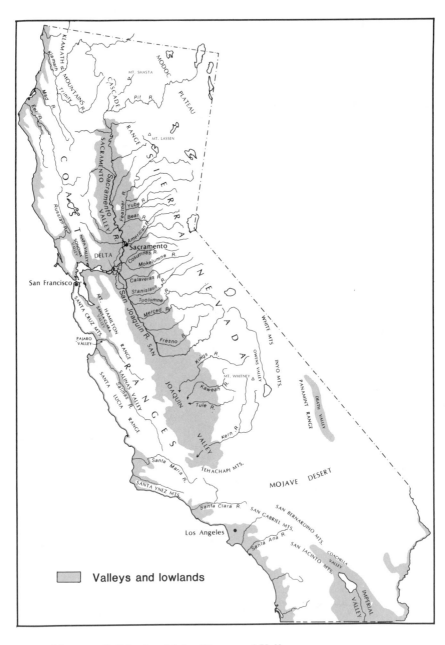

Map 1   California: Major Rivers and Valleys

Introduction

TODAY half a dozen or so Chinese American farmers
are well known in the Sacramento–San Joaquin Delta
of California as the owners of hundreds of acres. Their
farms are not too far from Locke, the last rural Chinatown in the United States, where several dozen ageing
Chinese Americans attempt to live out their last days
in peace and quiet despite the intrusive presence of tourists,
government bureaucrats, and community organizers. In another
agricultural region, around the southeastern shores of San Francisco Bay, some two hundred Chinese American chrysanthemum growers and nurserymen struggle to stay in business as
land values skyrocket with the establishment of an increasing
number of computer firms, turning much of the rich agricultural
Santa Clara Valley into "Silicon Valley." The Chinese American farmers in the Delta, the residents of Locke, and the Santa
Clara nurserymen are the last vestiges of an important but
relatively unknown segment of the Chinese American population—the thousands of Chinese pioneers who earned a living
as truck gardeners, farmers, commission merchants, fruit and
vegetable vendors and peddlers, fruit packers, and harvest
laborers in California during the latter part of the nineteenth
century. In that period, in the major agricultural regions, some
80 to 90 percent of the Chinese population earned a living by
growing and harvesting crops.

But numbers do not indicate their full significance. Though
the Chinese who earned a living in agriculture on a regular basis
never numbered more than six or seven thousand and represented only 10 to 15 percent of the total Chinese population in

the state even at the height of their prosperity in the last two decades of the nineteenth century, they played an important role in the development of California agriculture, and the nature of that involvement, in turn, influenced the evolution of Chinese communities in rural California. As the most stable members of these communities, the truck gardeners and tenant farmers served as the social nuclei around which several additional thousands of seasonal farmworkers gathered, while money earned in agriculture provided an economic foundation for Chinese immigrant communities outside the major urban centers. Profits earned in agriculture were a fairly reliable source of income as well as an important source of capital accumulation in Chinese American communities. The story of Chinese truck gardeners, farmers, and farmworkers in California has never been properly told. Students of Chinese American history read of them only in snatches of congressional testimony, brief newspaper accounts, and sections of books on the history of farm labor in California.

Chinese immigrants have been remembered mainly as gold miners, railroad builders, laborers, merchants, servants, laundrymen, and waiters. Gold miners from China were noticed because they challenged the white man's belief that gold was to be reserved for his own aggrandizement and not that of foreigners; railroad builders have entered the historical record because their feats were so spectacular that they could not be ignored; laborers have come down in history as the anathema of anti-Chinese agitators who decried Chinese "cheap labor" as the scourge of white civilization; merchants, though relatively few in number, have been repeatedly mentioned in the literature because they were considered to be the ruling elite in Chinese American communities; house servants, laundrymen, and waiters have been depicted as the stereotypical Chinese of the pre–World War II period. Chinese truck gardeners, farmers, and fruit growers, on the other hand, have not found their way into either the popular or the scholarly consciousness, because they were seldom the focus of anti-Chinese hostility.

The paucity of information on Chinese American agriculturalists reflects the nascent stage of development of Chinese American Studies. Many topics have not been covered, and available sources often present biased views of the Chinese.

Indeed, in few other areas of academic investigation has history itself so dominated historiography. In the past, citizens, politicians, journalists, and scholars have written about the Chinese mostly on those occasions when they were seen as a "problem" in American public life; consequently, many of the views expressed were vociferously anti-Chinese, though there is also a body of "pro-Chinese" literature. Even in the latter, however, the Chinese were defended mainly in terms of the American economy's need for cheap labor or on grounds that the presence of Chinese on American soil was part of God's providence because it facilitated the conversion of Chinese to Christianity.

Three topics—nineteenth-century Chinese immigration and the anti-Chinese movement, the social structure of urban Chinese immigrant communities, and the assimilation of Chinese Americans (or lack thereof)—have captured scholarly attention to the neglect of other aspects of Chinese American life. For the most part, historians have been concerned with the first topic, sociologists have been more interested in the second, and both have touched here and there on the third. Regardless of their disciplinary backgrounds, however, scholars have tried above all else to explain the causes of the anti-Chinese movement; Chinese immigration created such a furor in the United States that the Chinese became the first group of persons to be excluded from immigration on the basis of race.

Several theses have been advanced to account for the intense hostility toward the Chinese. The most common explanation locates the genesis of the anti-Chinese movement in the fears, perceptions, attitudes, and actions of white workers. The writings of Hubert Howe Bancroft, Lucile Eaves, Ira B. Cross, Elmer Clarence Sandmeyer, and Alexander Saxton explore different facets of what might be called the economic thesis, which argues that the fear of being undercut by Chinese "cheap labor" caused workingmen to oppose Chinese immigration.[1] These writers recognized, of course, that racial antagonism and perceptions of the Chinese as perpetual foreigners also played a part in the anti-Chinese outbursts. Mary Roberts Coolidge, though also cognizant of the role white workers played, was especially sensitive to the way the anti-Chinese movement fed into party politics at both the state and the national level. Thus, her work presents the most developed political explanation in the literature.[2] Finally, two kinds of cultural explanations have been given. Historians Gunther Barth, Stuart Creighton Miller, and Robert McClellan examined American perceptions; sociologists Rose Hum Lee and Stanford Lyman looked mainly at Chinese community life.[3] Barth argued that the Chinese, being sojourners and not true immigrants, were seen as obstacles to the realization of the most perfect form of American civilization "destined to culminate on the shore of the Pacific"; Miller showed that negative stereotypes of the Chinese were a national phenomenon and predated the arrival of the first Chinese on American soil; McClellan believed that negative images of the Chinese were based on the "private needs [of white Americans] and not upon the realities of Chinese life." Rose Hum Lee and Stanford Lyman both tried to dissect the structure of Chinese communities in America and, in the process, underlined their cultural and social peculiarities rather than whatever similarities they may have had to other immigrant communities.

In recent years, though many books on the Chinese in America have been published, the majority tend to be broad surveys written to meet the political and social needs of Chinese Americans who are rapidly forging themselves into a self-conscious ethnic group.[4] The best of these recent works have focused on Chinese communities in particular localities and include the study of the Chinese in Hawaii by Clarence E. Glick, the oral history of San Francisco's Chinatown by Victor G. and Brett de Bary Nee, the analysis of the Chinese in the Mississippi Delta by James W. Loewen, and the historico-anthropological study of the Chinese in the post–Civil War South by Lucy M. Cohen.[5] Though many of the recent works in the field have been written by Asian Americans, the only study so far published that has presented a Chinese American perspective based on a careful examination of Chinese-language primary sources is the work by Shih-shan Henry Tsai, who viewed Chinese immigration as a problem that shaped the history of U.S.–China relations.[6] Articles in scholarly journals and more popular publications, some of which are quite good, usually discuss single communities, analyze selected aspects of Chinese American life, or debate some contemporary community issue.[7]

In-depth monographic treatment of topics other than the anti-Chinese movement and the structure of Chinese immigrant communities is still to be done. Different facets of the economic history of Chinese American communities are especially in need of analysis, for economic interaction often spearheaded social and political contact and conflict. There are only two book-length studies that explore Chinese American economic history. One, *Chinese Immigration: Its Social and Economic Aspects*, by George Seward, who served as the United States minister to China just before his book was published in 1881, is essentially a compilation of excerpts from testimony given before the 1877 Congressional Joint Special Committe to Investigate Chinese Immigration. Seward pulled together the remarks of several large landowners who had used Chinese labor to reclaim and cultivate farm land and placed them in a chapter on farming and another on fruit culture. In the second half of the book, he assessed all the objections against the Chinese and dismissed all of them as unfounded except for the charges against Chinese prostitution and other vices.[8]

The other work, *Chinese Labor in California, 1850–1880*, by Ping Chiu, is the only in-depth study of Chinese economic life in nineteenth-century California. Chiu analyzed the role of the Chinese in mining, railroad building, agriculture, and manufacturing—particularly the woolen textile, clothing, shoe, cigar, and other "fringe" industries—within the context of the changing economy of California. He pointed out that the stands taken by white employers and employees on Chinese immigration were based on different assumptions: the former argued in terms of laissez-faire economics, whereas the latter were primarily concerned about the "human rights" of workers and blamed the depression of the 1870s on Chinese competition. The chapter on agriculture, based on information in the manuscript schedules of the censuses of population and agriculture and a small number of narrative sources, provides more information on the subject than any other work, but is nonetheless flawed because Chiu used random examples and did not deal systematically with regional variations in California's agricultural development.[9]

The most detailed studies of Chinese in California agriculture have been done not by historians of Chinese America but by three agricultural economists, a crusading journalist, a labor historian, and two sociologists. These works of Varden Fuller, Paul Taylor, Lloyd Fisher, Carey McWilliams, Cletus Daniel, and Linda and Theo Majka—all of which dealt with Chinese agriculturalists only in their capacity as farm laborers—will be discussed at length in chapter 8.[10]

This study attempts to examine an aspect of Chinese American economic history which has not been studied in depth before: the role played by Chinese in the growth of California agriculture. As individuals and companies, Chinese tenant farmers left a voluminous record of their activities in the form of official documents filed in the county archives of California. Together with information in the manuscript schedules of the censuses of population and agriculture, and glimpses given in contemporary periodicals, these records make it possible to reconstruct the history of Chinese agricultural activities all over the state. Though the subjective perceptions and attitudes of the Chinese themselves cannot be depicted by using such documents, still, whatever story that can be pieced together is well worth telling.

More important, that tale has a significance above and beyond itself: it is yet another variation on the pattern of economic interaction between Chinese and Westerners in the last century and a half. In China itself, some members of the emerging bourgeoisie struggled valiantly to survive in the face of competition from Western capitalists. A number not only survived but did quite well, among whom were middlemen who helped Westerners exploit their poorer Chinese brethren in China as well as overseas. In my view, the history of the overseas Chinese can quite properly be considered a chapter in the modern history of China because the question of why capitalism never developed fully in China cannot be answered adequately until one examines the economic activities of Chinese entrepreneurs abroad and the conditions in those lands that allowed them to flourish there more than at home. Chinese who farmed successfully in California can be grouped among the overseas entrepreneurs, even though farming is not often thought of as an entrepreneurial activity. But that is precisely what made the Chinese experience in California unique: given the organization of California agriculture along "industrial" lines, Chinese farmers in California were among the greatest entrepreneurs in the Chinese immigrant community in America.

ONE

The Chinese Diaspora

 THE CHINESE who came to California in the second half of the nineteenth century were but one branch of a much larger emigrant stream. Their migration was roughly contemporaneous with the infamous "coolie trade" to Peru and Cuba (1847–74) and the movement of Chinese from 1852 onward to join the gold rush in Australia. These nineteenth-century migrations represented a new stage in the long history of Chinese emigration.

Emigration as an aspect of Chinese history has frequently been overlooked because historians and the Chinese themselves have tended to view China as an inward-looking agrarian society. Students of the overseas Chinese, however, have been quite aware of the fact that certain groups of Chinese have been among the world's great migratory peoples. As a matter of fact, Chinese have emigrated to so many parts of the world that it is possible to speak of a Chinese diaspora, of which Chinese emigration to the United States was but one small segment.

Seaborne Chinese migration was a continuation of the overland movement over many centuries of the Han Chinese people who expanded from the Yellow River valley of northern China southward to populate contiguous territory.[1] One stream of migrants went in a southwesterly direction and settled in the area that became the provinces of Szechwan, Kweichou, and Yunnan, while a second, southeasterly movement of peoples brought what are now the provinces of Kwangsi, Kwangtung, and Fukien under Chinese jurisdiction. The latter came under firm Chinese control only during the T'ang dynasty (618–907); for that reason, people from southeastern China refer to themselves as T'ang-jen (T'ang people) rather than as Han-jen (Han people).

When one talks about pre–World War II Chinese emigration, one is really talking about the emigration of people from only five small regions in the two southeastern provinces of Fukien and Kwangtung and the island of Hainan. In the literature on the overseas Chinese, these major emigrant groups have usually been identified by the regional dialects they speak. Though there is only one script for written Chinese, there are many different Chinese dialects, some of which are so unintelligible to speakers of other dialects that they might almost be considered different languages. The dialect one speaks, therefore, often is the most important mark that separates an individual from members of other groups.

Before the 1940s, the five major groups of overseas Chinese were the Hokkien, Teochiu, Cantonese, Hakka, and Hailam. Hokkien is the pronunciation for Fukien in the Amoy dialect, and the term refers to people from Amoy and its environs in southern Fukien. Teochiu is the pronunciation for Ch'aochou in that dialect and is the name for people from Ch'aochou prefecture in the Han River delta of northeastern Kwangtung with Swatow as its chief port. Cantonese is the Anglicized term for people from the Kwangtung provincial capital, Canton, and its environs. Hakka or Kheh is the name for a group that migrated from north China to settle in an area stretching from the southwestern tip of Fukien to northern Kwangtung, interspersed among the Cantonese, with whom they engaged in sporadic feuds—Hakka being the Cantonese pronunciation, and Kheh the Hokkien and Teochiu pronunciation, for words meaning "guest families." Hailam is what people from Hainan Island are called, Hailam being the pronunciation for Hainan in the dialect of that place.

Chinese emigration can be divided broadly into two historical phases according to which groups emigrated and what motivated them to go abroad. Before the nineteenth century, Hokkien and Teochiu emigrants predominated, whereas from the early nineteenth century onward, Cantonese were more prominent. Hakka and Hailam emigrants never matched the Hokkien, Teochiu, and Cantonese in numbers or in social significance. The earliest emigrants were religious pilgrims, merchants, and artisans, but as time passed, more and more laborers joined the

emigrant stream. One group of emigrants who always stood out
were those who went abroad to mine for precious metals. Beginning in the fifteenth century, when Chinese went to mine for tin
in southern Thailand, aspiring miners have been an especially
dynamic element among their fellow emigrants. The tin miners
in southern Thailand and the ones who went later to Malaysia,
as well as those who went to mine for gold in Borneo, were pre-
cursors of the large number of Chinese who joined the gold
rushes in California, elsewhere in the United States, and in
Australia, British Columbia, and Alaska in the nineteenth
century.

Throughout the history of Chinese emigration, geography,
trade patterns, shipping routes, chain migration, and European
incursion each played a significant role in determining the pattern of overseas migration and settlement. Some of these factors
served to "push" Chinese out of their native places, others
served to "pull" them to particular locations abroad, while yet
others provided the means that made emigration possible.

The geography of Fukien and Kwangtung provided certain
natural conditions that created a propensity in the inhabitants
of those provinces to emigrate. Both provinces were relatively
isolated from the rest of China, being separated from central
China by mountains penetrated only by a small number of
passes. After Han Chinese settled this region, communication
with central and north China was easier by sea than by land.
Fukien, in particular, had little arable land, so the people there
depended a great deal on fishing for sustenance. Plying junks
along the coast to travel, fish, and trade allowed the residents of
these two coastal provinces to develop an early expertise in
seafaring—a factor that facilitated overseas migration.

Chinese imperial policy that confined maritime trade to a few
large ports in the provinces of Chekiang, Fukien, and Kwangtung also gave the people there the added advantage of gaining
experience in dealing with foreigners, enabling them to overcome
some of the psychological and social barriers posed by Chinese
culture against departure from home. Furthermore, the pattern
of trade at different ports in southeastern China helped to determine which groups went overseas in the largest numbers at particular points in time.

Chinese maritime trade, which began to flourish in the late

eighth century, was initially handled largely by foreign merchants who operated out of enclaves set aside for them in the major ports, but in the heyday of that trade during the Northern Sung, Southern Sung, and Yuan Dynasties (979–1127, 1127–1279, and 1279–1368), Chinese shipbuilders and -owners as well as sailors and merchants in the lower Yangtze valley and along the southern coast played an increasingly prominent part in it. Chinese officials confined the bulk of the foreign trade to several large ports to facilitate the collection of anchorage fees and custom duties. In the late T'ang and the Northern Sung periods, the city of Canton handled the bulk of the trade, but beginning in the Southern Sung dynasty, ports in Fukien also became ascendant. By 1760, however, the Chinese government had restricted foreign commerce solely to Canton, and foreign merchants were allowed to deal only with Chinese merchants of the *cohong*.[2]

Since Canton had been most consistently accessible to outsiders, Chinese merchants in Canton had to compete with Arab, Indian, Malay, European, and other merchants. Frequently, merchandise shipped out of Canton was carried in foreign bottoms. In the Fukien ports, which were not always open to foreigners, on the other hand, Chinese shipowners, sailors, and merchants controlled every phase of foreign commerce, so much so that before the nineteenth century, trade carried on in Chinese junks—be it Chinese coastal or foreign trade—became a near monopoly of Fukien merchants.[3] As a result, the Hokkien became the earliest group of Chinese to settle overseas in sizable numbers, particularly in the Philippines, Indonesia, the former Straits Settlements in Malaysia (Malacca, Penang, and Singapore), and southern Thailand.

Chinese junks of several hundred tons plied a number of trade routes: one went northward along the China coast and then eastward across to Japan; one followed a southwesterly direction, hugging the coast of Vietnam, Cambodia, Thailand, Malaya, and the east coast of Sumatra; a third proceeded across the South China Sea in a southeasterly direction to Manila, wound its way through the Philippine Islands to the Sulu archipelago at the southwestern tip, thence turned either southward to Borneo and Java or eastward to the Moluccas, the famed spice islands. Not surprisingly, these areas had some of the earliest Chinese overseas settlements.

Prior to the introduction of steamships, the monsoons dictated in which particular season trading junks could sail in what direction toward which destinations. During the winter months, when the northeast monsoons blew across the South China Sea, junks left Chinese ports in January or February to sail southward to Southeast Asia. These junks returned during the southwest monsoon season, usually departing from Southeast Asian ports for China in August. Chinese traders therefore had an eight-month trading season away from home, but some stayed away more than a year, giving rise to a practice that came to be called "double wintering." The natural elements had a further effect on the destination of these Chinese ships because varying degrees of turbulence encountered in different parts of the South China Sea caused captains to favor certain locations over others. Where sojourning traders landed, more permanent settlers tended to follow.

When steamships were introduced by Western firms in the second half of the nineteenth century, transportation and trade patterns changed. Junks had preferred sheltered upriver ports, but steamers required deep-water harbors and different docking facilities. Hong Kong, Canton, Amoy, and Swatow became the most important ocean ports on the southeastern China coast in the nineteenth century. For the passengers, the introduction of steamers meant that a far larger number of persons could be carried in each ship, fares became lower, passage was safer and quicker, and more distant destinations across oceans—rather than seas—became accessible. More important, migration increased dramatically between ports connected by direct steamship service. Regularly scheduled, direct steamer service between Amoy and Manila, and Hong Kong and Singapore, was established in the 1860s, facilitating the movement of Hokkien and Cantonese not only to those locations but to points beyond in Indonesia, the Malay peninsula, and Burma. In 1882 direct steamship service was opened between Swatow and Bangkok, enabling a large number of Teochiu and a smaller number of Hakka (to whom Swatow was the most accessible port) to migrate to central Thailand. Emigrants from Hainan continued to sail in small junks from their island to points along the coast of Indochina and into the Gulf of Siam.[4]

Once the initial migration patterns of the different regional groups had been established, the tendency of potential emigrants to go where they already had relatives or village mates created a chain migration that caused particular groups to cluster in certain localities and in a limited range of occupations. Old emigrants, known as "head guests," would return to their villages in China to recruit "new guests." Frequently, the old emigrants advanced the passage for the new emigrants, tying the latter to themselves through debt bondage, and made all arrangements for housing them on their arrival and finding employment for them. Common dialect and place of origin, therefore, not only became the main bases for associating with others overseas, but more often than not also helped to determine the emigrants' choice of occupation.

More than any of the above factors, the appearance of Europeans in Asia from the fifteenth century onward led to increased Chinese migration to Southeast Asia. In the early decades of the sixteenth century, the Portuguese set up trading posts stretching from the Iberian peninsula, around the coast of Africa, across the Indian Ocean to India, eastward to Southeast Asia, and on to China and Japan. The Spanish conquered the Philippines in the 1570s, and the Dutch showed up in the Indonesian archipelago soon afterward. After an initial effort to establish commercial hegemony in insular Southeast Asia, the British withdrew from the area to concentrate on building an empire in India, but they returned in force in the early nineteenth century to colonize Malaysia. The French colonial effort in Vietnam, Cambodia, and Laos was also a nineteenth-century phenomenon. European trading companies needed middlemen, and colonial settlements needed provisioners, craftsmen, and laborers. An increasing number of Chinese merchants, artisans, and workers came to earn a living in the newly established European colonies. European colonialism thus created powerful "pull" factors that lured Chinese in increasing numbers to different parts of Southeast Asia.

The presence or absence of European colonialists became a key factor in differentiating two different patterns of interaction between the Chinese and the Southeast Asian host societies. In the countries which were colonized by Europeans, the Chinese were treated with ambivalence because the colonial masters both needed and feared them. Consequently, the Chinese were alter-

nately courted and persecuted. In Thailand—the only Southeast
Asian country that was never colonized—on the other hand, the
Chinese were accepted more easily, and many became well inte-
grated into the host society.

European ambivalence toward the Chinese was perhaps most
clearly shown in the Philippines, the northern and central parts
of which the Spaniards had conquered by the early 1570s. Long
before the Spaniards appeared, Chinese had been traveling to
the Philippines. However, the presence of the Spaniards and the
Manila galleon trade they established created a new incentive for
Chinese to go to the archipelago. Hokkien merchants now came
regularly in their junks with the monsoons, bringing luxury
goods expressly for the galleon trade between Manila and
Acapulco. They also brought more ordinary household items,
such as pots, pans, and furniture, and foodstuffs, such as grain,
vegetables, fruit, poultry, and other livestock, to provision the
Spanish troops, administrators, and priests. Meanwhile, Chinese
artisans found work in building the new Spanish colonial capital
at Manila, offering skilled craftsmanship not obtainable locally.
As the Spaniards came to depend more and more on the Chinese,
however, they began to worry about their presence. By 1603
there were an estimated 20,000 Chinese and only 1,000 Spaniards in Manila. To control this growing population, the Spanish
colonial rulers used several (sometimes contradictory) means:
driving them out of Manila, restricting them to segregated
districts, limiting the maximum number of Chinese residents in
the country, imposing special taxes, attempting to convert them
to Catholicism, and periodically massacring them to forestall
incipient rebellions. In each of the bloody massacres of 1603
and 1639 some 20,000 Chinese were killed—almost the entire
Chinese population in the first episode and about two-thirds of it
in the second. By the end of the seventeenth century, Spanish
policy allowed a maximum of 6,000 Chinese to be resident in the
Philippines, and intermittent attempts to oust them continued
through the eighteenth century.[5]

The Chinese had also begun to go to Thailand from early
times when Chinese junks called at ports on the east coast of the
upper Malay peninsula, which constitutes southern Thailand.
By the fifteenth century, groups of Chinese were busily mining
tin in southern Thailand, where a large, independent Chinese

settlement soon appeared at Pattani, a flourishing port. Another
Chinese community was established in Songkla. In addition to
the Chinese who ventured to old Siam on their own, the Thai
rulers of the Ayuthia period (1350–1767) and the subsequent
Chakri dynasty (1782 to the present) encouraged Chinese immigration. It has been estimated that there were a minimum of
10,000 Chinese by the seventeenth century.

Unlike European colonizers, the Thai monarchs did not consider the Chinese to be suspect foreigners. When the Ayuthia
kings set up royal trading monopolies, they not only found the
Chinese to be the most experienced seamen and merchants, but
discovered the added advantage that employing Chinese traders
gave them access to all Chinese ports as well as to Japan, which
in 1536 had closed its ports to all foreigners except the Chinese.
In time, the Chinese came to manage all the king's maritime and
mercantile affairs, both in Thailand and abroad. Chinese merchants also traded on their own account, exporting rice and lumber from Thailand to Fukien. By the nineteenth century, Chinese
dominated rice milling and shipbuilding, and a small number
grew and refined sugar. In addition to playing their traditional
roles as traders and artisans, they also worked as royal entertain-
ers and physicians, and a few individuals even became high officials. Thus, although the Chinese in Thailand, as elsewhere in
Southeast Asia, served middlemen functions, they also gained
considerable political influence and became more integrated
socially and culturally into Thai society. Only from the
nineteenth century onward did European competition (under-
pinned by the terms of the 1855 Bowring Treaty) and an upsurge
of defensive Thai nationalism lead to anti-Chinese restrictive
measures.[6] However, despite their influence and amicable interaction with the Thais, it is doubtful that Chinese settlers in
Thailand ever considered themselves colonists.

Real Chinese colonization did take place closer to home on
Hainan and Taiwan: Chinese settlement on these two large
offshore islands led to their full incorporation into the Chinese
polity. Of the two, more is known about the settlement process
on Taiwan. It began in the seventh century when people from
Fukien province crossed the Taiwan Straits to fish and to settle
on the Penghu Islands and Taiwan. The initial movement was
sporadic and the number of persons involved was small. In the

eighth century, the Fukien emigrants were joined by people from Kwangtung who began to migrate westward into Kwangsi and eastward to southwestern Taiwan. According to Chen Ta, some of the emigrants to Taiwan were people fleeing from Japanese and Chinese pirates who periodically raided villages along China's central and southern coast. By the time the Dutch occupied Taiwan in 1624, some 25,000 Chinese emigrants resided on the island. The Dutch left Taiwan in 1662 when Cheng Ch'eng-kung (known to Westerners as Koxinga) ousted them and set up a base, where he and his descendants ruled for thirty-eight years, laying a firm foundation for Chinese settlement. The settlers cultivated rice, sugarcane, tea, and vegetables, fished, mined for coal, gathered camphor, and carried on both domestic and foreign trade. The Ch'ing rulers, who had tried strenuously to eliminate Koxinga, annexed Taiwan in 1683.[7]

Despite its long history, emigration was never something the central Chinese government approved of, much less promoted. In the 1640s, in the transition from the Ming to the Ch'ing dynasty, warfare and unstable social conditions led many to leave China from both Fukien and Kwangtung provinces. This outflow alarmed the Ch'ing government, which feared that Ming loyalists might foment anti-Ch'ing campaigns from overseas bases. Because of this worry, the Ch'ing government adopted the attitude of the Ming rulers (who had viewed Chinese emigrants as "deserters," "criminals," and "potential traitors")[8] and codified it into law. Between 1656 and 1729, the Ch'ing officials promulgated a series of edicts that prohibited voyaging overseas on pain of death. Reflecting the Ch'ing government's vacillating attitude, these imperial edicts first requested foreign governments to repatriate all Chinese residing abroad so that they might be executed, then stipulated that amnesty would be provided to those who returned to their homeland within a given period, and still later announced that those who failed to return would be barred forever from reentry.[9]

These prohibitory edicts notwithstanding, Chinese continued to go overseas to trade, work, and settle. In the middle of the nineteenth century, the forcible "opening" of China by European powers qualitatively changed the nature of Chinese emigration. Whereas in earlier centuries European activities served mainly to create the "pull" factors that lured Chinese overseas, in this new stage of Chinese international migrations, the actions of Europeans gave rise not only to the "pull" but also to the "push" factors. In addition, it was also Europeans who provided the means that enabled large numbers to sail to distant lands across not just seas but oceans.

Cantonese Emigration

In the century between the end of the Opium War in 1842 and the beginning of World War II, Chinese proceeded abroad to work and live in every continent on the earth. It has been estimated by Sing-wu Wang that in the last six decades of the nineteenth century, at least two and a half million Chinese went overseas.[10] Asian international migrations of this period were an integral part of Western economic development and imperialist expansion. Chinese emigrants left to escape poverty in China, which resulted from insufficient land and overpopulation but was exacerbated by the political, economic, and social disruptions caused by Western activities. They went to the Americas, the West Indies, Hawaii, Australia, New Zealand, Southeast Asia, and even Africa to take advantage of the economic opportunities created by the colonial development of some of these areas and by the discovery of gold in others. Thus, "push" and "pull" factors—both of which were linked to the dynamic growth of Western capitalism—appeared simultaneously to create conditions that stimulated emigration. At the same time, Western ships, eagerly waiting in Asian ports to take large numbers of destitute peasants to faraway lands to work, served as the means that made emigration possible.

The Cantonese emigrants to America originated in a small region stretching over some two dozen districts in the Pearl River Delta or on its periphery in Kwangtung province. Eight of the twenty-two districts were numerically quite important in terms of sending emigrants to North America. These eight can be subsumed under three groups: the first consists of Namhoi [Nanhai], Punyu [Panyu], and Shuntak [Shunte], which are collectively known as Sam Yup [San-i], or Three Districts; the

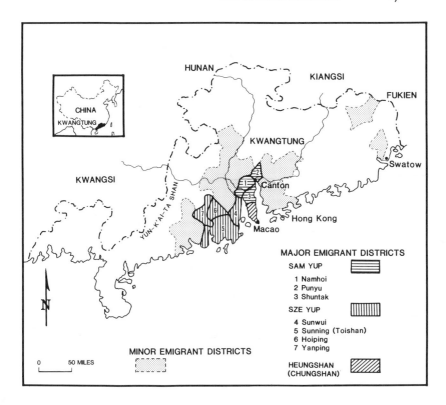

Map 2   *Kwangtung Province: Emigrant Districts*
          SOURCE: *Thomas Chinn, Him Mark Lai, and Phillip Choy,* A
          History of the Chinese in California: A Syllabus *(San*
          Francisco: Chinese Historical Society of America), 1969, p. 3.

second is the single district of Heungshan [Hsiangshan, now
known as Chungshan]; and the third consists of Sunwui
[Hsinhui], Sunning [Hsinning]—which was renamed T'oishan
[T'aishan] in 1914—Hoip'ing [K'aip'ing], and Yanp'ing [En-
p'ing], which are collectively known as Sze Yup [Szu-i], or Four
Districts.[11] The largest proportion of the nineteenth-century
emigrants to both North America and Australia came from
T'oishan. Different sources estimate that they constituted from
40 percent to over 50 percent of all the emigrants to America.[12]
The T'oishan people, together with their compatriots from the
other districts of Sze Yup, made up over 70 percent of the
Chinese immigrants in California.

Though most numerous, Sze Yup people were relatively poor. Most of them became laborers and domestic servants; those who went into business had laundries, restaurants, and small retail stores. The immigrants from Sam Yup, on the other hand, became some of the more important merchants in San Francisco's Chinatown, and many of the early sewing factories and butcher shops were owned by them. Chungshan people controlled the retail fish stores and ladies' garment factories in San Francisco.[13] Many also became truck gardeners and farmers in rural California. Their presence has been, and continues to be, notable in the floriculture and nursery business in San Mateo and Santa Clara counties; at the turn of the present century, they virtually dominated large-scale tenant farming in the Sacramento–San Joaquin Delta.[14]

Although these eight districts are contiguous and are all clustered around the provincial capital, Canton, they in fact have rather different geographic conditions. The main difference lies between Sam Yup and Chungshan, which are in the Pearl River Delta proper, and Sze Yup, which is on the southwestern periphery of the delta. The former districts have relatively productive land, whereas the latter ones are hilly and barren, with only 10 percent of the land arable.

The Pearl River Delta, stretching approximately seventy miles from north to south and fifty miles from east to west, is a collection of islands, ever-changing natural channels, and man-made canals. It is the economic heartland of Kwangtung province. Formed by the confluence of three rivers, the Sikiang [Hsichiang], or West River, the Peikiang [Peichiang], or North River, and the Tungkiang [Tungchiang], or East River, it is not a single flat plain but a fragmented area of alluvium intersected by hills and wide streams. The Tropic of Cancer bisects it, and depending on local terrain, both subtropical and tropical crops can be grown. Temperatures vary from the mid-50s to the mid-80s Fahrenheit, and the region generally does not suffer from frost. The annual rainfall is about eighty inches, most of which falls in the summer.

Much of the farmland has been reclaimed and is protected by dikes lined with trees. Crops can be grown the year around, with rice as the dominant crop. The bulk of the rice is grown as freshwater paddy, which produces two crops a year; smaller amounts of brackish-water paddy and dry upland rice are also grown. Kwangtung province is famous throughout China as a cornucopia of fruit and vegetables, including oranges, tangerines, pomelos, lichee, longan, papayas, guavas, persimmons, bananas, and pineapples. Other cash crops include mulberry (for feeding silkworms), peanuts, sugarcane, indigo, tea, tobacco, and rape (an oil-bearing seed). Major farm animals are hogs, chickens, and ducks. Extensive freshwater and saltwater fishing is done, and the upper delta is notable for its fish-breeding ponds.[15]

This image of abundance must be juxtaposed against the extraordinary population density in the lowland areas of the delta. In the eighteenth and nineteenth centuries, the delta had accommodated two thousand or more persons per square mile, forcing peasants to cultivate miniscule plots with meticulousness. It has been estimated that in the early nineteenth century, the ratio of land to population in Kwangtung was 1.67 *mou* per person (one mou equals one-sixth acre). In the poorer districts, as little as one mou had to suffice to produce enough food for the sustenance of each person, when it has been estimated that three mou per person were needed for an adequate diet. In south China, the rate of tenancy was high because a large proportion of the arable land was held by clans rather than by individual proprietors.[16] Worse, for many peasants no land at all was available. Those who could not afford to rent any land eked out a living as landless laborers. Supplementary income from a wide variety of handicrafts and employment in personal service and common labor enabled many peasant households to survive, though barely. In addition, from the time when emigration began in earnest, monetary remittances from household members abroad have sustained many a peasant household.

If the peasants in the relatively well-off Pearl River Delta had a hard time making ends meet, their counterparts in T'oishan district (the source of most of the emigrants to America) had an even more difficult time surviving. The area is hilly, rocky, and barren. It was bad enough that only 10 percent of the land could be cultivated, but of this area, a large proportion adjoining the South China Sea had been ruined for farming by the intrusion of salt water. To make matters worse, the land farther inland, which did not suffer from salinity, was vulnerable to drought because the hilly terrain made it impossible to construct an irrigation system. Surrounded by mountains on three sides and the ocean on the other, the district's population found it difficult to engage in overland commerce. Their only outlet was the sea. In the latter half of the nineteenth century, the district produced only one-quarter of the grain needed to feed its people, and there were no compensating industries to provide supplementary income.[17]

Compounding the problems caused by meager natural resources and population pressure, the emigrant region suffered from a series of man-made devastations from the 1840s through the late 1860s, including the Opium Wars, the T'aip'ing Rebellion, secret society uprisings, and clan warfare. The effects of the Opium Wars will be described below. Although the T'aip'ing Rebellion wreaked its greatest devastation in central and not south China, still, peasants were conscripted and fields were trampled upon, and local brigands used the passing of the T'aip'ing armies through the region as a pretext to prey upon the populace. Greater damage was caused by local warfare. In 1854 several uprisings involving thousands of clan and secret-society members wearing red headbands erupted in the environs of Canton. One group captured the city of Fatshan to the west of Canton, while others pressed toward the provincial capital from the north and farther to the west. Imperial troops as well as local militia were mobilized to combat the rebels, and areas around Canton were burned to deny cover to them, while the city's merchants raised funds to help suppress the outbreaks. Namhoi, Shuntak, and Punyu in Sam Yup, Hoip'ing in Sze Yup and Chungshan districts all suffered from the Red Turbans uprisings.[18] Though the rebels were suppressed within the year, rural pacification continued for a decade.

Meanwhile, the Punti-Hakka [Penti-K'echia] feuds erupted, causing severe damage to the Sze Yup area, where T'oishan was a major battlefield. Harboring age-old hatred for each other, in the wake of the devastations caused by larger conflicts, the two groups found new expression for their traditional antagonism in a bloody war that lasted over a decade.[19] Some of the Chinese sold to coolie traders were allegedly prisoners taken during the Punti-Hakka war. Given these simultaneous multiple disasters, emigration overseas to find work seemed an absolute necessity for families that wished to survive.

The emigrants of this period can be divided between those taken overseas in the coolie trade[20] and those who left more voluntarily as free or semifree emigrants. Of the latter, individuals who paid for their own fares were free, whereas those whose passage was financed by the so-called credit-ticket system can be classified as semifree. Though contemporary observers saw little difference between the Chinese shipped abroad in the coolie traffic and the free and semifree emigrants, the Chinese themselves, as well as British and American consuls stationed in China in the 1850s and 1860s, distinguished between the two groups: free emigrants to California and Australia—a large number of whom initially were aspiring gold miners—were usually referred to as "passengers," while those taken to Latin America were called "coolies."[21]

The coolie trade had begun in response to the termination of the African slave trade and of slavery as a form of labor in European colonies and former colonies.[22] Owners of plantations that had depended on slave labor now had to look mainly to two new sources of human muscle power: Indians and Chinese. The global division of coerced labor followed the geographic boundaries of the European empires. Indians were taken mainly to regions within the British Empire, although in the last decades of the nineteenth century Britain allowed France to import Indian laborers into the island of Réunion, a French "sugar island" in the southwestern Indian Ocean, and Denmark was permitted to import them into St. Croix, its "sugar island" in the West Indies.[23] Chinese were taken mainly to Peru and Cuba and in smaller numbers to the three Guianas, Jamaica, and Trinidad in Latin America.[24] Javanese were also recruited to work abroad, though on a far smaller scale, when the Dutch shipped some of their colonized subjects from the densely populated island of Java to their former Cape Colony in South Africa—where the Javanese became known as "Cape Malays"—and to Surinam (Dutch Guiana).[25]

Most writers have identified the first shipment of Chinese coolies to Peru in 1847 as the beginning of the Chinese coolie trade, but in fact the transportation of Chinese by Westerners to lands outside of the Asian continent had begun many years earlier when several hundred tea growers were taken to Brazil by the Portuguese in 1810.[26] Another precedent was set in 1843 when a shipload of laborers was recruited from the Straits Settlements (most probably Singapore) to work on sugar plantations in Mauritius.[27] The coolie traffic did not become heavy, however, until the third quarter of the nineteenth century. In those decades, the major ports handling the coolie as well as the passenger traffic were Hong Kong, Canton, Amoy, Swatow, and Macao.

That so many Chinese were available to venture overseas was a result of the upheavals that European encroachment had caused in China. In the middle of the nineteenth century, Britain had taken the lead in "opening" China to foreign trade, Christian proselytization, and political control. The terms of the Treaty of Nanking that ended the Opium War (1839–42) opened five Chinese ports—Canton, Amoy, Foochow, Ningpo, and Shanghai—to foreign commerce, thereby undercutting the monopolistic position the city of Canton had occupied in that trade for the preceding eight decades. The *cohong* system was abolished, and a limit was placed on the amount of customs duty the Chinese could charge. The island of Hong Kong was ceded to Great Britain, and foreigners were granted the right of extraterritoriality under which they could only be tried by their own consuls, making them immune to Chinese law. Other Western nations, which had not participated in the Opium War, received the same privileges as Great Britain under the most-favored-nation clause. Diplomatic etiquette that had irked the Europeans—such as the *kowtow*—was summarily rejected, and Christian missionaries were now allowed to proselytize with greater freedom.[28]

The common people were greatly affected by the Opium War and its consequences. Many dockhands, as well as laborers who transported merchandise inland, were thrown out of work. Peasants also lost a traditional source of supplementary income when imported foreign goods—particularly British textiles—offered native-made piece goods stiff competition. The continued influx of opium, the ever-increasing number of addicts among the Chinese population, and the increased taxation of peasants to help pay for the indemnity of 21 million silver dollars imposed by Great Britain on China—all made the peasants' existence a precarious one.

The Treaty of Tientsin ending the Second Opium War (1856–

60) wrung further concessions from the Ch'ing government. More ports, many on waterways in the interior of China, were opened; opium was legalized; a further indemnity was imposed; and Kowloon—territory on the Chinese mainland opposite Hong Kong island—was ceded to Great Britain. British troops that had been stationed in Canton in 1858 remained as an occupation force until 1861. During those years the city was governed jointly by foreigners and Chinese.[29] This fact had an important effect on Cantonese emigration, since both local Chinese officials and the Allied occupation commanders—for different reasons—took the opportunity to chip away at the Ch'ing government's prohibition against emigration, thereby greatly increasing the number of Chinese who went overseas.

Chinese administrators in Kwangtung and Fukien were concerned about the notorious coolie trade, but they recognized that the Ch'ing government could not regulate this traffic or other forms of emigration, because it considered such exodus illegal. Thus they tried to legalize emigration on their own authority as a precondition for regulating the traffic. The British government, eager to ship Chinese laborers abroad but wary of the moral dilemma involved, tried to ameliorate some of the horrible conditions surrounding the coolie traffic by passing a series of acts, the most important of which was the British Passengers Act of 1855, which imposed certain space allocation and health standards on ships departing from Hong Kong. For their part, American consuls and ministers on the scene who were also concerned about the trade could not do very much to stop American participation in it, because the United States had no clear policy on the legality of American involvement. Existing laws that stipulated minimal sanitary conditions on passenger ships landing in American ports did not apply to ships engaged in the coolie trade, because the latter sailed from one foreign port to another.[30]

Despite attempts to control it by various parties, the trade in coolies flourished because it was profitable and because what regulations existed were easy to circumvent. According to anecdotal information, in the 1860s and 1870s, it sometimes cost the coolie brokers from $120 to $170 to secure a coolie and ship him to Latin America; upon arrival, the same coolie, if alive, was sold for $350 to $400. Those desiring the services of the coolies bought

them outright, often at auctions.[31] In China itself, regulations
were easy to circumvent because within the foreign enclaves
carved out in the treaty ports, Chinese law did not apply. Thus,
not only were foreigners outside Chinese jurisdiction, but their
Chinese employees who violated Chinese laws could also find
refuge in the foreign "settlements," as the areas enjoying extraterritoriality were called. Extraterritoriality enabled labor
recruiters and kidnappers of all nationalities to operate freely,
spiriting away hapless Chinese youths to hard labor overseas.
Potential free emigrants often also preferred to buy their tickets
from Western "passage brokers" because dealing with them
rather than with Chinese brokers provided greater protection.[32]
It was relatively easy for aspiring emigrants to proceed from
their villages to Hong Kong or another port to buy passage on
foreign ships without interference from Chinese government
officials.

Sporadic efforts to regulate the coolie traffic culminated in a
"Convention to Regulate the Engagement of Chinese Emigrants
by British and French Subjects," which contained twenty-two
articles spelling out detailed regulations for each step in the recruitment, export, employment, and return of Chinese laborers,
and was signed by Sir Rutherford Alcock, British minister in
China, Henri de Bellonet, French chargé d'affaires in China, and
Prince Kung, head of the Chinese Tsungli Yamen, in 1866. The
American and Prussian ministers in China also signed the convention even though they had not been party to its negotiation.
Unfortunately, though British and French officials had participated actively in the discussions that brought the convention
about, the British and French governments declined to ratify it.
Two years later, these two governments proposed a revised convention that deleted all references to the length of service (set at
five years in the 1866 draft), free passage for those laborers who
had fulfilled their contracts and who wished to return to their
homeland, and several other conditions protecting the rights of
the coolies. Prince Kung refused to accept the watered-down
convention, and the Chinese government continued to deal
with emigration according to the terms contained in the 1866
version.[33] In the late 1860s and early 1870s, wishing to escape
Chinese jurisdiction, labor recruiters and coolie ships began to
operate primarily out of Macao, a Portuguese colony, which at

one point had dozens of "barracoons"—prison-like detention centers to hold coolies awaiting shipment. The coolie traffic flourished in Macao until pressure from other governments finally forced the Portuguese to declare an end to it in 1874.[34]

In the United States, after a decade of correspondence from American diplomats in China, imploring the United States government to clarify its policy regarding the coolie traffic and American participation in this despicable enterprise, Congress finally passed a law in 1862 forbidding American ships to engage in the traffic, but some American captains and their ships continued to do so by sailing under foreign flags.[35] In the mid-1860s the United States desired the immigration of nonindentured Chinese laborers, however, and formalized it under the terms of the Burlingame-Seward Treaty of 1868.[36]

While most of the Chinese coolies were destined for Latin America, the majority of the free and semifree Cantonese emigrants proceeded to California and Australia. In California some of the earliest Chinese immigrants were merchants who came of their own accord. However, once news of the gold discovery reached China, boatloads of aspiring miners outnumbered all other classes of emigrants. A similar situation prevailed in Australia, where the earliest Chinese immigrants had been a thousand or so indentured laborers taken there prior to the discovery of gold; but after news of the latter event also reached China, thousands again climbed aboard whatever means of transportation were available to reach the gold fields "Down Under." That gold—everywhere it was found—was the most powerful lodestone attracting large numbers of free and semifree Chinese voyagers is clearly shown by the fact that Chinese miners were among the first on the scene in all the major gold rushes of the second half of the nineteenth century; they clambered to the Fraser River strike of British Columbia, the Klondike strike in Alaska, and the less spectacular rushes in the Pacific Northwest. They showed up also when silver was discovered in the Comstock Lode in Nevada, but the vehement opposition of white miners there kept them away from the mines. Only after all hopes of striking it rich in gold mining had faded did Chinese succumb to the necessity of laboring for low wages.

Though there were conflicting opinions regarding the status of the emigrants to California and Australia, there is sufficient evidence to show that most of them either paid their own passage or secured it under the credit-ticket system. Under this system, various persons—returned emigrants, Chinese merchants, Western labor recruiters, and ship captains—provided the emigrants with tickets on credit for the voyage abroad. The indebted emigrants were obligated to repay the fares out of their future earnings upon arrival at their destinations. Interest was extremely high, ranging from 4 to 8 percent a month.[37] Two methods of repayment existed. Those who went to a nearby place such as Singapore—where the fare was relatively cheap—were turned over upon arrival by their passage brokers to employers who wished to hire them. The employers, having paid the brokers the cost of passage, then had legal claim to the labor of the immigrants. Passage to Australia or North America, on the other hand, was ten to twenty times more expensive. Few employers were willing to "buy" the newly arrived immigrants, who continued to be indebted to the passage brokers until they had paid them off fully.[38] The obligation incurred by those Chinese immigrants who came under the credit-ticket system was thus one of debt bondage, and not contract labor, but anti-Chinese groups argued that there was in fact no difference between the two. There is no reliable evidence to indicate how many immigrants to the United States paid their own passage and were therefore completely free, and how many came under the credit-ticket system; in all probability, the latter far outnumbered the former.

The Journey

Little is known of the ships that carried Chinese to California in the first two decades of Chinese emigration, except that many were British and all were overcrowded. If conditions on these ships were at all similar to those found on ships engaged in the coolie trade, then it can be assumed that the mortality on board was high—perhaps 5 to 10 percent—and the steerage passengers were hardly, if ever, allowed on deck.[39] Conditions encountered on the crossing greatly improved as steamships were introduced. After 1867 the great majority of trans-Pacific voyagers traveled in ships of the Pacific Mail Steamship Company, which became, in time, one of the means facilitating Chinese immigration into the United States. These ships also made it possible for both Chinese and American merchants to carry on a vigorous trans-Pacific trade. Though a wide variety of merchandise was carried,[40] the passenger traffic was always one of the chief aspects of that trade.

The Pacific Mail Steamship Company had been established in 1848 to carry mail from the Atlantic seaboard to Oregon with a subsidy from Congress. Upon the discovery of gold in California, the company's coastal steamers, which had been running between Panama and Oregon, entered the lucrative business of carrying passengers from the Isthmus of Panama to San Francisco. In the mid-1860s the United States Post Office offered the line an additional annual subsidy of half a million dollars to carry mail between San Francisco and Hong Kong on twelve round trips per year. The company built four of the largest wooden side-wheel steamships ever manufactured, at one million dollars each, for the purpose. Larger than British steamships, these giants measured 460 feet in length, had paddle wheels 40 feet in diameter, had a gross displacement of 4,000 tons, and burned 45 tons of coal a day. Later, vessels with iron screw-propellers and hulls were introduced. In 1872 the company's subsidy from the United States government was doubled to one million dollars a year in return for carrying mail to Asia in two crossings, instead of one, per month.[41]

Each ship had room for 250 cabin and 1,200 steerage passengers, the latter generally composed entirely of Chinese, although occasionally some white passengers also traveled in steerage. Steerage passengers had to carry their own bedding, which was placed on wooden frames; the latter were cleared in the daytime for sitting and eating on. It was said that even well-to-do Chinese often preferred steerage to cabin accommodations because they could eat Chinese food in the former but not the latter class. The thousand or more steerage passengers crowded into each ship actually violated legal limits, but with the exception of an occasional suit brought by U.S. government officials against it, the company was able to get away with the practice most of the time. Depending on the season and the route chosen, trans-Pacific journeys ranged from 4,500 to 5,000 miles and took an average of thirty-three or thirty-four days.

The company not only transported Chinese passengers but

began to employ Chinese sailors in increasing numbers, although the ships' officers were always white. With the exception of the Chinese cooks, the wages paid the Chinese crew members, regardless of their occupational classification, were lower than those offered to the most lowly paid white employee.[42] Charging each steerage passenger $50 to $55 from Hong Kong to San Francisco and $40 going the other way, and paying its Chinese employees low wages, the company profited greatly from Chinese immigration. Between 1867 and 1875 Pacific Mail steamships carried some 124,800 Chinese across the Pacific (in both directions), receiving $5,800,000 in steerage fares alone. In the early 1870s, the company was using over forty ships in this traffic. However, the profit was not lucrative enough for competitors to attempt to enter the business until two decades later.[43] Chinese did not come out completely on the short end, however, because steamships were cheaper, faster, and safer than clipper ships.

The company had complex and far-ranging dealings with Chinese communities in both the United States and China: Chinese merchants depended on the ships to carry the merchandise traded in their import-export businesses; hundreds of ticket agents worked for the company in Chinese and American ports and even in towns and cities in the interior of the United States; and company officials frequently defended Chinese immigration. During the half century (1867–1917) that the Pacific Mail Steamship Company was involved in the Chinese immigration business, anti-Chinese groups repeatedly attacked it as one of the chief culprits that brought Chinese "cheap labor" to the United States, and a number of attempts were made in the late 1870s to burn its wharves, offices, and holding shacks in San Francisco.[44]

Though they were paid lower wages than white workers, Chinese immigrants were by no means the dregs of Chinese society. There is, unfortunately, no reliable information on the social origins of the Chinese emigrants of this period. It is not known what the relative distribution of peasants, laborers, artisans, merchants, or gentry among the emigrants was, and how the distribution varied for groups going to different destinations. There is little evidence that members of gentry families went abroad except for individuals sent to foreign countries for higher studies, but it is certain that among the first immigrants to California, a considerable number were merchants, because California newspapers reported their presence.[45] A number of artisans must have gone also, because some of the construction workers in gold-rush San Francisco—such as those who built the famous Parrott Building at the corner of California and Montgomery Streets with precut granite imported from China in 1852—were Chinese, and there were reports by journalists of other prefabricated houses that came from China and were put up by Chinese construction crews.[46] Logic alone would indicate that the majority of the emigrants must have been peasants and laborers, given the agrarian nature of Chinese society and the fact that the existence of these groups, in any society, has always been the most precarious.

Regardless of their class background, it is probable that family members left behind depended on those who went abroad to send sufficient money home that they might live, but this hope must have been tinged with realism, for centuries of experience with emigration must have acquainted Chinese villagers with the fact that while some lucky ones indeed might succeed economically in foreign lands, many others would face destitution or even death. It can be assumed that the members chosen to go abroad were among the strongest and bravest because their families knew that physical stamina, ingenuity, and the willingness and ability to provide mutual aid and protection to each other would determine how well the emigrants could cope with hostile treatment in alien lands.

One small bit of evidence suggests that many emigrants were in fact second-generation migrants. In the 1880 manuscript schedules of the population census for Union Township, San Joaquin County, California, the census taker went to the trouble of putting down the birthplace of each Chinese—usually the name of a district or the name of the district association to which the individual belonged[47]—unlike his peers who usually only wrote "China." More than half of the 536 Chinese listed in this township were born in places that differed from the birthplaces of their parents. Although many of the listed individuals had been born in Canton, Whampoa, or a relatively large urban center, their parents had originated from smaller, more rural settings. This meant that the families had already moved once from a rural area to a town or a city, and the young men's emigrations overseas were but continuations of journeys that had begun a generation earlier.

In the 1850s, California, the "Old Gold Mountain," and Australia, the "New Gold Mountain," were the most desired destinations even though they were farther away than the more familiar lands of Southeast Asia and passage was accordingly more expensive. Traveling abroad to mine for precious metals was, after all, nothing new, so it can be assumed that aspiring gold miners to California and Australia were an intrepid lot whose movement across the Pacific was based on rational calculations. Migration being a gamble and an investment, the possibility of finding gold made emigration to California and Australia the best bet available.

At least, that was their expectation. What they did not know was that they would be the first large groups of Chinese to enter white men's countries. Though the lands of Southeast Asia they had long traveled to had been colonized by Europeans, those were colonies of trade and exploitation acquired for the sake of gaining access to desired tropical agricultural products and minerals. The number of Europeans who actually settled there was never large. In contrast, the United States, Canada, Australia, New Zealand, and southern Africa became colonies of settlement to which millions of Europeans emigrated. Chinese would be treated in these temperate countries quite differently from the way they had been received in the tropical colonies, where, despite sporadic persecution, they had been tolerated because they were needed as middlemen and artisans. In the latter, by dint of perseverance and hard work, it was possible for some of them to succeed economically, while more than a few even acquired considerable social standing. In the temperate countries of European settlement, however, Chinese—and other peoples of color—historically have been wanted only as cheap labor. Any attempt by them to rise above the status of laborer would be met with resistance and retaliation. Persecution now came not from the ruling elite, as in pre-nineteenth-century colonial Southeast Asia, but from white workers.

As a matter of fact, Chinese immigration and opposition to it helped to consolidate the white labor movement in California and probably elsewhere too.[48] In many Western industrializing countries, attempts were made to organize workers in the late nineteenth century in response to the growing power of those who owned and managed large industries and businesses. Although collective action did give white workers strength and a modicum of political power, the newly acquired power was always tenuous, so that white workers remained vigilant against the potential threat posed by cheaper and more exploitable and exploited foreign (mainly Asian) labor. Whenever they perceived an "invasion" of "alien" workers to be imminent, they rose up to repel it. Regardless of whether the Chinese who came to these countries were free or contract laborers, white workers indiscriminately called them coolies and argued that white societies were imperiled by the influx of hordes of people able to survive on very thin margins of subsistence.[49] Politicians and some employers eventually also climbed on board the anti-Chinese bandwagon because racism overrode considerations of class interest. In the United States, Canada, Australia, and New Zealand, anti-Chinese movements ended Chinese immigration by the end of the nineteenth century.

Thus, the tragic irony of the Chinese experience in America lies in the fact that although the Chinese had come filled with hope and were imbued with a strong desire for economic success—desire fueled by the burden of being responsible for the survival of their extended families—many were never allowed to become much more than mudsills in white men's societies.

TWO

Trans-Pacific Pioneers

CANTONESE EMIGRANTS who embarked on Western ships for long voyages across the Pacific Ocean to California left behind a peasant society in turmoil and entered a frontier environment equally in flux. Their sense of uprootedness must have been similar to the alienation experienced by European immigrants as that feeling has been so poetically evoked by Oscar Handlin.[1] After weeks of being confined below deck, weak from eating poor food and drinking foul water, often suffering from seasickness and diarrhea, they disembarked in San Francisco, walking down the gangplank single file, each man carrying his worldly belongings in two bundles or baskets hung one at each end of a bamboo pole slung across his shoulder. The arrivals were met by a small number of their fellow countrymen—usually merchants or their agents—who immediately took them to their lodgings. There, they were allowed to rest a few days and were then outfitted for the gold fields and sent on their way.

Some took steamers across San Francisco Bay and up the Sacramento River, landing at Sutter's Fort, which blossomed overnight into Sacramento City. From there they proceeded, usually on foot, into the northern portion of the mining country in the foothills of the Sierra Nevada. Others walked southward from San Francisco to San Jose, then eastward toward Stockton, whence they continued into the southern mining region. Along the way they encountered all manner of non-Chinese people traveling on horseback, in stage coaches and carts, or on foot. They also met heavily loaded mule trains, many of which were run by Latino packers, and saw herds of untended cattle roaming loose. By the mid-1860s, however, instead of live cattle, more often than not they saw carcasses or piles of bones littered in many places throughout the Central Valley because torrential rains followed by several years of severe drought in the early 1860s had decimated California's herds. The Central Valley, which they had to traverse to get to the mining country, was rather barren, with few large boulders and few trees except for occasional clumps of black oak and stands of willows alongside streams. From May through October, the air was hot and dry, becoming cooler only as they approached the foothills where the ground was covered with chaparral. Miners in canvas tents and makeshift lean-tos could be seen camped along streams and rivers, many of which were rich in gold placers. Ascending higher, the Chinese found groves of majestic coniferous trees, which soon thickened into forests. On the horizon, towering mountain ranges loomed.

Had its gold never been discovered, California would have remained, in all likelihood, a frontier for several more decades. Located at the edge of a continent, flanked by an ocean, ranges of snow-capped mountains, and a desert, it was a self-contained physical entity extremely difficult to reach from the settled portions of the United States except by a long sea voyage around stormy Cape Horn, or across the Isthmus of Panama, or overland across the semiarid Great Plains and the forbidding ranges of the Rocky and Sierra Nevada mountains. Many tribes of Native Americans originally inhabited California, but because they were small and scattered, they found it difficult to resist the intrusion of white colonists.[2] Spanish priests, soldiers, and a small number of colonists came northward from Mexico between 1769 and 1821 to build a string of twenty-one missions along the coast from San Diego to Sonoma, and also the pueblos and presidios of Los Angeles, Monterey, Branciforte (Santa Cruz), San Jose, and Yerba Buena (San Francisco). In this northernmost outpost of the Spanish empire, thousands of Native Americans were taken into the mission compounds and made into peons and Christian neophytes. The mission gardens became well known for the great variety of crops they raised. Spanish and—after 1821, when Mexico became independent—Mexican colonists (who called themselves Californios) were granted large tracts of land with hazily defined boundaries, to be used as ranchos for cattle raising.[3]

During the Spanish and Mexican periods, Americans and Europeans had become conscious of California when a small number of individuals drifted there in search of pelts. Sea captains involved in the China trade came to the California coast to hunt otters, whose pelts were desired items of trade; Boston merchants captured the hide and tallow trade of the Pacific Coast, carrying thousands of California cowhides away each year; fur trappers came overland to obtain beaver skins. Other Yankee traders, of whom William G. Dana, Abel Stearns, and Thomas O. Larkin became the best known, came to trade and to settle. Early settlers in the interior of California included the American John Marsh and the Swiss German Johan August Suter who soon anglicized his name to John Sutter. Sutter asked the Mexican government for, and was granted, eleven leagues of land at the confluence of the Sacramento and American rivers, where he built a fort. In the ensuing years, Sutter's Fort became a favorite gathering place for European and American immigrants arriving in California. In the early 1840s wagon-train loads of immigrants began to arrive, increasing the number of foreigners in California to such an extent that the Mexican authorities became alarmed. Their uneasiness increased when John C. Fremont, a lieutenant in the United States Army Corps of Topographical Engineers and son-in-law of the influential Missouri senator Thomas Hart Benton, carried out three well-publicized explorations of the valleys and mountains of Alta California with an eye to American westward expansion.

In the years immediately preceding the gold rush, American political interest in California became quite clear.[4] Fearing British designs on the Pacific Coast, ships of the American navy routinely patrolled the offshore waters, their officers eager for action. In May 1846 the United States declared war on Mexico in connection with events in Texas. A month later, out in California, a motley crew of mountain men and runaway sailors, acting on their own authority, took possession of Mariano Vallejo's home and compound in Sonoma and established the Bear Flag Republic. Meanwhile, on July 2 of the same year, Commodore John D. Sloat of the Pacific squadron, who had raced up the coast from Baja California upon receiving news that Mexico and the United States were indeed at war, took Monterey. Sloat proclaimed to the native Californian population that he had come in peace and ordered his men to act with strict discipline. However, being in ill health, he soon turned his command over to Commodore Robert F. Stockton, who viewed the American occupation in a quite different light. Stockton accepted Frémont's forces into United States military service and marched south to pacify the rest of California. Meeting unexpected resistance in Los Angeles and elsewhere, the American forces did not fully conquer southern California until the beginning of 1847.

Pending the establishment of a civil government, a military commander was put in charge of the newly acquired territory, but his authority was nominal at best. At the local level, the Mexican system of alcaldes continued to provide whatever governance that existed. Americans began to gain control of the existing local government when they were first appointed and then elected to the post of alcalde in Monterey and San Francisco. Then in 1848, as a result of Mexico's defeat in the Mexican-American War, California—along with what is now most of the southwestern United States—was ceded by Mexico to the United States under the terms of the Treaty of Guadalupe-Hidalgo. Gold was discovered on the banks of the American River just as the war was coming to an end; this fact gave urgency to the question of whether California should become a United States territory or enter the union directly as a state. The sentiment was overwhelmingly in favor of immediate statehood, but heated debate over whether California should enter the union as a free or a slave state delayed a decision. Too impatient to wait for congressional action, Americans in California, joined by a number of native Californian leaders, convened a constitutional convention, wrote a constitution, and declared California a state before Congress did so.

In 1848 some 6,000 aspiring miners appeared, but the following year an estimated 100,000 Americans and foreign-born immigrants from Europe, Latin America, Asia, and Australia converged on California. Since California land had not yet been surveyed and no legislation regarding its possession or use was in effect, the gold-rushers surging into California in 1849 took matters into their own hands: miners formed mining districts, each of which established its own regulations. When disputes erupted, brute force was often used to settle them.

Native Americans and Californios suffered the greatest losses from this demographic and political invasion. Large numbers of Native Americans perished from the loss of their land and consequently of their traditional means of subsistence, from diseases introduced by whites, and from massacres. The Californios also suffered from a loss of land and, more important, from a loss of political power and social status.[5] Although they were granted American citizenship by the terms of the Treaty of Guadalupe-Hidalgo, the Californios became, at best, second-class citizens, the majority eventually being reduced in status to laborers. By the 1870s both groups had become insignificant in the eyes of Americans except as a source of cheap labor.

The manner in which California was populated led to the establishment of extraordinarily complex racial and ethnic relations in the thirty-first state. The Californios, though themselves often the offspring of interracial marriages, retained many of the arrogant attitudes that Spanish colonizers of the New World had held toward Native Americans. Old-stock white Americans, especially the large numbers who originated from the South, brought with them a legacy of black slavery and elaborate beliefs to justify the existence of this institution that contradicted the noble ideals of the American creed. Protestants held strong prejudices against Catholics, particularly the Irish. European immigrants, many of whose native countries were imperialist powers, also came with negative perceptions of, and attitudes toward, peoples of color.

The Chinese who came, for their part, were by no means free of ethnocentrism and racial prejudice. For centuries, Han Chinese had discriminated against non-Han peoples within China's borders and had considered themselves superior to all foreign barbarians. They had repeatedly dealt with foreigners—even invaders such as the Mongols and Manchus—by assimilating them. Though the Cantonese were probably more cosmopolitan than other Chinese, having traded with Arabs, Indians, Southeast Asians, Japanese, and Europeans for centuries, still, they were quite aware of the fact that they were children of the Middle Kingdom—a land frequently encroached upon, but never, in their minds, to be vanquished.

Immigration

Though some Chinese had reached Mexico in the seventeenth century, and individual sailors had visited ports on the east coast of the United States in the late eighteenth, the Chinese who arrived in California during the first years of the gold rush were the true vanguard of more than 200,000 of their countrymen who came to the "Old Gold Mountain" in the nineteenth century. Some merchants had arrived by 1850, but a real influx began when more than 20,000 Chinese entered the port of San Francisco in 1852. The following year, however, though 4,270 came, 4,421 departed from California, because of the reimposition of a Foreign Miners' Tax in 1852 and the rush of aspiring gold miners from China to Australia where gold had been discovered in 1851. News of the finds in New South Wales and Victoria reached China in 1852; Chinese began going to the "New Gold Mountain" that year, building up such a steady movement that by mid-1855 there were some 17,000 Chinese in Australia. This development reduced the number destined for the "Old Gold Mountain," though California's drawing power was still strong: in 1854 over 16,000 entered, but the flow decreased again the year after that—this time because the more accessible surface placer claims had already been exhausted and the imposition of a capitation tax of fifty dollars on each Chinese passenger who landed in California discouraged ship owners and captains from bringing Chinese to San Francisco. From 1855 onward, 2,000 to 8,000 Chinese came each year until 1868, when over 11,000 entered. In 1869 almost 15,000 arrived, and in 1870 another 11,000 came.

The factors that promoted Chinese trans-Pacific migration changed over time. In the 1850s and early 1860s, while gold pulled Chinese to California, the political, social, and economic turmoil caused by the two Opium Wars, the T'aip'ing Rebellion, the Red Turbans uprisings, and the Punti-Hakka feuds pushed Cantonese out of their homeland. By the late 1860s, however, a new set of forces emerged to boost Chinese emigration to the United States: the initiation of regular steamship service by the Pacific Mail Steamship Company from Hong Kong and Shanghai to San Francisco in 1867; the active recruitment of laborers by the Central Pacific Railroad Company, which was building

the western half of the first transcontinental railroad in the
United States; the signing of the Burlingame-Seward Treaty in
1868, formalizing the Chinese central government's recognition
of the legality of Chinese emigration to the United States; and
the growth of the California economy. The role of the Pacific Mail
Steamship Company has already been described briefly, so the
only point that needs to be repeated here is that without direct
steamship service, it is unlikely that as many Chinese could
have traveled to California as did so.

The Central Pacific Railroad began to employ Chinese on an
experimental basis in 1865; within a year, Chinese constituted
four-fifths of its entire work force. At one point there were 10,000
to 11,000 Chinese workers on the line. A firm, Sisson and Wallace, and a Dutch broker, Cornelius Koopmanschap, played key
roles in importing laborers directly from China when numbers
already in California proved insufficient.[6] The Central Pacific
discovered, as did other employers later, that hiring large
numbers of Chinese laborers to work in gangs under Chinese
foremen—called "China bosses"—was extremely convenient, for
the labor contractors and foremen assumed all supervisory and
managerial responsibilities over the men, including provisioning
them. The Central Pacific Railroad's large-scale employment of
Chinese workers crystallized an image that Americans had
held—as a result both of the coolie trade to Latin America and of
conflicts between white and Chinese miners in the 1850s—that
Chinese were exploitable and exploited cheap labor. The use of
so many Chinese as gang labor, along with the decline of placer
mining, quickly changed the social status of many Chinese in
America from that of independent gold miners to that of
laborers.

Though it cannot be ascertained how many Chinese were
actually encouraged to immigrate because of the Burlingame-
Seward Treaty, it is historically important because through it the
Chinese central government finally sanctioned emigration to the
United States—which had in fact been going on for two decades.
The treaty contained eight articles, which affirmed China's con-
tinued right of eminent domain over the territorial concessions
she had granted to foreign powers; reiterated the right of the
Chinese government to regulate commerce; gave China the right
to appoint consuls "who shall enjoy the same privileges and im-

munities as those ... enjoyed by ... the Consuls of Great Britain
and Russia" and who were to be stationed in American ports;
protected citizens of either nation from persecution on the basis
of religion; recognized the "inalienable right of man to change
his home and allegiance" and outlawed the coolie trade; gave
each contracting party most-favored-nation status, but dis-
allowed Chinese in the United States the right of naturalization;
opened the public educational institutions of each country to the
citizens of the other; and disclaimed any desire by the United
States to interfere in the internal administration of China.[7] Article V, which recognized the "inalienable right of man to change
his home and allegiance," ensured a steady supply of Chinese
laborers for those California entrepreneurs who desired to use
them. Within less than a dozen years of the treaty's ratification,
however, the United States government, giving in to the de-
mands of the anti-Chinese movement, felt compelled to negotiate
a new treaty to repudiate it.

Regardless of the conditions in China that forced people to
emigrate, or the stimulus that direct steamship service, recruitment for railroad labor, and the Burlingame-Seward Treaty
provided to lure Chinese to the United States, in the final
analysis, economic conditions and the anti-Chinese movement
in California were the most important determinants of the number of Chinese in the state, especially after the completion of
the transcontinental railroad in 1869. Before that date, the
California economy had been relatively autonomous because
bulky merchandise could only be exported to, or imported from,
the Atlantic Coast by ship around Cape Horn or to and from
Asia across the Pacific. Beginning in the early 1870s, and more
so after the 1880s when railroad freight rates were reduced, the
state's economy became increasingly linked to the national one,
so that the latter's recurrent business cycles began to affect it.
Economic depressions fueled anti-Chinese hostilities, which
tended to reduce the number of Chinese in the state.[8]

Anti-Chinese hostility took many forms. It first manifested
itself in the mining regions when Chinese miners were harassed
by tax collectors, bandits, and white miners who drove them
away from good claims. Many Chinese miners were robbed and
others were murdered. A different kind of anti-Chinese activity
emerged in San Francisco where, as the Chinese population

burgeoned in the late 1860s, anti-coolie clubs were formed in increasing numbers and anti-Chinese mass meetings were held. Soon, politicians began to take up the "Chinese question" in their preelection oratory, and the issue of Chinese immigration was joined to contending positions over Reconstruction, with the Fourteenth and Fifteenth Amendments cited in support of civil rights for the Chinese or condemned for granting those rights to nonwhites. Labor unions, such as the shoemakers' Knights of Saint Crispin, and the plumbers and carpenters' Eight Hour League, sponsored mass rallies in 1870 during which the Burlingame-Seward Treaty was castigated. A statewide anti-Chinese convention held in August also called for the abrogation of the treaty. In that same year, the California School Law made children of African and Indian descent attend segregated schools, and a later court decision as well as statutory enactment also relegated Chinese children to an Oriental school. Several municipal ordinances passed in San Francisco in the 1870s harassed the Chinese for living under crowded and unhealthy conditions. Furthermore, discriminatory license fees affected the manner in which they conducted their businesses. The mayor vetoed some of these measures, others were later declared unconstitutional, but a few took effect. Violence broke out in 1871 when buildings in the Chinatown of Los Angeles were burned, property was looted, and people were killed. Sporadic instances of anti-Chinese violence occurred throughout the 1870s.

Anti-Chinese sentiment spread throughout the nation when an investigation conducted in 1876 by a specially appointed committee of the California legislature, and the hearings held in California the following year by a joint special committee of the United States Congress, called public attention to the presence of the Chinese in the United States. The report of the 1876 state committee, which focused on the moral effects of continued Chinese presence, was used effectively as an anti-Chinese weapon. Copies of the report and a memorial to Congress based on it were sent to all the leading newspapers in the country, five copies were sent to each member of Congress, ten copies to each state governor, and ten thousand copies were distributed to the general public. Anti-Chinese forces, led by Denis Kearney and his Workingmen's Party of California, achieved considerable success in California's 1878–79 Constitutional Convention when anti-Chinese clauses were placed in the state's second constitution. These included a prohibition against the employment of Chinese by corporations, and municipal, county, and state government. Other clauses, such as the legalization of residential and business segregation, the prohibition of land purchase by foreigners ineligible for naturalization, and denial of the franchise, were proposed but not adopted.

The effect of anti-Chinese activities depended on economic conditions. Poor conditions enhanced, whereas prosperity dulled, the effectiveness of such activities, which in turn influenced the volume of immigration. From 1869 onward, when thousands of discharged Chinese railroad workers returned from Utah to Nevada and California and flooded the labor market, the temporary high unemployment rate among the Chinese already here and a flurry of anti-Chinese activities reduced Chinese immigration in 1871–72. However, after the discharged railroad builders found work in other areas, Chinese immigration resumed, and there was a five-year spurt from 1873 to 1877, when an average of 18,000 entered each year. This was a period when land reclamation and large construction projects requiring gangs of laborers created a demand for workers. Also, grain cultivation was giving way to more labor-intensive horticulture. Thus, in the mid-1870s economic development exerted a stronger influence than the anti-Chinese movement on the volume of immigration. A few years later, however, with the onset of national depression and a more organized marshaling of the anti-Chinese forces, immigration began to decrease again to about 7,000 arrivals a year—a number more or less equal to departures. The trend was reversed with a large final influx when knowledge of the impending passage of a Chinese Exclusion Act brought over 50,000 Chinese into the United States in 1881 and 1882.

Chinese immigration after 1882 did not cease, but it declined drastically, being restricted to persons in four "exempt classes": merchants, students, diplomats, and temporary travelers. An unknown number of illegal entrants also slipped in. After 1890 the Chinese population in the United States grew smaller with each census until the slow birth of a second generation caused the numbers to climb again in the fourth decade of the twentieth century.

Important as it was for the western states, in the national context Chinese immigration represented only a small fraction of the total immigration into the United States during the latter half of the nineteenth century. To place the numerical significance of Chinese immigration in perspective, in the 1861–70 decade only 2.7 percent of the total number of immigrants into the United States were Chinese. Only 4.4 percent were Chinese in the 1871–80 decade, 1.2 percent in the 1881–90 period, and a miniscule 0.4 percent in the last decade of the nineteenth century.[9] Yet, because their presence was so controversial, the Chinese who came to the United States have borne a heavy burden of many wrongs against them.

Settlement

The Chinese presence in America was concentrated almost entirely on the Pacific Coast, especially in California, during the first three decades of Chinese immigration. Only when they discovered that persecution against them did not cease even after the passage of the Chinese Exclusion Law in 1882, did an increasing number of those who had chosen to remain the United States begin to disperse to the rest of the country. Many settled in the larger metropolitan areas of the East Coast. To place the significance of the Chinese involvement in California agriculture in perspective, it is necessary first to describe the geographic distribution of the Chinese, and then to look at how their occupational distribution pattern changed over time.[10]

There were already Chinese in California when the 1850 federal census was taken, and there were even more Chinese by 1852 when a special state census was taken. These early counts were not accurate, however, given the extraordinary mobility of the gold rush population and the fact that a part of the returns for the 1850 census was lost. The 1860 census, which was more complete and reliable, counted 34,933 Chinese in California. (Chinese in other states were not listed discretely in the published census reports.) As table 1 shows, at that time 84 percent of the Chinese in California were in the three mining regions: the Southern Mines (El Dorado, Amador, Calaveras, Tuolumne,

Table 1 Distribution of the Chinese in California by Geographic Region and Year, 1860–1900

	1860		1870		1880		1890		1900	
	Number	Percent	Number	Percent	Number	Percent	Number	Percent	Number	Percent
Mining Regions	29,335	84.0	22,285	45.2	23,664	31.5	9,464	13.1	5,609	12.3
Central Valley	2,519	7.2	8,473	17.2	13,372	17.8	16,296	22.5	12,124	26.5
San Francisco	2,719	7.8	12,022	24.4	21,745	28.9	25,833	35.6	13,954	30.5
Other Bay Area	295	0.8	5,240	10.6	11,445	15.2	9,882	13.6	6,511	14.2
Other California	65	0.2	1,257	2.6	4,906	6.5	10,996	15.2	7,555	16.5
Total Calif.	34,933	100.0	49,277	100.0	75,132	100.0	72,472	100.0	45,753	100.0
Total U.S.A.	34,933[a]		63,199		105,465		107,488		89,863	
Percent in California		100.0[a]		78.0		71.2		67.4		50.9

SOURCE: Computed from figures for individual counties given in U.S. Bureau of the Census, *Census of U.S. Population, 1870*, p. 15; *Census of U.S. Population, 1900*, p. 565.

[a] There were some Chinese outside of California, but they were not listed as "Chinese" under the category "race" except in California. Without going through the manuscript schedules of the census for the states outside of California, it is not possible to state exactly how many Chinese there were in addition to the 34,933 given in the published census.

and Mariposa counties), the Northern Mines (Plumas, Butte, Sierra, Yuba, Nevada, and Placer counties), and the Klamath/ Trinity Mines (Del Norte, Klamath, Siskiyou, Trinity, and Shasta counties). As might be expected, the great majority of them mined for a living.

The 2,719 Chinese found in San Francisco represented less than 8 percent of the Chinese population in the United States in 1860. The third area where Chinese could be found was in the Central Valley—the northern half of which is known as the Sacramento Valley, the southern half as the San Joaquin Valley. The former contained 1,866 Chinese, 1,731 of whom were in Sacramento County, while the latter had only 653. The bulk of the Chinese population in the valley counties lived in towns located at the confluence of major rivers that drain through the Central Valley, where they plied various trades; but even here—particularly in Sacramento, Stanislaus, and Fresno counties, where mining was possible in the foothills—a considerable number were miners. With the exception of some truck gardeners, Chinese had not yet begun to farm in the Central Valley. There were virtually no Chinese anywhere else: the rest of the counties in California taken together had fewer than 30 Chinese in 1860.

The first Chinese to move out of California were those who went to seek their fortunes in newly discovered gold deposits in the Pacific Northwest, both north and south of the forty-ninth parallel. As Rodman Paul has shown, the mining frontier in western America moved eastward—in the opposite direction from the main stream of western settlement.[11] Chinese were part of this eastward countermovement. By 1870 Oregon had 3,330 Chinese, Idaho 4,274, Montana 1,949, and Nevada 3,152. There were few Chinese in any other part of the United States in 1870: 99.4 percent of all the Chinese in the country were in the "western division," the census designation for the Pacific Coast and Rocky Mountain states.

In California itself, the Chinese population in 1870 was concentrated in the same three areas as a decade earlier, but the relative distribution was now more even. Over 22,000 of the 49,277 of the Chinese population in the state (45 percent) still resided in the three mining regions, though the number of Chinese miners had fallen from over 25,000 in 1860 to approximately 16,000 in 1870. San Francisco had over 12,000 Chinese, and the other Bay Area counties over 5,000—almost a sixfold increase for the area since 1860. The population in the Sacramento and San Joaquin valleys had more than tripled from about 2,500 in 1860 to almost 7,500 in 1870. Chinese here now worked as farm laborers, farmers, truck gardeners, fishermen, common laborers, merchants, professionals, laundrymen, cooks, servants, and prostitutes.

By 1880, 3.2 percent of the Chinese in the United States had moved to areas outside of the western division. The largest gain was in the North Atlantic division, where some 900 Chinese now resided in New York. Massachusetts and Illinois each had a little over 200, Boston and Chicago having established nascent Chinatowns. Over a hundred Chinese were found in Texas. These men were railroad builders who had worked on the Houston and Texas Central and the Texas and Pacific railroads in the 1870s and who, after their contracts expired, settled in the Lone Star state.

Approximately three-quarters of the Chinese in the western division—75,132 of them—were still in California. The mining regions, San Francisco and the Bay Area, and the Sacramento and San Joaquin valleys continued to hold the bulk of them, but a southward movement of the Chinese population had become apparent, with 753, 702, and 1,169 Chinese counted in Fresno, Kern, and Los Angeles counties, respectively. The percentage in the mining regions had declined to 31.5 percent of the state's total Chinese population, though the absolute number of Chinese there actually had increased by about 1,400 over the 1870 figure. The growing availability of nonmining occupations made such a rise possible.

San Francisco and the other bay counties made the most spectacular gain in Chinese population—together they now contained over 34,000, constituting 44 percent of the Chinese population in California, of whom almost 22,000 were in the city. The continued growth of manufacturing and the increasing use of Chinese servants, cooks, laundrymen, and laborers by white employers provided the Bay Area Chinese with means of livelihood. The most notable change was an increase in the number as well as the variety of Chinese factory workers in San Francisco.

Like San Francisco, valley counties made substantial gains,

their Chinese population increasing from about 8,500 in 1870 to over 13,000 in 1880, almost 18 percent of the state's Chinese. Agricultural development and the need for laborers to clear land, plant new fields and orchards, harvest and pack crops, build roads, bridges, and levees, and dig ditches and drains drew the Chinese there.

Most scholars believe that the Chinese population in the United States reached a peak in the early 1880s, although it will never be known what the actual number was. Immigration records indicate that in 1881 and 1882, 57,271 entered and 26,788 departed, leaving a net increase of 30,483. If that figure is added to the 1880 census count of 105,465, then the probable number of Chinese in the United States in 1883 was about 136,000. This was probably the nineteenth-century peak.

The 1880s witnessed a gradual movement of Chinese away from the western United States, so that by 1890, 9.9 percent of the 107,488 Chinese in the country lived outside of the western division. Three-quarters of the 96,844 in the western division— that is, slightly over two-thirds of the Chinese population in the United States—were still California residents. Major gains were made by the North Atlantic division, which now had 5.7 percent of the Chinese population in the country, with almost 3,000 in New York, over 1,000 in Pennsylvania, and almost 1,000 in Massachusetts. In the rest of the country, 1.3 percent were in the south central states, 2.2 percent in the north central states, and 0.6 percent in the South Atlantic states.

The same decade witnessed a real exodus of Chinese from the three mining regions in California, where their numbers fell from almost 24,000 in 1880 to only about 9,500 in 1890. The population in the Sacramento Valley remained about the same at over 8,600, while that in the San Joaquin Valley increased by almost 3,000 to 7,657. San Francisco picked up over 4,000, so that its 25,833 Chinese formed almost 36 percent of the total Chinese population in the state, but it also became a community under siege, serving as a sanctuary for Chinese escaping persecution elsewhere. The movement to the southern part of the state became even more noticeable with the Chinese population in Los Angeles quadrupling from 1,169 in 1880 to 4,424 in 1890, and that in Fresno increasing three and a half times from 753 to 2,736. The rapid development of irrigated agriculture and a

lesser degree of anti-Chinese hostility in the southern part of
the state drew Chinese there in those years. The other Bay Area
counties suffered a decline from about 11,500 in 1880 to less than
10,000 in 1890.

The last decade of the nineteenth century was one of
demographic, social, and economic depression for the Chinese
in America. Of the 89,863 Chinese counted in the United States
(excluding Alaska and Hawaii) in the 1900 census, only 67,729
(75.4 percent) were still in the western division. Of these, 67.6
percent were in California, which meant that only half of the
Chinese in the United States at the turn of the century were
still in that state. Further gains had been made by the North
Atlantic states, whose Chinese population had more than doubled
from 6,177 in 1890 to 14,693 in 1900. The midwestern and southern states also showed modest increases.

Within California, San Francisco had lost almost half of its
Chinese population in the 1890s, and every other area also
suffered sizable lossess. Between 1890 and 1900, the Chinese
population increased in only two counties—San Joaquin and
Contra Costa, portions of which are located in the San Joaquin
Delta, an area where Chinese tenant farming thrived in the
1890s. Chinese grew huge crops of potatoes, beans, onions, and
fruit there.

The relative demographic importance of the Chinese in
nineteenth-century California may be seen in table 2, which
shows the percentage of Chinese among the total population in
selected counties from 1860 to 1900. The Chinese presence was
most significant, as might be expected, in the mining counties,
where from 1860 through 1880 they ranged from one-tenth to
almost a quarter of the total population. Although there were
many Chinese in the San Francisco Bay Area, they represented a
far smaller percentage there, the peak being reached in the 1880s
when 7 to 9 percent of the population in some of the bay counties
were Chinese. Their presence was less significant in the Sacramento and San Joaquin valleys where, with the exception of
Sacramento County, Chinese seldom exceeded one-tenth of the
total population. By the turn of the century, even in those counties with the largest number of Chinese, they averaged only
about 5 percent of the total population.

The geographic movement of the Chinese in California is

Table 2   The Chinese Population as a Percentage of the Total Population by County, Selected Counties, 1860–1900

	1860			1870		
	Chinese Population	Total Population	Percent Chinese	Chinese Population	Total Population	Percent Chinese
MINING COUNTIES						
Butte	2,177	12,106	18.0	2,082	11,403	18.3
Sierra	2,208	11,387	19.4	810	5,619	14.4
Yuba	1,781	13,668	13.0	2,337	10,851	21.5
Nevada	2,147	16,446	13.1	2,627	19,134	13.7
Placer	2,392	13,270	18.0	2,410	11,357	21.2
El Dorado	4,762	20,562	23.2	1,560	10,309	15.1
Calaveras	3,657	16,299	22.4	1,441	8,895	16.2
Tuolumne	1,962	16,229	12.1	1,524	8,150	18.7
SACRAMENTO VALLEY						
Shasta	415	4,360	9.5	574	4,173	13.8
Tehama	104	4,044	2.6	294	3,587	8.2
Sutter	2	3,390	0.1	208	5,030	4.1
Sacramento	1,731	24,142	7.2	3,195	26,830	11.9
Yolo	6	4,716	0.1	395	9,899	4.0
Solano	14	7,169	0.2	920	16,878	5.5
SAN FRANCISCO BAY AREA						
Alameda	193	8,927	2.2	1,939	24,237	8.0
San Francisco	2,719	56,802	4.8	12,022	149,473	8.0
Santa Clara	22	11,912	0.2	1,525	26,246	5.8
SAN JOAQUIN VALLEY						
San Joaquin	139	9,435	1.5	1,626	21,050	7.7
Fresno	309	4,605	6.7	427	6,336	6.7
Tulare	13	4,638	0.3	99	4,533	2.2
Kern	–	–	–	143	2,925	4.9
COASTAL COUNTIES						
Monterey	6	4,739	0.1	230	9,876	2.3
Los Angeles	11	11,333	0.1	234	15,309	1.5
TOTAL CALIFORNIA	34,933	379,994	9.2	49,310	560,247	8.8

SOURCE: Computed from U.S. Bureau of the Census, *Census of Population, 1880*, pp. 51 and 382; and *Census of Population, 1900*, pp. 11 and 75–80.

NOTE: The figures for the state's total are *not* the same as the aggregate of the counties listed in this table, because many counties are not included here.

Dashes in this and the following tables indicate that no information was given.

	1880			1900	
Chinese Population	Total Population	Percent Chinese	Chinese Population	Total Population	Percent Chinese
3,793	18,721	20.3	712	17,117	4.2
1,252	6,623	18.9	309	4,017	7.7
2,146	11,284	19.0	719	8,620	8.3
3,003	20,823	14.4	632	17,789	3.6
2,190	14,232	15.4	1,050	15,786	6.7
1,484	10,683	13.9	206	8,986	2.3
1,037	9,094	11.4	148	11,200	1.3
805	7,848	10.3	158	11,166	1.4
1,334	9,492	14.1	102	17,318	0.6
774	9,301	8.3	729	10,996	6.6
266	5,159	5.2	274	5,886	4.7
4,892	34,390	14.2	3,254	45,915	7.1
608	11,772	5.2	346	13,618	2.5
993	18,475	5.4	903	24,143	3.7
4,386	62,976	7.0	2,211	130,197	1.7
21,745	233,959	9.3	13,954	342,782	4.1
2,695	35,039	7.7	1,738	60,216	2.9
1,997	24,349	8.2	1,875	35,452	5.3
753	9,478	7.9	1,775	37,862	4.7
324	11,281	2.9	370	18,375	2.0
702	5,601	12.5	906	16,480	5.5
372	11,302	3.3	857	19,380	4.4
1,169	33,381	3.5	3,209	170,298	1.9
75,218	864,694	8.7	45,753	1,485,053	3.1

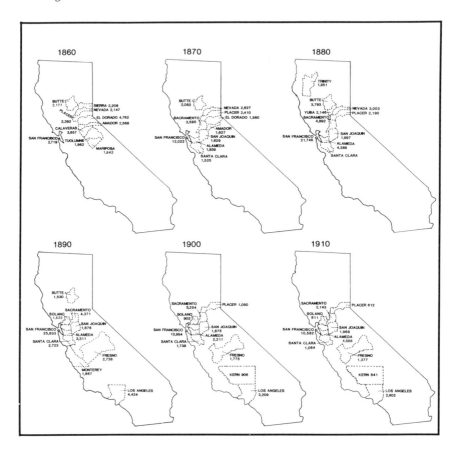

Map 3 California: The Ten Counties with the Largest Chinese Popula-
tion, 1860–1910
SOURCE: Based on U.S. Bureau of the Census, Census of the
U.S. Population, 1860, 1870, 1880, 1890, 1900, 1910.

depicted in map 3, which shows the ten counties with the largest
number of Chinese from 1860 through 1910. The six inset maps
show in graphic form how the Chinese presence shifted from the
mining regions to the Sacramento Valley and the San Francisco
Bay Area and thence southward to the San Joaquin Valley and
Los Angeles County in this fifty-year period. These movements
coincided with changes in the California economy, which was
based first on mining and then on agriculture and manufac-

turing. That the geographic shifts of the Chinese population followed so closely the developments in California's economy underscores the fact that economic factors exerted the most important influence on Chinese settlement patterns in the state.

Livelihood

Since the bulk of the Chinese population lived in California in the nineteenth century, an understanding of how economic changes affected Chinese livelihood can best be gained through a detailed analysis of the Chinese occupational distribution in the major geographic-economic regions of the golden state. Raw data on this topic amenable to systematic analysis are available in only one source: the manuscript schedules of the United States censuses of population. Tallying the information from these schedules, which have been preserved in hundreds of reels of microfilm, is a most time-consuming and labor-intensive task, which no scholar has yet fully undertaken. So far, I have counted only the Chinese in the major regions of California, and the results, as summarized in tables 3, 4, 5, and 6, will be discussed below.[12]

As for other areas of the state and the rest of the country, only sketchy information in the published census for certain years (but not for others) is available, so an accurate picture of the statistical distribution of Chinese occupations in those other places has not yet been put together.

There are many problems associated with using both the published and the manuscript censuses. Without doubt, the census undercounted the number of Chinese, some of whom probably hid from census takers, having experienced persecution or extortion from collectors of the Foreign Miners' Tax or immigration and customs officials. Inaccuracies also crept in because census takers could not speak Chinese and communication probably depended on the few words of pidgin English that some of the Chinese commanded. The published censuses for 1880, 1890, 1900, and 1910 are not very useful for a study of Chinese occupations because they provided no discrete occupational information for the Chinese, who were lumped together with "Civilized

Indians," Japanese, and Blacks as "Coloreds." Thus statistics
must be compiled from scratch from the manuscript schedules,
which are not easy to use, because the handwriting of the census
enumerators is often difficult to read, and the ink on many pages
is so faint that it is virtually impossible to decipher the words.
There are also many torn or smeared pages, as well as incomplete data for entire townships and counties. Finally, in comparing data compiled from the manuscript schedules with the
figures in the published censuses, I discovered many arithmetical errors in the latter, especially for the earlier census years
when adding machines were not yet available. For all the above
reasons, my tallies often do not agree with the published figures.
Despite these difficulties, the manuscript census is an indispensable source which no serious student of Chinese American
economic history can ignore.

The development of the Chinese occupational pattern in
nineteenth-century California may be divided into four stages:
an initial period from 1850 to 1865, when the Chinese worked
mainly as miners and traders; a period of growth and development from 1865 to the late 1870s, when they branched into agriculture, light manufacturing, and common labor; a period of
consolidation from the late 1870s to the late 1880s, when they
competed successfully with others in a wide variety of occupations; and a period of decline from the late 1880s to the turn of
the century, when they were forced to abandon many occupations. Data tallied from the manuscript schedules of the 1860
census of population provide a glimpse into the first period, data
from the 1870 census show the pattern in the years of development, data from the 1880 census give a picture of the period of
consolidation, whereas the years of decline are difficult to document because the manuscript schedules of the 1890 census were
lost in a fire, so that there is a twenty-year gap between the 1880
and the 1900 data.

The occupational pattern of the Chinese differed considerably
among the mining regions, the city of San Francisco, and the
agricultural counties, reflecting differences in the structure of the
three sectors of the California economy—mining, manufacturing, and agriculture—and how Chinese fitted into them. Overall, in the 1850–80 period, mining occupied more Chinese than
any other economic activity. Even more important, the presence

Chinese miner panning for gold, Mongolian Flat on the American River, California, ca. 1851. (Photograph by Eadweard Muybridge, courtesy of the California Historical Society, San Francisco)

Table 3   The Occupational Distribution of the Chinese in California by
Economic Sector and by Region, 1860

	Southern Mines[a]	
	Number	**Percent**
PRIMARY PRODUCERS AND EXTRACTORS	13,062	88.7
Agriculturalists	161	1.1
Agric. laborers	3	0
Fishermen	0	0
Miners	12,898	87.6
PROFESSIONALS AND SKILLED ARTISANS	174	1.2
ENTREPRENEURS AND THEIR ASSISTANTS	614	4.2
Merchants	447	3.0
Clerks and shop assistants	53	0.4
Owners of recreational vices	28	0.2
Laundrymen/women	86	0.6
Factory owners	0	0
Factory workers	0	0
PROVIDERS OF PERSONAL SERVICES	640	4.3
Cooks	153	1.0
Servants	29	0.2
Prostitutes (listed)	139	0.9
Prostitutes (probable)	291	2.0
Others	28	0.2
NONAGRICULTURAL LABORERS	112	0.8
MISCELLANEOUS AND NO OCCUPATION	126	0.9
TOTAL	14,728	100.0
(Published Total)	(14,792)	

SOURCE: My tally from U.S. National Archives, Record Group 29, "Census of U.S. Population" (manuscript), 1860.
[a] Includes El Dorado, Amador, Calaveras, Tuolumne, and Mariposa counties.
[b] Includes Plumas, Butte, Sierra, Yuba, Nevada, and Placer counties.
[c] Includes Del Norte, Siskiyou, Shasta, Trinity, and Klamath counties.

Northern Mines[b]		Klamath/Trinity Mines[c]		San Francisco	
Number	Percent	Number	Percent	Number	Percent
9,195	83.8	3,027	88.1	470	17.3
126	1.1	12	0.3	31	1.1
6	0.1	0	0	13	0.5
2	0	0	0	418	15.4
9,061	82.6	3,015	87.7	8	0.3
96	0.9	27	0.8	197	7.2
554	5.0	76	2.2	575	21.1
270	2.5	49	1.4	147	5.4
50	0.5	7	0.2	10	0.4
20	0.2	1	0	274	10.1
214	2.0	19	0.6	144	5.3
0	0	0	0	0	0
0	0	0	0	0	0
639	5.8	107	3.1	723	26.6
261	2.4	48	1.4	67	2.5
73	0.7	12	0.3	11	0.4
58	0.5	40	1.2	0	0
199	1.8	5	0.1	636	23.4
48	0.4	2	0.1	9	0.3
418	3.8	43	1.3	650	23.9
71	0.6	157	4.6	104	3.8
10,973	100.0	3,437	100.0	2,719	99.9
(11,104)		(3,439)		(2,719)	

or absence of mining in an area strongly affected what other occupations the Chinese entered because there was an inverse relationship between the percentage of the Chinese population who mined and those who earned a living as laborers or providers of personal services. Where mining was available, few became laborers or providers of personal services, but as mining waned from the 1860s onward, an increasing number of Chinese became laundrymen, laborers, servants, and cooks. Where mining was not available, the first Chinese to enter those areas had to take menial jobs from the beginning. The proletarianization of the Chinese population was partially arrested in the early 1870s when many Chinese became independent entrepreneurs in light manufacturing and agriculture. While the Chinese engaged in light manufacturing (primarily in San Francisco) for only about a quarter of a century, they farmed all over California for six or seven decades. Therefore, agriculture succeeded mining in providing an economic foundation for Chinese communities in rural California. However, agriculture never dominated the livelihood of rural Chinese in quite the same way that mining had done.

As is shown in table 3, a large proportion of the Chinese in the three mining regions in 1860 were miners: seven-eighths of those in the Southern Mines, more than four-fifths of those in the Northern Mines, and seven-eighths of the ones in the Klamath/ Trinity mining region. In California as a whole, miners constituted over 70 percent of all gainfully employed Chinese above age fifteen; this concentration was more than double the 32 percent found among the non-Chinese population in 1860. Chinese miners were strung out along rivers and streams, since most of them engaged only in placer mining. Few Chinese worked in hardrock areas.

All other occupations were subsidiary to and were sustained by the income from mining. In the major mining camps, a small number of Chinese merchants, grocers, truck gardeners, cooks, servants, laundrymen, barbers, herbalists, prostitutes, professional gamblers, and even a fortune-teller or two served the subsistence and recreational needs of their fellow countrymen. In each of the three mining regions, professionals made up about 1 percent of the Chinese population, while merchants ranged from 1.5 percent of the Chinese population in the mining camps of the remote Klamath/Trinity mountains to 3.0 percent in the

Chinese miners using pick and rocker, California, ca. 1850s. (Photograph by Eadweard Muybridge, courtesy of the California Historical Society, San Francisco)

prosperous Southern Mines. In these mining camps and mountain towns, Chinatown took up a few blocks or part of a block in
the central district; here, Chinese grocery and general merchandise stores and recreational facilities were clustered. The owners
of these enterprises usually lived in the back or the upstairs of
their establishments.

The Chinese miners in 1860 were men who had survived a
decade of sporadic violence against them. In the 1850s, Chinese
had been frequently robbed, beaten, driven away from their
diggings, or even killed.[13] As early as 1849, before the arrival
of large numbers of Chinese miners, white miners in Tuolumne
County were already paranoid enough to pass a resolution to
prohibit Chinese from working claims.[14] Chinese had little
recourse to justice because an 1850 law that barred Indian
and Negro testimony in court was extended to apply to them in
1854.[15] Once in a while, however, some white men out of a
sense of fair play defended the Chinese. For example, the *Shasta
Republican* reported in late 1856 that Francis Blair was the first
white man to be hanged for murdering Chinese.[16] Also, though
Chinese miners were expelled from many localities, there were
some spots where they managed to work in relative peace for
many years consecutively.[17] Chinese miners had also persisted
despite a discriminatory Foreign Miners' Tax, first set at twenty
dollars a month but later reduced to either three or four dollars.
Originally intended to drive Mexican and Chileno miners away,
it soon came to be collected only from Chinese miners. The tax
was reduced because too large an amount drove even the
Chinese away, and their departure deprived merchants of trade,
and county governments of their chief source of revenue. Ping
Chiu has calculated that the receipts from this tax provided over
half of the total revenue for the mining counties in the 1850s.[18]

From the beginning of Chinese immigration, the Chinese in
San Francisco developed an occupational pattern that was quite
distinct from that found in the mining regions.[19] Though many
of the Chinese counted in the San Francisco census were transients, there was a stable core of people in the city who handled
the immigration traffic, San Francisco being the chief port of
entry for Chinese immigrants to America. Merchants had
already established themselves as representatives of the Chinese
community by 1852 when large numbers of aspiring miners

came. They had built or rented buildings to house the new arrivals, and soon made money by provisioning them and finding them employment. Even more lucrative were the establishments that served the miners' recreational needs. By 1860 over 10 percent of San Francisco's gainfully employed Chinese males worked in one of several branches of the vice industry: gambling, the sale of opium, and prostitution. In addition, 23.4 percent of the city's Chinese were prostitutes. (Only nonagricultural laborers rivaled the prostitutes in number.) Thus, fully one-third of the gainfully employed Chinese in the city were engaged in providing recreational vices—an unfortunate fact which gave rise to strong negative images of the Chinese. Another notable occupation that San Francisco Chinese engaged in was fishing— an industry that the Chinese dominated in those days—which provided a living for 15 percent of the Chinese population in the county.[20] Merchants not engaged in the vice industry and laundrymen each made up about 5 percent of the city's Chinese population. Most of the merchants and professionals served their fellow countrymen, whereas servants, cooks, laundrymen, laborers, and some prostitutes served whites.

In the late 1860s and early 1870s, the employment of Chinese workers by the Central Pacific Railroad, the decline of placer mining, and the development of manufacturing and agriculture helped to change the occupational distribution of the Chinese in California. The employment of approximately 10,000 Chinese workers by the Central Pacific Railroad confirmed the belief of many Americans that Chinese could be employed in gangs to carry out large construction projects. After the transcontinental railroad was completed, several hundred Chinese were kept on by the Central Pacific for maintaining the right of way and to build branch lines. Thousands more, however, were discharged, and many sought work as common laborers in agriculture and various construction projects, while others became cooks, servants, and laundrymen. This development served to harden the lines of division in California's racial hierarchy, which eventually relegated all Chinese—regardless of their standing within the Chinese immigrant community—to a position well below the lowest class of whites.

Data compiled from the manuscript schedules of the 1870 census reveal the changes taking place in this period of change

Chinese fishermen, Monterey County, California, 1875. (Photograph courtesy of the California State Library, Sacramento)

Table 4   The Occupational Distribution of the Chinese in California by Economic Sector and by Region, 1870

	Southern Mines[a]		Northern Mines[b]	
	Number	Percent	Number	Percent
PRIMARY PRODUCERS AND EXTRACTORS	5,593	77.1	6,076	54.5
Agriculturalists	56	0.8	253	2.3
Agric. laborers	66	0.9	240	2.2
Fishermen	0	0	€	0
Miners	5,471	75.4	5,577	50.0
PROFESSIONALS AND SKILLED ARTISANS	43	0.6	287	2.6
ENTREPRENEURS AND THEIR ASSISTANTS	414	5.7	972	8.7
Merchants	149	2.1	374	3.4
Clerks and shop assistants	36	0.5	41	0.4
Owners of recreational vices	132	1.8	171	1.5
Laundrymen/women	18	0.2	292	2.6
Factory owners	0	0	3	0
Factory workers	79	1.1	91	0.8
PROVIDERS OF PERSONAL SERVICES	364	5.0	1,366	12.3
Cooks	162	2.2	873	7.8
Servants	23	0.3	101	0.9
Prostitutes (listed)	148	2.0	306	2.7
Prostitutes (probable)	16	0.2	11	0.1
Others	15	0.2	75	6.7
NONAGRICULTURAL LABORERS	474	6.5	2,205	19.8
MISCELLANEOUS AND NO OCCUPATION	370	5.1	239	2.1
TOTAL	7,258	100.0	11,145	100.0
(Published Total)	(7,236)		(11,177)	

SOURCE: My tally from U.S. National Archives, Record Group 29, "Census of U.S. Population" (manuscript), 1870.
a Includes El Dorado, Amador, Calaveras, Tuolumne, and Mariposa counties.
b Includes Plumas, Butte, Sierra, Yuba, Nevada, and Placer counties.
c Includes Del Norte, Siskiyou, Shasta, Trinity, and Klamath counties.
d The figures in this column come from U.S. Bureau of the Census, *Census of U.S. Population, 1870*, p. 799, Table XXXII, and not from my tally of the manuscript schedules. (The given figures are for the city only; figures have been rearranged from the published table.)
e Includes Sonoma, Napa, Marin, Contra Costa, Alameda, Santa Clara, and San Mateo counties; *excludes* San Francisco.
f Includes Tehama, Colusa, Sutter, Sacramento, Yolo, and Solano counties.
g Estimate
h Not given in published table.

Klamath/Trinity[c]		San Francisco[d]		Other Bay Area[e]		Sacramento Valley[f]	
Number	Percent	Number	Percent	Number	Percent	Number	Percent
3,420	88.5	547	4.9	1,502	29.5	2,015	36.4
34	0.9	6	0	417	8.2	150	2.7
21	0.5	53	0.5	934	18.4	971	17.6
0	0	145	1.3	135	2.7	7	0.1
3,365	87.1	343	3.1	16	0.3	887	16.0
32	0.8	565	5.1	81	1.6	125	2.3
117	3.0	4,394	39.5	708	13.9	679	12.3
45	1.2	518	4.7	84	1.6	101	1.8
13	0.3	97	0.9	0	0	25	0.5
30	0.8	—[h]	–	1	0	50	0.9
28	0.7	1,333	12.0	471	9.3	450	8.1
0	0	—[h]	–	2	0	35	0.6
1	0	2,446	22.0	150	3.0	18	0.3
187	4.8	2,414	21.7	883	17.4	924	16.7
114	3.0	—[h]	–	415	8.2	395	7.1
20	0.5	1,256	11.3	305	6.0	252	4.6
44	1.1	—[h]	–	91	1.8	60	1.1
2	0.1	1,000[g]	9.0[g]	15	0.3	175	3.2
7	0.2	158	1.4	57	1.1	42	0.8
35	0.9	2,210	19.9	1,645	32.4	1,665	30.1
73	1.9	1,000[g]	9.0[g]	264	5.2	122	2.2
3,864	99.9	11,130	100.0	5,083	100.0	5,530	100.0
(3,872)		(12,022)		(5,240)		(5,683)	

and development.[21] As table 4 reveals, overall, mining still dominated Chinese livelihood. In the Southern, Northern, and Klamath/Trinity mines, 75.4, 50, and 87.1 percent of the Chinese, respectively, still mined for a living. The 16,000 or so remaining Chinese miners constituted about 35 percent of the gainfully employed Chinese in the state, whereas only about 11 percent of non-Chinese still mined for a living. More interesting than the Chinese persistence in mining, however, was the fact that developments in the Southern Mines differed from those in the Northern Mines, while the situation in the remote Klamath/Trinity mountains remained more or less the same. The Southern Mines had lost more than half of its Chinese population in the 1860–70 decade as placer mining in that region gave out and other avenues of employment did not open up. In contrast, in the Northern Mines where the Chinese population remained about the same as ten years earlier, half remained as miners while the other half managed to find other means of livelihood, primarily as nonagricultural laborers, whose ranks had swelled from 418 in 1860 to 2,205 in 1870. Providers of various personal services more than doubled from 639 in 1860 to 1,368 in 1870. Such work was available because the western portions of four of the counties in the Northern Mines lay in the Sacramento Valley—an area that provided employment opportunities other than mining.

Though Chinese managed to make a living, from both their own point of view as well as that of white Americans, working as common laborers, servants, and cooks represented a considerable descent in social status from that of independent miners. While they cannot be called an industrial proletariat, these laborers and providers of personal services did constitute an emergent Chinese working class in rural California. Meanwhile, a small number of Chinese had begun to work in factories, woolen mills, and sawmills in the more prosperous mining counties—Placer, El Dorado, and Mariposa. Such individuals were the forerunners of what might have become a true Chinese industrialized proletariat in California. (No such development ever materialized, because of Chinese exclusion.) Finally, as both the absolute number and the percentage of merchants in all three mining regions increased, they began to consolidate their position as the socioeconomic elite in Chinese communities in rural California, just as they had in San Francisco.

Considerable changes had also occurred in San Francisco, which was blossoming into a manufacturing, trade, and service center for Chinese as well as white Californians.[22] Fully 22 percent of the some 12,000 Chinese in the city in 1870 worked in factories, making cigars, shoes and boots, woolen textiles, and clothing. The Civil War had boosted manufacturing in California because it had disrupted the importation of goods from the East Coast. After the war ended, importation resumed, and the lower-priced eastern goods threatened the survival of California's nascent industries. Only those establishments that had low enough production costs to compete against eastern imports managed to survive, and Chinese sweatshops were among them.[23]

As was true of white-owned manufacturing enterprises during this transitional period of industrial development, Chinese production took two forms: artisans, working alone or with a small number of partners and assistants, made most of the manufactured items, but in selected industries, factory production employing wage workers had also begun. For example, shoemakers produced shoes, slippers, and boots in tiny workshops, but boot and shoe factories had also come into being. Similarly, cigar makers worked either in small storefronts or in larger establishments. Tailors and seamstresses had their own shops, but there was also an increasing number of garment workers making underwear, shirts, pants, and overalls in Chinese-owned sewing factories. Thus, from census data alone, it is difficult to determine how many of these individuals can be regarded as wage workers and how many as independent producers.

In addition to these factory workers and independent producers, there were over 1,000 Chinese professionals, artisans, and merchants in the city, along with over 1,300 laundrymen and laundresses. The growth of manufacturing, trade, and services thus greatly enlarged the economic base of San Francisco's Chinatown, diversifying it, and moving it well beyond its former role as a provider of imported goods and recreational vices.

The other Bay Area counties, though close to San Francisco, did not share in its industrial development. The small valleys they contained were becoming productive agricultural areas, which offered work in farming and fishing to about 30 percent of the Chinese there. Santa Clara County had over 250 Chinese strawberry-growers and over 100 truck gardeners, and Alameda County had over 100 truck gardeners. Chinese had also begun to work as seasonal farm laborers in these counties: Marin, Solano, and Santa Clara counties each had over 100 farm laborers, San Mateo County had almost 200, and Alameda County had over 300. The bulk of the other Chinese found work as laundrymen (9.3 percent), cooks (8.2 percent), servants (6 percent), and common laborers (32.4 percent). Factory workers included 49 who made bags in Alameda County, 64 who made shoes in San Mateo County, and 24 who made woolen textiles in Santa Clara County. The 165 professionals, artisans, and merchants together made up only a little over 3 percent of the Chinese population, and there were virtually no prostitutes, gamblers, or opium dealers—perhaps a reflection of the fact that Chinese in these areas preferred to go to the much better stocked Chinatown in San Francisco for their shopping and recreational needs.

The occupational distribution pattern in the Sacramento Valley—another prime agricultural region—was very similar to that in the Bay Area counties, except for the 887 miners found in the Sierra Nevada foothills. Farmers, truck gardeners, and poultry raisers made up 2.7 percent of the Chinese population, another 17.6 percent worked as farm laborers, and 0.1 percent fished, bringing the total in agriculture and fishing to 20.4 percent. Just over 30 percent earned a living as nonagricultural laborers, while laundrymen, cooks, and servants made up 8.1, 7.1, and 4.6 percent of the Chinese population, respectively. The number of merchants, professionals, and artisans was similarly small, with the cities of Sacramento and Marysville containing the largest clusters. The 100-plus truck gardeners were distributed among all the counties of the valley, but the 32 farmers were concentrated in the Sacramento Delta, where Chinese had helped to reclaim the land and had begun to lease it for farming. Though Chinese had engaged in truck gardening almost from the beginning of their immigration into California, their entry into larger-scale farming was a phenomenon associated with these transitional years.

The period from the mid-1870s to the end of the following decade saw a consolidation of the patterns that had begun to emerge in the preceding period.[24] In 1880, even though mining itself was on the decline except for quartz mining in Nevada and

Chinese engaged in placer mining, California. (Photograph by Charles Bierstadt, courtesy of the California Historical Society, San Francisco)

Table 5   The Occupational Distribution of the Chinese in California by Economic Sector and by Region, 1880

	Southern Mines[a]		Northern Mines[b]		Klamath/Trinity[c]	
	Number	Percent	Number	Percent	Number	Percent
PRIMARY PRODUCERS AND EXTRACTORS	3,795	74.4	7,954	61.0	3,799	72.6
Agriculturalists	87	1.7	436	3.3	77	1.5
Agric. laborers	82	1.6	784	6.0	22	0.4
Fishermen	1	0	1	0	0	0
Miners	3,625	71.1	6,733	51.6	3,700	70.7
PROFESSIONALS AND SKILLED ARTISANS	63	1.2	215	1.6	48	0.9
ENTREPRENEURS AND THEIR ASSISTANTS	305	6.0	1,020	7.8	190	3.6
Merchants	151	3.0	337	2.6	91	1.7
Clerks and shop assistants	38	0.7	76	0.6	14	0.3
Owners of recreational vices	66	1.3	178	1.4	51	1.0
Laundrymen/women	49	1.0	367	2.8	28	0.5
Factory owners	1	0	4	0	1	0
Factory workers	0	0	58	0.4	5	0.1
PROVIDERS OF PERSONAL SERVICES	328	6.4	1,298	10.0	313	6.0
Cooks	157	3.1	1,014	7.8	182	3.5
Servants	26	0.5	48	0.4	25	0.5
Prostitutes (listed)	84	1.6	108	0.8	38	0.7
Prostitutes (probable)	22	0.4	29	0.2	38	0.7
Others	39	0.8	99	0.8	30	0.6
NONAGRICULTURAL LABORERS	322	6.3	2,136	16.4	790	15.1
MISCELLANEOUS AND NO OCCUPATION	285	5.6	420	3.2	93	1.8
TOTAL	5,098	99.9	13,043	100.0	5,233	100.0
(Published Total)	(5,120)		(13,255)		(5,289)	

SOURCE: My tally from U.S. National Archives, Record Group 29, "Census of U.S. Population" (manuscript), 1880.
[a] Includes El Dorado, Amador, Calaveras, Tuolumne, and Mariposa counties.
[b] Includes Plumas, Butte, Sierra, Yuba, Nevada, and Placer counties.
[c] Includes Del Norte, Siskiyou, Shasta, and Trinity counties. (Klamath County was dissolved in 1874 and its territory was divided up between Siskiyou and Humboldt counties.)
[d] Includes Sonoma, Napa, Marin, Contra Costa, Alameda, Santa Clara, and San Mateo counties; excludes San Francisco.
[e] Includes Tehama, Colusa, Sutter, Sacramento, Yolo, and Solano counties.
[f] Includes San Joaquin, Stanislaus, Merced, Fresno, Tulare, and Kern counties.

San Francisco		Other Bay Area[d]		Sacramento Valley[e]		San Joaquin Valley[f]	
Number	Percent	Number	Percent	Number	Percent	Number	Percent
660	3.1	3,009	26.6	3,035	35.1	1,417	29.3
46	0.2	286	2.5	956	11.1	332	6.9
187	0.9	1,975	17.4	982	11.4	741	15.3
371	1.7	670	5.9	15	0.2	95	2.0
56	0.3	78	0.7	1,082	12.5	249	5.2
2,186	10.1	188	1.7	182	2.1	116	2.4
10,091	46.8	1,505	13.3	1,241	14.4	649	13.4
2,230	10.3	306	2.7	276	3.2	139	2.9
331	1.5	7	0.1	3	0	46	1.0
95	0.4	12	0.1	16	0.2	30	0.6
2,148	10.0	1,036	9.1	722	8.4	335	6.9
1,023	4.7	4	0	35	0.4	1	0
4,264	19.8	140	1.2	189	2.2	98	2.0
4,986	23.1	2,192	19.4	1,490	17.2	1,076	22.3
857	4.0	958	8.5	1,019	11.8	721	14.9
2,443	11.3	1,104	9.7	376	4.4	191	4.0
432	2.0	16	0.1	10	0.1	78	1.6
600	2.8	27	0.2	12	0.1	43	0.9
654	3.0	87	0.8	73	0.8	43	0.9
2,336	10.8	4,147	36.6	2,280	26.4	1,179	24.4
1,289	6.0	285	2.5	412	4.8	396	8.2
21,548	99.9	11,326	100.1	8,640	100.0	4,833	100.0
(21,745)		(11,445)		(8,503)		(4,869)	

Amador counties, there were still about 15,000 Chinese miners, one-fifth of the Chinese population in the state. In comparison, only about 7 percent of the gainfully employed non-Chinese still mined for a living. As is shown in table 5, though the Chinese population in the Southern Mines had decreased by some 2,000, there were still over 3,500 Chinese miners there, forming 71 percent of the Chinese population. In the Northern Mines, where the Chinese population had increased by over 2,000 since 1870, over 6,700 of the 13,000-plus Chinese, or 51.6 percent, were miners. The number of Chinese miners in this region, therefore, had increased by almost 1,200 over that of ten years earlier, reflecting the general prosperity brought about by the boom in hardrock mining in Nevada County in the late 1870s. In the Klamath/Trinity region, 3,700, or 70.7 percent, of the over 5,000 Chinese dug gold for a living. Over a thousand additional Chinese miners were found outside of the major mining regions, mostly in Sacramento County. So far as the Chinese were concerned, the gold rush was not yet over.

The differences noted in the occupational distribution of Chinese in the three mining regions in 1870 were even more prominent in 1880. The Northern Mines, which had a more diversified economic base, were able to continue to sustain a sizable Chinese population because nonmining work was available. Almost 1,300 provided personal services, while some 2,100 earned a living as common laborers. These individuals were no doubt disappointed miners forced to accept menial jobs for survival. Similar work was apparently not available in the Southern Mines or the Klamath/Trinity region. The larger population in the Northern Mines also enabled a larger absolute number and percentage of Chinese to flourish as merchants, professionals, and artisans.

In San Francisco the most notable change was an increase in the number as well as the variety of Chinese factory workers who were now employed in broom, candle, cigar, coffee, collar, cotton, fuse, gum, match, pen, sack, sewing, shoe, soap, and sugar factories; flour, jute, paper, saw, and woolen mills; electrical, powder (ammunition), and salt works, as well as tanneries and packing houses. The inconsistent manner in which occupations were listed in the manuscript census makes it difficult to state exactly how many Chinese were factory workers and how many were independent producers. While all those listed as working in a factory were counted as factory workers, there were also others, listed as "makers" of cigars, shoes, slippers, boots, candles, or matches, who might have worked in factories, but who could also have been independent artisans. Moreover, it is not certain whether persons listed as "manufacturers" were owners of factories or workers in them. Only those specifically listed as "proprietors" of factories can be considered owners, but these were few in number. If all persons listed as working in factories, all manufacturers, and all makers of goods amenable to factory production are included, then over 6,000— some 30 percent of San Francisco's gainfully employed Chinese, or 28 percent of its total Chinese population—were engaged in manufacturing.

The 2,230 merchants and their 331 employees made up 11.8 percent of the Chinese in the city, while an additional 95 individuals ran brothels, gambling halls, and opium joints. The 2,148 laundrymen constituted 10 percent of the Chinese population. San Francisco's Chinatown was not only a thriving center of Chinese-owned and operated factories and sweatshops but also the home of many independent, skilled craftsmen, 1,614 of whom engaged in a wide array of crafts, including shoemaking, tailoring, carpentry, and the making of beds, bricks, brooms, cages, candles, chairs, coffins, fishnets, harnesses, ivory carvings, lace, lanterns, locks, roofs, ropes, silverware, tents, umbrellas, watches, and whiplashes. There were also Chinese printers, machinists, painters, brick and stone masons, as well as bakers, candy makers, bean-cake makers, and butchers. No doubt a great deal of the Chinese-manufactured goods was sold to the general public, but many trades in this city obviously specialized in the production of items intended for the Chinese ethnic market. Professionals in the city included actors and musicians, herbalists and physicians, linguists and interpreters, teachers, priests and missionaries, fortune-tellers, and even artists, photographers, letter writers, lawyers, newspapermen, nurses, and secretaries. The status of San Francisco as the metropolis of Chinese America is best gleaned from the fact that professionals, artisans, entrepreneurs of various sorts, and their assistants made up 56.9 percent of the city's total Chinese population.

Providers of personal services, including barbers, cooks, dishwashers, drivers, errand boys, janitors, porters, prostitutes, servants, waiters, and night watchmen, made up 23.1 percent, while nonagricultural laborers formed another 10.8 percent of the city's Chinese population. There were dozens of sailors, steamer stewards, mess boys, coal heavers, and firemen among the nonagricultural laborers, San Francisco being a port for both ocean liners and river steamers. A small number of Chinese also worked as ragpickers and scavengers.

Finally, the city and county had 371 Chinese fishermen, 37 truck gardeners, 2 farmers, 7 poultry raisers, 66 woodchoppers, 56 miners, and 121 farm laborers—the last two groups obviously transients visiting the city.

The other Bay Area counties continued to grow as agricultural areas, providing employment for few Chinese farmers but for more than twice as many farm laborers as a decade earlier. In the Santa Clara and Sonoma valleys, four-fifths of the Chinese earned a living in agriculture while in the Vaca Valley of Solano County, about three-fifths worked in various agricultural pursuits. Individuals providing personal services and nonagricultural laborers each increased two and a half times in number, the former growing from 883 in 1870 to 2,192 in 1880, the latter jumping from 1,645 to 4,147 in the same decade. The Bay Area outside of San Francisco thus had the largest proportion of working-class Chinese of any of the regions under consideration.

Between 1870 and 1880, the number of Chinese earning a living in agriculture had increased in the Central Valley. The almost 1,000 Chinese farm laborers and woodchoppers in the Sacramento Valley made up 11.4 percent of the Chinese population, the bulk of whom, over 500, were in Sacramento County. Unlike the Sacramento Valley, where rainfall and water from streams provided sufficient water to grow a wide variety of crops without irrigation, the more arid San Joaquin Valley was still devoted to ranching and wheat growing, the sole exceptions being that portion of San Joaquin County which forms part of the San Joaquin Delta, and an area in Tulare County around the fans of the Kings and Kaweah rivers. Though ranching and wheat growing were less labor-intensive, nonetheless, 741 Chinese farm laborers were found in the San Joaquin Valley in 1880, forming 15.3 percent of the Chinese population, the largest number being in San Joaquin County.

The inverse relationship between mining and menial labor can most clearly be seen by comparing the occupational distribution of Chinese in the mining regions with those in the agricultural ones. In the mining regions, the percentage of nonagricultural laborers in the total Chinese population varied from 2.8 percent in Shasta County, where 82 percent of the Chinese were miners, to 27 percent in Butte County, where only 46 percent of the Chinese were miners. Nonagricultural laborers were more numerous in the Sacramento and San Joaquin valleys, which had mining only in the eastern flanks of a few of their counties: here, laborers constituted 26.4 and 24.4 percent of the Chinese population, respectively. Not only were there more of them, but the work Chinese laborers did was also more diverse, with hundreds of brick makers, soap-root diggers, woodchoppers, lumber sawyers and pilers, as well as many engaged in other forms of semiskilled labor.

Similarly, the percentage of Chinese who worked as cooks in the various regions of rural California also varied inversely with the percentage who were miners. Only about 3 percent of the Chinese in the Southern and Klamath/Trininty mines, which still had a high percentage of miners in 1880, were cooks, while almost 8 percent of the Chinese in the Northern Mines were. In the Sacramento and San Joaquin valleys, 11.8 and 14.9 percent of the Chinese, respectively, earned a living as cooks. Among these, at least one-third were farm cooks who resided in the households of farmers, and many more may have been, but there is no way to ascertain the exact number because a large proportion of the cooks listed in the manuscript schedules of the census lived in their own households. The percentage of Chinese servants in different regions also varied inversely with the number of miners. These statistics indicate clearly that as Chinese left mining, the more enterprising ones became manufacturers or merchants in the urban areas and farmers or truck gardeners in the rural and suburban areas, while the less fortunate everywhere became laborers, cooks, and servants.

The 1880–90 decade brought many important changes for Chinese communities in the United States because the 1882 Chinese Exclusion Law not only cut off the major source of addition to the Chinese population but made it quite clear to the Chinese who chose to remain that they were a despised and unwanted minority. Unfortunately, since the 1890 published census

Table 6   The Occupational Distribution of the Chinese in California by
Economic Sector and by Region, 1900

	Northern Mines[a]		San Francisco	
	Number	Percent	Number	Percent
PRIMARY PRODUCERS AND EXTRACTORS	1,698	47.9	286	2.0
Agriculturalists	203	5.7	61	0.4
Agric. laborers	489	13.8	165	1.2
Fishermen	0	0	38	0.3
Miners	1,006	28.4	22	0.2
PROFESSIONALS AND SKILLED ARTISANS	56	1.6	1,669	11.7
ENTREPRENEURS AND THEIR ASSISTANTS	541	15.3	5,455	38.1
Merchants	344	9.7	1,440	10.1
Clerks and shop assistants	36	1.0	259	1.8
Owners of recreational vices	6	0.2	54	0.4
Laundrymen/women	155	4.4	1,924	13.4
Factory owners	0	0	84	0.6
Factory workers	0	0	1,694	11.8
PROVIDERS OF PERSONAL SERVICES	479	13.5	2,603	18.2
Cooks	357	10.1	1,197	8.4
Servants	18	0.5	659	4.6
Prostitutes (listed)	5	0.1	269	1.9
Prostitutes (probable)	76	2.1	0	0
Others	23	0.6	478	3.3
NONAGRICULTURAL LABORERS	669	18.9	3,534	24.7
MISCELLANEOUS AND NO OCCUPATION	104	2.9	771	5.4
TOTAL	3,547	100.1	14,318	100.1
(Published Total)	(3,614)		(13,954)	

SOURCE: My tally from U.S. National Archives, Record Group 29, "Census of U.S. Population" (manuscript), 1900.
[a] See note [b] in table 5.
[b] See note [d] in table 5.
[c] See note [e] in table 5.
[d] See note [f] in table 5.

Other Bay Area[b]		Sacramento Valley[c]		San Joaquin Valley[d]	
Number	Percent	Number	Percent	Number	Percent
1,704	31.2	2,353	43.5	2,295	40.6
302	5.5	502	9.3	338	6.0
1,135	20.8	1,815	33.5	1,942	34.3
158	2.9	1	0	0	0
109	2.0	35	0.6	15	0.3
38	0.7	170	3.1	133	2.4
1,471	27.0	806	14.9	765	13.5
268	4.9	316	5.8	436	7.7
8	0.1	51	0.9	32	0.6
1	0	1	0	0	0
821	15.1	346	6.4	294	5.2
1	0	0	0	1	0
372	6.8	92	1.7	2	0
1,173	21.5	1,015	18.7	1,222	21.6
815	14.9	666	12.3	905	16.0
263	4.8	174	3.2	43	0.8
0	0	14	0.3	5	0.1
38	0.7	97	1.8	180	3.2
57	1.0	64	1.2	89	1.6
754	13.8	933	17.2	1,001	17.7
314	5.8	138	2.5	238	4.2
5,454	100.0	5,415	99.9	5,654	100.2
(6,511)		(5,959)		(6,165)	

provided no information on the occupations of the Chinese population, and the manuscript schedules are no longer available, almost nothing can be said about the immediate effects of these legal restrictions on the kind of work the Chinese did.

It has been noted that an increasing number of Chinese moved to other sections of the country after the passage of the Chinese Exclusion Act in 1882. What needs to be added is that the ratio of urban to rural Chinese in California as well as the nation increased substantially during the last two decades of the nineteenth century. This urbanizing population, however, had fewer and fewer means of livelihood open to its members because Chinese were being driven out of manufacturing from the mid-1880s onward. Boycotts of Chinese-made goods by white consumers reduced the number of Chinese factory owners and workers in San Francisco from 1,023 and 4,264, respectively, in 1880 to only 84 and 1,694, respectively, in 1900. By 1920 there were virtually no Chinese cigar or shoe and boot makers left. At the same time, it must be remembered that manufacturing in the United States had long left its artisanal stage behind, so that only those with large amounts of capital could afford to engage in manufacturing industries. Thus, even had there been no discrimination, it is not certain how many Chinese could have remained in manufacturing by the early twentieth century.

In 1900 there were barely any Chinese left in the Southern and Klamath/Trinity mining regions of California, but of the 3,547 counted in the Northern Mines, over a thousand were still doggedly mining.[25] As table 6 shows, the other Chinese in the latter region were more or less evenly distributed as agriculturalists and agricultural laborers, entrepreneurs, providers of personal services, and nonagricultural laborers. In San Francisco, although the percentage in business and manufacturing was still high (38.1 percent), the most notable change between 1880 and 1900 was the increase in the percentage engaged in common labor. Agriculture absorbed an ever larger percentage of the Chinese population in the Bay Area counties outside of San Francisco and in the Sacramento and San Joaquin valleys. In particular, not only the percentage but the absolute number in farming increased in the San Joaquin Valley.

Chinese farmers, farmworkers, fruit packers, and commission merchants were most concentrated in the Sacramento–San Joaquin Delta, where some 95 percent of the Chinese population earned a living in agriculture at the turn of the century. The percentage of those who survived through farming and farm work was almost as high in other major agricultural areas, such as the Sonoma and Vaca valleys and the foothills of Placer County. Elsewhere in California, from one-quarter to over one-half of the Chinese depended on agriculture for a living. Rather than having been driven out of rural California, as some writers have alleged, Chinese agriculturalists in fact flourished during these decades when the Chinese population as a whole, paradoxically, was experiencing a general demographic and economic decline.

Sizable numbers continued in farming until the end of World War I. Except for the Sacramento–San Joaquin Delta, where a settled Chinese community persisted, the Chinese exodus from agriculture finally took place in the late 1910s and 1920s. They left because they were growing old, their children preferred other work, they could not compete against the more aggressive Japanese immigrants who now greatly outnumbered them, and a great drop in agricultural commodity prices in the mid-1920s caused a farm recession, which made it hard for farmers to make ends meet.

Throughout the nation wherever Chinese had settled, more and more retreated to urban centers to earn a living in trade and common labor during the early decades of the twentieth century. By 1920 fully 48 percent of all the Chinese in the United States were in small businesses, while 27 percent provided personal services. Only 11 percent were in agricultural occupations, 9 percent in factory work and the skilled crafts, 2 percent in transportation, 1 percent in the professions, and 2 percent in white-collar work. There were only 151 Chinese miners left in the United States, but it is not known where they were located.[26]

The extraordinary increase in the number of merchants and small shopkeepers resulted partly from the implementation of the various Chinese exclusion laws. After the initial 1882 law came into effect, additional laws were enacted in the following two decades which not only made it impossible for all but a handful of Chinese to enter legally, but caused those who were already here great hardship. Merchants, belonging to one of the "exempted" classes, became the main source of new blood in Chinese American communities. Since few Chinese women were in the United States at the time exclusion was imposed, and even fewer came in the ensuing six decades, the Chinese population was unable to replenish itself easily through natural increase. The only addition to the community, therefore, came from immigration. Although many of the immigrants probably were not merchants originally, they very likely bought shares in existing businesses in order to qualify for entry as merchants. Those who were not fortunate enough to become businessmen eked out a living as cooks, servants, and laborers. Seven decades of rapid economic change, anti-Chinese discrimination, and finally Chinese exclusion—rather than any inherent racial or cultural characteristics—had made the Chinese in the United States into an urban mercantile and servile population by the early twentieth century.

THREE

Feeding The Miners

THOUGH the Chinese who came to California were peasants, they had not come to farm: they took up cultivation only because they needed to feed themselves. Some eventually became full-time farmers because they discovered—like some of the white immigrants who came in the same period—that farming could often provide a more steady income than mining for gold. In the period 1860–1920 many Chinese earned a living as truck gardeners, tenant farmers, owner-operators, commission merchants, fruit packers, harvest laborers, and farm cooks. Finding a ready market for their produce, much of which they distributed and sold themselves, they soon learned to treat agricultural production and marketing not simply as a source of subsistence but as a commerical enterprise. Thus, the Chinese may have arrived as peasants, but they quickly became capitalist farmers and agricultural wage laborers.

The Chinese truck gardeners and farmers in the New World faced many obstacles that had to be surmounted. Though they brought with them copious agricultural knowledge accumulated by their ancestors through centuries of experience, the agricultural techniques they knew had to be adapted to California's semiarid conditions. They had to learn to grow crops in a land of summer drought and moderate winter rains—the average annual rainfall in the Sierra Nevada foothills and the Sacramento Valley being less than one-third of what they were used to at home. California's virgin soil, though rich, was different from the worn-out soil of China, which had been cultivated lovingly, fastidiously, for centuries. At the same time, they had to learn to lease or buy land and to deal with landlords who spoke an alien tongue. Those who grew crops for sale had to discover which crops white Americans would buy. Most important of all, growing crops for sale meant that the Chinese immigrant peasants had either to create, or to learn to participate in, marketing networks based on business practices with which they were unfamiliar.

Chinese immigrant farmers in California did not follow the upward mobility path implied in the concept of the "agricultural ladder."[1] The term refers to the occupational sequence followed by most landless persons aspiring to be farm owners who must work first as farmhands, then as tenants, and finally as owner-operators. Chinese agriculturalists in California followed a reverse sequence: only some two decades after they began to grow crops as owner-operators and tenants did many of their members become agricultural laborers. The reverse mobility path followed by Chinese agriculturalists in California was one manifestation of the proletarianization of the Chinese working population in the United States in the nineteenth century, but at the same time, it reflected some of the peculiarities of California agriculture—which has been dominated by large landholdings used to produce specialty crops for a world market, although smaller family farms have also existed. Because of these characteristics, tenancy has had a different cast in California than elsewhere in the United States.

Generally speaking, three patterns of tenancy have prevailed in American agriculture. In the eastern and midwestern United States where family farms predominated, those farm boys who did not leave their parents' farms for the cities aspired to become owner-operators of their own farms. Here, according to the agrarian myth, tenancy was but a temporary step up the agricultural ladder.[2] In the plantation South, however, tenancy—both black and white—was a more permanent condition. Slaves who became sharecroppers faced almost insurmountable odds: they were held to the land by their former masters through various devices, including debt bondage and vagrancy laws, and found few alternative opportunities to earn a living.[3] White tenants in the South were not much better off, struggling against some of the same economic odds, though they were not encumbered by the color of their skin.

In California, tenancy has been associated with yet another set of social conditions. Here, where the scale of operations more than the farmer's tenure status affects farm income, tenancy is often an economic choice. As R. L. Adams and W. H. Smith have stated, there are four groups of California farmers who prefer to rent rather than to own land for cultivation. First are newcomers to California, unfamiliar with its soil and climatic conditions, who wish to test the suitability of certain areas for particular crops before purchasing any land. Second are those who use tenancy as a means to increase their farm size and consequently their farm income; not possessing sufficient capital to buy as much land as they desire, they resort to leasing part of the land they cultivate. Third are farmers who specialize in particular crops. Since certain crops draw specific nutrients from the soil, good agronomy requires the farmer to practice crop rotation if he wishes to cultivate the same land year after year without letting it lie fallow. If a grower prefers to plant the same crops all the time, to prevent soil depletion he must then move from farm to farm. Thus farm rotation rather than crop rotation is practiced; leasing makes farm rotation possible. Fourth are large grower-shippers who use tenancy to maximize the flexibility of their overall financial operations.[4] Tenancy, therefore, is not necessarily an undesirable status—it could even be a mark of success. Chinese became tenant farmers in California for all of the above reasons.

Provisioning the Miners

To understand why the earliest Chinese farmers in California were truck gardeners who grew vegetables and small fruit, it is necessary to consider the Chinese concept of food, and the logistical problems of sending provisions to the mining camps during the gold rush. Chinese divide food into two categories: *fan* (rice or some other kind of grain) and *choy* [*ts'ai*], which literally means leafy vegetables but is the generic term for anything eaten along with the staples to provide taste. Rice is the preferred staple, but in north China, wheat, millet, and *kaoliang* are also eaten. Poor peasants who cannot afford rice or other grain eat root crops. Fresh or dried vegetables are the most commonly eaten *choy*. In Kwangtung province, in particular, the people are used to eating a wide variety of vegetables, including several kinds of cabbage, mustard greens, melons, squash, string beans, green onions, and different kinds of tubers. With the exception of fowl and fish, which may be served whole, meat (also considered *choy*) is generally not served in large chunks, but is used in combination with vegetables in a great variety of dishes.

Given the importance of vegetables in the diet of the Chinese—especially those from southeastern China—it is not surprising that almost everywhere Chinese have gone, they have grown vegetables both for family consumption and for sale. In various Southeast Asian countries, in Australia and Canada, as well as in the United States, Chinese immigrants have become truck gardeners in considerable numbers. Growing vegetables requires only small plots of ground and few tools; moreover, vegetables mature in only a few months, so even persons who stay in a locality for only a short time may engage in truck gardening, a relatively easy occupation to enter for those without much capital.

Chinese miners in California had to procure both their traditional staples and *choy* by importation and by local production. In the nineteenth century, rice and sweet potatoes were the main staples in the diet of the Chinese immigrants, but it is not known whether they attempted to grow rice in California. If they tried, they did not succeed, because rice was not grown on a commercial scale in California until the second decade of the twentieth century. Rice purchased by Chinese on the Pacific Coast was imported primarily from China and secondarily from Hawaii. In the latter locality, most of the rice farmers were Chinese immigrants. In 1874, of the 34,586,287 pounds of rice imported into California, 31,838,116 came from China. In 1875, 46,270,365 out of 54,231,961 pounds imported came from China, and in the following year, 35,885,761 out of 42,203,501 did.[5] Sweet potatoes, the secondary staple, were locally grown.[6]

Rice was a highly valued item in the Chinese mining camps. A bill of sale drawn up in the mid-1860s in Camanche, Calaveras County, provides a glimpse of the value of rice relative to the value of different merchandise in those days. In 1865 Chew Lung sold a house and lot located on Main Street in Camanche to

Table 7 Merchandise Stocked in a Chinese Store, Camanche, Calaveras County, 1865

House and Furnishings		Provisions and Medicine	
House and lot	$40.00	30 sacks of rice	180.00
1 counter	2.00	38 gallons of gin	57.00
2 tables	2.00	1 basket of tea	13.25
6 chairs	2.00	25 lb. of China sugar	2.50
		25 lb. of China beans	2.50
Horse and Feed		20 lb. of China peas	1.00
		35 lb. of China peas	1.50
1 horse	30.00	50 lb. of salt	3.50
3 bales of hay	4.50	25 lb. of ginger	3.00
		15 gallons of oil	22.50
Utensils and Dry Goods		65 lb. of salt fish	5.00
1 pair of China scales	1.00	4 sacks of flour	6.00
4 one-gallon kegs	2.50	China medicine	1.75
3 two-gallon kegs	2.00	20 bottles of	
2 China pans	2.00	China peppermint	1.00
2 dippers	.75	2,000 China cigars	2.00
1 teapot, 1 oilpot			
and 2 lamps	1.00	(Items with Chinese names only):	
1 dozen China bottles	3.00	10 lb. China cam chum	2.00
30 bowls	1.00	(dried stalks of	
2 China bowls	2.00	plant used in cooking)	
5 China bowls	1.00	1 jar cam chung	2.00
1 pair of China shoes	2.00	(same as preceding item?)	
3 pairs of China shoes	3.00	1 lb. hung sen	.50
2 boxes of candles	8.00	(medicinal herb?)	
20 dozen matches	9.00	6 moy chung	1.50
20 (piles?) of China tobacco		(medicinal herb?)	
paper	8.00	20 pieces of lin hung	2.00
800 pieces of China paper	3.00	(medicinal herb?)	
China paper	1.00	20 lb. pue tung	2.00
letter paper	2.50	(medicinal herb?)	
5,000 firecrackers	2.50		
(Illegible items):			
32 lb. _____	7.25		
6 Chorok _____			
(Cherook brand cigars?)	1.75		
1 piece China _____	.87		

SOURCE: Calaveras County, "Bill of Sales," Book C, p. 129.
NOTE: During the nineteenth century, the word "China" was used as an adjective in many English-language writings about the Chinese in America. The items from the bill of sale have been rearranged for this table.

Chung Hop for $468.87. The building housed a store with all kinds of imported Chinese goods, which were included in the sale. Table 7 reproduces this bill of sale, which shows that rice, tea, gin, and oil were among the most expensive items. Rice cost $6.00 a sack, gin cost $1.50 a gallon, as did oil, and one basket of tea leaves cost $13.25. Flour was only $1.50 per sack, or one-quarter as much as a sack of rice. Of the other food items, salted fish was quite cheap—sixty-five pounds cost only $5.00. This fish might have been imported from China or might have been dried and salted by Chinese fishermen around San Francisco Bay or in the Sacramento–San Joaquin Delta. On the other hand, the Chinese beans and peas, as well as important cooking ingredients such as ginger and *cam chum* (the dried stalks of lilies used in Chinese cooking), were definitely imported.

When Chinese immigrants could not afford to eat rice, they ate sweet potatoes, which were grown along with green leafy vegetables in some of the earliest Chinese truck gardens in the mining counties. The sweet potato originated in the New World. Spanish traders brought it to the Philippines, and from there it was introduced into China via Fukien province toward the end of the sixteenth century. The crop began to be grown all over south China, and Kwangtung province was one of the main producers.[7] Sweet potatoes became an important staple for peasants who were too poor to eat rice. The diffusion process came full circle when Chinese truck gardeners in gold rush California began to grow sweet potatoes and became their main consumers. Interestingly, while Chinese treat the sweet potato as a staple, they consider the Irish potato to be *choy*, and it is most commonly used as an ingredient in curried meat dishes by overseas Chinese.

For *choy*, fresh or salted vegetables were eaten on ordinary days, and pork and fish were eaten usually on festive occasions such as Chinese New Year. Several sources provide information on the great variety of food Chinese imported into California. Robert Spier examined the records of the Customs House at San Francisco and found that as early as 1852, shipments of food arrived from Hong Kong consigned to Chinese firms in San Francisco. Items listed on the invoices included fresh oranges and pomelos, dried fruits, dried oysters, dried shrimps, cuttle fish, salted fish, mushrooms, salted beans, dried bean curd, bam-

boo shoots, dried green vegetables, yams, ginger, sausages, dried
duck, duck liver, salted eggs, and sweetmeats, besides staples
such as rice, noodles, sugar, tea, and vinegar.[8] (It should be
noted that although dried seafood was imported from China in
the early 1850s, after the Chinese began to fish in California
waters, they started to export dried marine products to China.)[9]
Charles Nordhoff, in recording his visit to a Chinese railroad-
construction camp in Merced County in the early 1870s, noted
an equally impressive array of imported food items.[10]

The account books of Chung Tai and Company of North San
Juan, and Kwong Tai Wo of Grass Valley, both of Nevada
County, together with archaeological evidence, indicate that the
Chinese also ate locally grown meat, especially pork and chicken
that they butchered themselves, plus lamb and beef that they
purchased from American butchers. (Since Chinese used and
still use cleavers, which leave distinctive marks on the bones, it
has been relatively easy for archaeologists examining bones
found at various sites to determine which items were butchered
by the Chinese themselves.) Companies of Chinese truck gardeners supplied fresh vegetables to Chung Tai and Company in bulk
once a week, and Chinese local butchers supplied pork.[11] Sar-
dine cans found at archaeological sites show that the Chinese ate
American canned food also.[12] Finally, the manuscript population census listed Chinese duck-raisers, so fresh duck must have
been available, too.[13] In addition to Chinese food and various
dry goods, such as bowls, paper, writing brushes, ink, matches,
firecrackers, joss sticks, clothing, and textiles, Chinese stores in
inland California also stocked items for recreational consumption, such as opium, and Western cigarettes, brandy, and whis-
key. Both regular opium and "opium shit" (the scrapings from
opium pipes) were available for sale.[14]

All the imported food items which came from China in sacks,
earthen jars, glass bottles, and woven baskets were apparently
transported into the mining camps and valley towns in their original containers. In the early days, the merchandise was taken
by river steamer as far as the head of navigation on the major
rivers and then by wagon or pack mule to the mining camps.
(The Chinese used the Wells Fargo Express Company to send
gold dust to San Francisco, but it is not known whether they
used the same company for transporting provisions inland.)[15]

Although some contemporary observers believed that Chinese could not handle horses and other American draft animals, the manuscript population census provides evidence that there *were* Chinese immigrants who became teamsters. The 1860 manuscript population census for Stockton listed three Chinese whose occupation was given as freighter. John Wokee, aged thirty-eight, Ah Ling, aged twenty-four, and La Sing, aged twenty-three, lived in the same household as a thirty-year-old Chileno named Juan who was also a freighter. (Perhaps the Chinese had learned freighting from Juan.) The 1860 manuscript census also listed Chinese mule-packers in the other counties of the Southern Mines: thirteen in El Dorado County, one in Amador County, seven in Calaveras County, and one in Tuolumne County.[16] After railroads were built, Chinese merchants used them, but the railway lines did not run to the more remote foothill settlements. That is why as late as 1880 there were still Chinese freighters and mule-packers listed in the manuscript census. A number of freight bills for goods sent on the Southern Pacific Railroad by a San Francisco Chinese firm to Wing On Wo and Company in Dutch Flat, Placer County, in the 1880s listed "mats," tins, boxes, packages, cases, and sacks of various "China goods" as well as American canned ham, oysters, tomatoes, lard, and soda crackers. Live chickens from Auburn and Wheatland were shipped in coops to the same firm.[17]

For fresh food, the Chinese began to grow vegetables and to raise hogs and poultry from the 1850s on. The need to provision such a large number of miners—over 25,000 in 1860, and over 15,000 in 1870 and 1880—made truck gardening a viable means of livelihood, and it, in turn, provided other Chinese with an entree into larger-scale commercial farming. Understandably, Chinese truck gardeners were initially concentrated in the mining counties and in developing urban centers, such as the towns around San Francisco Bay and along major rivers in the Sacramento Valley. There were Chinese truck gardeners in scattered localities throughout the San Joaquin Valley also. They cultivated crops on land they had purchased or leased; hence, the first Chinese to engage in agriculture in California were independent owner-operators or small-scale tenants and not farm laborers. Although truck gardening employed fewer than 50 Chinese in each county (with the exception of Sacramento County, which had 125 of them), in some localities they were overrepresented. Passing references to Chinese vegetable-peddlers in local newspapers, and a separate account book kept by Chung Tai and Company for its white customers, indicate that Chinese sold to white as well as Chinese clients.

Chinese truck gardeners were suburban agriculturalists who formed an important link in the food supply chain. They played a far more important social and economic role than either the value of their products or their numbers would imply, for they combined production with merchandising. Peripatetic Chinese vendors functioned as California's earliest group of retail distributors of fresh produce. Some peddlers traveled great distances, making round trips that measured hundreds of miles. For example, an eyewitness remembers a Chinese vegetable-peddler, Tu Charley, a well-known character in the Yuba River basin in the 1890s, who carried fresh cucumbers, tomatoes, beans, melons, and other produce in a horse-drawn wagon, traveling periodically between Marysville and Sierra City, a distance of over a hundred miles in hilly terrain.[18] Both the short-distance vegetable-peddler, who carried his produce in baskets suspended on a bamboo pole balanced across his shoulder, and the long-distance vegetable-peddler, who used a horse-drawn cart or pack mules, were adapting an ancient Chinese practice to the California environment. By combining the economic roles of producer and trader, they were among the first farmers in California to make use of vertical integration in their enterprises.

Chinese truck gardeners flourished for many decades, though the locations where they were found shifted over time. Concentrated in the mining regions in the 1850s and 1860s, their relative numbers there had declined by 1870—a reflection of the fact that there were now some 10,000 fewer Chinese miners than a decade earlier. However, in certain localities, such as Grass Valley, Nevada City, and Placerville, clusters of Chinese truck gardeners remained. The largest concentrations of truck gardeners were now in the growing urban areas around San Francisco Bay and in the larger towns of the Sacramento Valley, such as Sacramento and Marysville. The 1870 manuscript census listed over

a hundred Chinese truck gardeners in San Francisco, and over
two hundred farmed plots along the stretch of land between
Oakland and San Jose. According to Ping Chiu, by the 1870s
Chinese vegetable-vendors were purchasing vegetables from
white wholesalers as well as from Chinese, so they were
distributing produce grown by gardeners of many different
ethnicities.[19]

By 1880 Chinese truck gardeners had achieved prominence in
several new localities: Los Angeles in the south, and the upper
Sacramento Valley (Tehama and Shasta counties) in the north.
In Los Angeles, Chinese truck gardeners made up almost 90 percent of all the market gardeners in the county.[20] In cities such as
San Francisco, Sacramento, Marysville, Yuba City, and Chico,
Chinese-grown fresh produce supplied the consumption needs of
the growing population. There was agitation against the presence of Chinese vegetable-plots in a number of cities, but somehow the Chinese truck gardeners managed to continue in their
business.[21] The most likely reason was that the Chinese habit of
peddling their fresh produce from door to door was simply such a
great convenience for housewives that whatever anti-Chinese
measures were passed were ignored by all parties. In 1880,
though the number of white truck gardeners had also increased,
widely dispersed Chinese truck gardeners made up more than
one-third of all the truck gardeners in the state.

Truck gardening and vegetable peddling not only were the
two agricultural roles Chinese immigrants took up earliest,
but were also the ones in which they have remained the
longest. Long after the anti-Chinese movement drove Chinese
immigrants and their American-born children out of diverse
occupations, small numbers of Chinese market gardeners and
produce sellers remained. Though they almost disappeared from
the mining regions as mining waned, their importance grew in
centers of urban settlement throughout California even after the
Chinese Exclusion Act of 1882 went into effect. Chinese truck
gardeners and produce merchants have persisted because their
lack of fluency in the English language was not as great a
handicap here as it might have been in other lines of work, their
produce was always fresh and attractive, and the marketing
networks they early established enabled them to remain
competitive.

Chinese peddler of garden truck, California. (Photograph from the Pierce Collection, courtesy of The Huntington Library, San Marino)

Truck Gardening in the Mining Regions

Chinese truck gardeners in California acquired land in different ways. The pioneers laid out vegetable plots on some corner of the grounds they had staked, purchased, or leased for mining. In the early 1850s, many mining districts passed laws to prohibit Chinese from staking claims, though this ban was by no means universal—county records in Calaveras, Yuba, and Sierra counties contain many instances of claims staked by Chinese. Far more frequently, though, Chinese miners bought claims from white miners. Having acquired land for mining, it cost nothing additional to plant vegetables on some portion of it. Moreover, the latter activity provided not only food but an extra source of income.

Successful vegetable-growing requires an adequate water supply and fertilizers. In the hot summer months, especially, vegetables have to be watered daily. Securing water was no problem, since almost all the Chinese miners worked placer claims, and these were either in riverbeds or along sparkling mountain streams. Later, as the river claims were worked out, they built wing dams, ditches, and other waterworks to conduct water to dry ground. In Yuba County, for example, Chinese miners were well known for the waterwheels they built to raise water from streams to higher ground.[22] Water procured for mining could also be used to irrigate vegetable plots, thus saving the miners-turned-gardeners an immense amount of labor. (In China, one of the most strenuous tasks facing vegetable growers was carrying water in buckets or cans from wells or rivers to their fields.) Fertilizer was also locally available, as the Chinese used "night soil."

After the Chinese realized that truck gardening could be a profitable enterprise, they began to buy or lease land for the express purpose of growing crops. A decade after their arrival, they had accumulated sufficient experience in land transactions to buy and lease farms or ranches that ranged from 5 to 160 acres in size. Since 5 acres (or 30 *mou*) would have been considered a huge parcel in China, the purchasers of even such small farms must have felt very successful. Being a landowner was a much desired status in China, but in addition, for several reasons Chinese immigrants found it attractive to purchase agricultural land. The Foreign Miners' Tax, technically applicable to all noncitizens but in practice collected primarily or solely from the Chinese, made mining less than lucrative. In contrast, no special tax had to be paid by truck gardeners or farmers. Moreover, in this period when white men did not seem as determined to guard their right to farm as jealously as they guarded their right to mine for gold, it is likely that the Chinese who retreated to outlying farms to earn a living encountered less harassment.

Chinese immigrants not only bought and leased land but knew how to make their transactions official. More than half of the leases and deeds for land they acquired were recorded at county offices at their own request, that is, not through an agent. They apparently knew how to hire someone to draw up the required documents and understood that recording these gave them some legal protection. Where Chinese buyers or lessees themselves did not bring their documents in for recording, a real estate agent, a notary public, or a lawyer did it for them. Seldom did a landlord or a seller perform this task in the mining counties, in contrast to agricultural counties where leases were most frequently recorded at the request of landlords.

County archival documents contain many records of vegetable gardens located on mining claims. In Calaveras County in 1861, two groups of Chinese miners purchased claims that already contained vegetable gardens. Ah Ty and Company purchased a claim with sluices, pumps, a cabin, and a vegetable garden along Esperanza Creek for $500 from John Caster.[23] Ah Yee and Company purchased a mining claim on the east bank of Chile Gulch, together with a ditch three-quarters of a mile long, sluices, mining tools, a house, and a vegetable garden with fruit trees, for $375 from Elisha Payne.[24] Chile Gulch had been so named because of the presence of a large number of Chileno miners there in the first years of the gold rush before Latin American miners were brutally driven out of the mines. In 1866 Ah Loy bought a plot of ground called "the Chinese Garden," on the banks of San Antonio Creek, from two Chinese, Ah Wo and Wa In, who had located it as a mining claim three years earlier. Ah Loy paid them $60 in cash for the ground, a house, some sheds, and tools.[25] Given the nomenclature of this piece of land, it is apparent that although mining was carried out there, it was more famous as a truck garden.

In Amador County there were four identifiable Chinese vegetable-gardens located on mining claims in the 1860s. In 1861 Ah Fook purchased two acres from Attoria Guidoni for $110 for the purpose of mining and vegetable growing.[26] The following year Ah Ping bought one acre, known as "Ah Chut and Company's Garden," from Ah Chut for $175.[27] The year after that, the partners of Yin Noo Company paid the large sum of $1,650 for a lot measuring 476 feet by 200 feet on the North Fork of Jackson Creek, which they used for mining as well as vegetable cultivation.[28] Since the price for this piece of land was high compared with other transactions during this period, the buyers obviously expected to find a considerable amount of gold on the premises. In 1864 Con Sin and Company purchased a small plot one mile northeast of Jackson—county seat of Amador County—for $80 from S. C. Combs.[29] The vegetables they grew were probably sold in Jackson, which had a sizable Chinatown where, between 1855 and 1880, a total of twenty-three town lots on both the east and the west side of Main Street were purchased by Chinese. Many of the Chinese-owned lots were contiguous, but interspersed were some lots owned by non-Chinese. The lots fronting on Main Street stretched back all the way to the North Fork of Jackson Creek, so it is possible that some of them were used for truck gardening.[30]

Aside from these small gardens on the corners of mining claims and in town lots, Chinese residents in Amador County also grew vegetables on much larger plots they purchased. In 1861 Ah Tim, Ah Loke, Ah Chow, and Ah Lick bought 100 acres, which were part of the Indian Creek Store Ranch, from Lucian Guinand. Located three miles west of Fiddletown, which had a large Chinese population, this ranch was sold four years later to G. E. Wallich.[31] In 1867 J. H. Callison of the State of Nevada sold 10 acres known as the Bob Callison Ranch, located along the Fiddletown–Volcano Road, to Ah Hay, Ah King, Ah Poy, Ah Cow, Ah Yan, Ah Kook, and Ah Lu for $300.00.[32] The following year, Ah Yoak purchased the 80-acre Cummings and Batchelders Ranch, one mile southeast of Jackson, at a public auction for $801.90—the highest bid offered.[33] In the 1870s and 1880s, Chinese continued to buy agricultural land in Amador County, but in generally smaller tracts measuring about 5 acres.[34] In the same period, they leased farms 20 to 80 acres in size.[35]

Agricultural land leased by Chinese immigrants in El Dorado County was also larger than garden size. In 1856 Chic Ah San leased approximately 5 acres in the eastern corner of the Lloyd and Smalley Ranch from its owners for $25.00 a year. The lease period was four years; should the tenants wish to extend it, the rent would be $37.50 a year.[36] In 1862 Quen Yun and Ah Woe leased a farm located within the city limits of Georgetown from John F. Gates for three years at $16.00 a month. Fruit trees were already growing on this parcel, and the tenants were given the right to use the fruit so long as they took care not to injure the trees.[37]

The archives of El Dorado County also contain records of Chinese purchases of agricultural land. In 1867 Ah Lung and Company bought 5 acres located within the city limits of Placerville from Charles W. Brewster for $500;[38] three years later they sold the tract for only $50 to George W. Wray.[39] In 1868 Ke Chin, a resident of Auburn, Placer County, bought 25 acres in Greenwood Township for $1,250.[40] Twenty years later, he sold it to Ma Wing Quong, another Chinese resident of Auburn, for $1,300.[41] In the mid-1870s, four other pieces of property, ranging from 4.5 acres to 160 acres, one of which adjoined an orchard, were purchased by Chinese immigrants in the county.[42] Although not every lease or deed document provided information on the kind of crops grown, episodic evidence suggests that the Chinese used these farms to grow vegetables, sweet potatoes, different kinds of berries, and deciduous fruit.

Besides growing fruit and vegetables, the Chinese in the counties of the Southern Mines contributed to the food supply in two other ways. There were a large number of Chinese butchers in this region, and there is evidence that at least some of the meat they cut up was supplied by Chinese animal-breeders. The 1860 manuscript agriculture census listed twenty-three Chinese hog-raisers in Calaveras County, and since pork was the most commonly eaten meat among Chinese, it is not surprising that they would raise hogs. In 1859–60 they slaughtered $42,650 worth of pigs, which amounted to over one-fifth of the total value of animals slaughtered in the four townships where Chinese hog-raisers resided.[43] Pigs were quite valuable, each worth from $10 to $20. If the average price of a pig was approximately $15, then in 1859–60 Chinese hog-raisers in Calaveras County slaughtered and sold some 3,000 pigs. Each of the hog-raisers owned at least one or two horses, and occasionally a mule, as draft animals. The manuscript agriculture census for Amador County for the same year listed thirteen Chinese households which together owned 21 horses, plus pigs and dairy cows. In Kwangtung province, the most common draft animal was the water buffalo; the fact that Chinese owned and used horses and mules in California showed that they were flexible in adapting to their new environment.

Residents of Kwangtung province who dwelled along the sea-coast and in a delta full of marine life were expert saltwater and freshwater fishermen and fish breeders. Even in the interior of California, Chinese engaged in fish breeding. In Amador County in 1885, just outside the town of Ione, Ah Guan and Doc Wah leased 80 acres for three years from William P. Hays for $225 per year. In addition to an existing vegetable garden, the farm contained a large fishpond. The tenants were told they could take any fish heavier than half a pound out so long as they left enough large fish in the pond for reproduction.[44] These bits of evidence, together with records of Chinese imported-food items, show that Chinese immigrants ate relatively well even when they lived and worked hundreds of miles away from San Francisco.

In the 1860s the Chinese demographic center shifted north-ward from the Southern to the Northern Mines. Between 1860 and 1870 the Chinese population in the Southern Mines declined by more than 7,000 persons, but increased slightly in the North-ern Mines. In the following decade the mining population moved even farther north into the Klamath and Trinity mountains. Chinese truck gardeners followed this population shift.

Chinese truck gardening activities in the northern mining counties differed from those in the southern mining counties in several ways. First, although the absolute number of Chinese truck gardeners was smaller in the northern counties, their relative percentage was larger because there were fewer truck gardeners in the northern counties as a whole. Second, fewer of the Chinese bought agricultural land. Given their later entry into the northern mining region, by the time they arrived anti-Chinese attitudes had hardened into clearly discriminatory practices, and the immigrants probably found it more difficult to purchase farms. Third, truck gardeners in the north were more geographically concentrated. Instead of being scattered over many localities, as in the southern region, Chinese truck gardeners in the northern region were assembled in major towns in the Sacramento Valley, where the climate was warmer and more arable land was available.

Two of the earliest recorded truck gardens in the northern region were in Nevada County. In 1859 Ah Kim paid James Lamar, a resident of Bridgeport, $800 for the five-acre Hargrave Ranch on the south side of the Middle Yuba River. The sale included a house, several outhouses, two stoves, and miscellaneous chairs, tables, and bedsteads.[45] Two years later, Chung, who had paid Manuel Francis $370 for several mining claims stretching for 1,500 feet along Rush Creek, was given permission to fence in a plot on the south side of the creek for gardening purposes. The deed specifically stated that the transaction included water rights for irrigating the garden.[46]

Between 1863 and 1881 in Nevada County, fourteen farm leases made by Chinese were recorded.[47] Most of the parcels were located in Grass Valley and Nevada townships where a small but verdant valley lay. Almost all the leases stated that the tenant was to provide the landlord with vegetables for the latter's family use. Nevada County landlords seemed to have a special concern for maintaining the fertility of the soil: several leases specified that the tenants had to use manure.[48]

In the 1860s and 1870s almost all the agricultural land used by Chinese truck gardeners in the northern counties was leased, and many of these already had growing gardens and orchards. The documentary records also indicate that by this time Chinese had learned to grow strawberries and grapes—two crops not grown in the Pearl River delta of China. One lease signed in 1869 by Chinese tenants was for a vineyard;[49] another document indicated that muscat grapes were growing on the land being leased.[50]

Truck Gardening in Sacramento Valley Towns

Some of the larger towns in the Sacramento Valley became major centers of Chinese truck gardening. While truck gardeners in the mining regions combined vegetable growing with mining, those in towns such as Oroville, Chico, Marysville, and Sacramento were full-time gardeners; if they had a joint or subsidiary occupation, it was most likely to be vegetable peddling. In the mining regions the most common work group among Chinese truck gardeners consisted of two partners, but in the valley towns larger partnerships were the rule.

Oroville, situated along the Feather River in Butte County, had the earliest recorded Chinese truck garden within city limits in the Sacramento Valley. Chinese appeared in Butte County very early because the mines there were more easily accessible than the ones in the foothills. Eager gold-seekers could reach the mines in the vicinity of Oroville directly by boat instead of having to trek hundreds of miles on foot or by wagon. Boats departing from San Francisco crossed the bay, stopped at either Benicia or Martinez—located on opposite banks at the mouth of the Sacramento River just before it pours into the upper bay through the Carquinez Straits—proceeded upriver to Sacramento City, and thence sailed on to Marysville and Oroville. The journey took approximately a week and cost sixteen dollars.[51]

A bluff in Oroville served as a natural levee fronting the river. By 1852 a flourishing Chinatown, which eventually occupied three full blocks on Broderick Street immediately behind the bluff, had already been established. A "China Garden" slightly larger than two acres was located to the west of Chinatown, where a plot had been purchased for $700 by Ah Jim and Company in 1854 from Jack Wesson, who already had a garden there.[52] This Chinese vegetable-plot was in use in Oroville for more than half a century; today its grounds are a city park in a quiet residential neighborhood. Several blocks from the park still stands a temple—now a state historical monument—built by Oroville's Chinese residents.

Originally named Ophir City, Oroville and its environs in 1860 had almost 1,000 Chinese residents, most of whom were located in two mining camps, named Bagdad and Lava Beds, south and west, respectively, of the city. Chinese worked mostly the tailings left over by white miners;[53] by the mid- and late 1870s, Lava Beds was mined almost exclusively by Chinese.[54] These miners provided a large and ready market for the produce grown by Oroville's Chinese truck gardeners.

If one wanted to go to a region farther west than Oroville in

Butte County, one could travel up Butte Creek. Rich gold deposits were found in the Butte Creek canyon as well as on Paradise Ridge, a promontory lying between Butte Creek and the Feather River. Chinese mining camps existed at Diamondville, Centerville, and Helltown along the canyon, and in Paradise, Dogtown (also known as Magalia), and Kimshew on Paradise Ridge. Several Chinese vegetable-gardens were located along Little Butte Creek about eight miles from Paradise below the Honey Run grade. Fresh vegetables grown there were brought by Chinese peddlers to be marketed in Magalia about once a week.[55]

Another center of Chinese vegetable-cultivation in the Sacramento Valley was the city of Marysville, located at the confluence of the Feather and Yuba rivers in Yuba County. There were thirty-four Chinese truck gardeners in Yuba County in 1860, twenty of whom resided in Marysville. Numerically, Chinese represented more than half the truck gardeners in the county. According to the 1860 manuscript agriculture census, Marysville produced $23,210 worth of market garden crops; of this amount, the Chinese had grown $4,260 worth, or 18.4 percent of the total.[56] White truck gardeners cultivated much larger plots; the largest, a farm and not a garden, had 183 acres.[57]

The Chinese truck gardens in or around Marysville were scattered. Several were located right in the heart of downtown Marysville; others were found either at the outer fringes of the city or outside the levee that surrounded and protected the city. Among the larger plots was one leased by four Chinese, Gem, Mang, Wang, and Wa, in 1863 from John and Mary Williams of Marysville. The 1860 manuscript agriculture census shows that John Williams owned 100 acres of land on the south bank of the Yuba River just outside the city levee, on which he produced 900 bushels of wheat, 250 bushels of oats, and 60 tons of hay. He also raised livestock and slaughtered $2,500 worth of animals in 1859–60.[58] Williams leased to the four Chinese tenants 5.3 acres of his farm for $210 a year for five years, an annual rent of $40 per acre.[59] The total value of Williams's farm as given in the census was $4,000, which means that his farm was worth $40 per acre. The annual rent he charged was exactly the same as the average per acre value of his farm. Being a landlord was thus quite lucrative.

It is not possible to determine how many other landlords in the early 1860s pegged the amount of rent they charged to the assessed value of their farms, because few Chinese leases were recorded at this early date. In the 1870s and 1880s, in general, in those cases where names in county records can be matched against those in the manuscript census, the annual rent per acre paid by Chinese tenants varied from one-quarter to one-half of the assessed value of the land they leased.

Some of the gardens leased within Marysville already had growing orchards in them. The Chinese planted vegetables on the ground between the rows of fruit trees, but they had to take care not to injure the roots of the trees when they plowed. In addition to paying rent, the Chinese tenants had to look after the orchards, so the landlords reaped a double benefit.

Early on, Marysville became famous for its Chinese bitter melons and string beans. According to Joe Kim, an old resident of Marysville whose grandfather farmed in the city, it was possible to grow certain Chinese vegetables in Marysville better than elsewhere because warm weather arrived there earlier in the spring. Since it was difficult to grow bitter melons in the foggy San Francisco Bay Area, Marysville-grown produce was shipped to San Francisco for sale.[60] Given its strategic location, Marysville supplied the consumption needs of residents in three locations: in the city and its environs, in the Yuba River basin stretching over 100 miles eastward, and in San Francisco more than 120 miles away to the west. Chinese-grown fresh produce was distributed by vegetable peddlers on foot around the city, was taken to the mining camps by horse-drawn wagon, and was shipped to San Francisco by river steamer and later by railroad.

The presence of vegetable gardens within the city limits of Marysville eventually became a point of friction between the city's white and Chinese residents. The issue of "China gardens" emerged as a topic for political debate in April 1876 when the Sanitary Committee of the Marysville City Council recommended passing an ordinance to prohibit "China gardens" on Eighth Street, in the section between K Street and Yuba Alley. Council members recommended that the garden be moved to Sacramento Square, which was at the northern edge of the city, but it was discovered that the square had already been rented to a Mr. Beach.[61] Six months later in a different location in the city, residents on H Street between Fifth and Sixth streets complained that the Chinese were creating a nuisance in the neighborhood because they had planted a garden on a lot owned by a Mr. Sheehan. The Sanitary Committee was asked to warn the Chinese about this complaint, but no official action was taken against them.[62] Although council minutes do not mention what kind of fertilizer the Chinese were using, it is possible that their presence was considered a nuisance because they used night soil, which creates a bad odor.

The objection that white Americans had to the use of night soil by Chinese gardeners began soon after Chinese gardens became prominent in different localities in California. The *Auburn Stars and Stripes* commented in 1866:

It is a well-established fact that Auburn, which up to 1853—about the time when Chinese gardens were first allowed to be cultivated within the limits of the town— was one of the healthiest inland towns in the State.... But ever since the establishment of the Chinese gardens, the residents of this place, and more especially those living in certain localities peculiarly exposed to the miasmata aris- ing from the gardens, have been subject to endemic and epidemic diseases, and the rate of mortality, particularly among children, has been absolutely fearful to contem- plate.... The evil consists mainly in the Chinese mode of cultivation, which is filthy and disgusting in the extreme. Their gardens are made on low grounds, and the soil is stimulated to rank productiveness by the application of the most offensive manures. Large holes are excavated in the ground, which are filled with human ordure, dead animals, and every imaginable kind of filth, water is added, and the feculent mass is left to thoroughly decompose, when it is ladled out and scattered broadcast over the garden. In addition to this, urine is kept in large earthen jars until it has acquired the proper degree of offensiveness, when it is poured over the growing cabbages and other vegetables, which grow rapidly under such treatment, and acquire a richness of flavor grateful to Chinese stomachs, but intoler- able to most white palates. Incredible as it may seem, we are assured that some white families actually use vegetables

cultivated in this mode! ... Our citizens, after enduring the
evil already too long, have determined that these and kindred
nuisances shall be abated.... As the Chinese gardens are a
considerable source of revenue to the owners of the soil, it is
not wonderful that some opposition should be made to their
summary abatement ... the Chinese gardeners will no
longer be tolerated at the expense of life and health.[63]

An editorial in the *Wheatland Graphic* of January 30, 1886, also
alluded to the Chinese use of night soil. It was said that the
Chinese gardeners in Wheatland were cutting prices by 25 percent in order to retain their customers in the face of impending
competition from two or three white gardeners who were preparing a plot of ground to grow all kinds of vegetables. Readers were
told that "lovers of cleanliness should avoid patronizing hea-
thens because of the filth used by them in their cultivation."[64]
The writer demurred by saying he would not give details, but
"those who wish to know should ask Officers Wadell, Bevan or
Kesner." It was said that once they had been told, their
knowledge would "give them such a disgust of Chinese
vegetables that they will never use them again."[65]

Despite public censure, however, the Chinese managed to
continue cultivating vegetable gardens within Marysville's city
limits. In January 1890 their presence once again became an
issue when a "large number of residents and tax payers" sent a
petition to the City Council declaring the "China gardens" west
of E Street to be a nuisance and a health menace. At a meeting
of the council in March, it was said that the problem included
drainage and sewage, and the Board of Health was asked to
investigate. Dr. Powell, president of the board, agreed that the
gardens were indeed a menace, but stated that the board had
"no power to compel the abatement of the nuisance." He said
that only the City Council had the requisite power; but the
council members took no action.[66] In November of the same
year, Dr. Powell, after visiting all the Chinese vegetable-gardens
in the city, declared they were "in excellent sanitary condition."
He said the Chinese had been "unusually particular the past
season in keeping everything cleaned up and are obeying the
Health Officer's instructions in every respect." Powell said that
if they continued thus, people should "take no exception to
them."[67]

Three years later, however, the issue surfaced again. Citizens living on Fifth Street petitioned to have all gardens removed from the area south of Ninth Street—the heart of downtown Marysville. The City Council compromised by drawing the line at Eighth Street, but later passed an ordinance to prohibit the cultivation of market gardens within the limits of the city altogether. This time, the mayor vetoed the proposed ordinance; unfortunately, council minutes do not state the reason he gave for his veto.[68] Chinese truck gardens probably survived because in Marysville, from the beginning, Chinese vegetable-growers had a virtual monopoly of truck gardening in the city—a situation that did not exist in any other California city. If the truck gardeners had been driven away, the people of Marysville would have been deprived of their source of fresh vegetables. So despite the disgruntlement of those who resided next to the smelly Chinese vegetable-plots, the dependence of the rest of the city on Chinese vegetable-growers enabled the latter to continue to ply their trade.

The most important urban center of Chinese truck gardening in the valley was the city of Sacramento, which had 110 Chinese vegetable-cultivators in 1860. In addition, fifteen Chinese gardeners were listed in the townships outside of city limits. Together, they constituted 58.1 percent of the 215 gardeners in the county. Lying at the confluence of the Sacramento and American rivers, Sacramento was the major supply post for the Northern Mines and the entire Sacramento Valley, and produce grown there fed both the city's residents and people in the hinterland. Within the city, Chinese gardeners made up 75.3 percent of the gardening population in 1860, greatly outnumbering gardeners of other ethnicities. Of the 36 non-Chinese gardeners in the city, 27 had been born in Europe: 5 in England and Scotland, 7 in Ireland, 8 in what eventually became modern Germany, and 7 in other European countries. There was also a lone truck gardener from Chile. American-born gardeners came mostly from New England and the Middle Atlantic states.[69]

In the 1860s Sacramento was divided into four wards. Ward One, the smallest and most densely populated, occupied the northwest section of the city and was bounded on the north by the American River, on the west by the Sacramento River, on the south by K Street, and on the east by Fifth Street. Chinatown was located there. There were 70 Chinese truck gardeners living

in Ward One, 4 of whom had their wives with them; 40 Chinese
gardeners and 4 of their wives lived in Ward Three in the northeast portion of the city; but no Chinese truck gardeners were
found in Wards Two and Four, the southern half of the city.
White gardeners, however, were listed in the latter locations.[70]
The Chinese had vegetable plots along the bottomland of the
Sacramento and American rivers—land that was fertile but was
subject to unpredictable flooding. Since no Chinese leases were
recorded in the 1850s, it is likely that the truck gardeners were
squatters, since much of the swampy bottomland they used still
belonged to the federal government, which had few officials on
the scene, so the Chinese may not have bothered to lease these
plots officially.

Chinese truck gardeners moved their operations from place to
place as land reclamation was carried out. Beginning in 1850 the
citizens of the state capital raised enormous sums of tax money to
pay for the construction of levees because floods constantly inundated the city. The early attempts at reclamation and flood control were not successful, for the city was flooded in 1850, 1852,
1853, 1862, and 1867. It was not until a railroad embankment
was built along the south bank of the American River in the
1870s that a levee large and strong enough to prevent flooding
became a reality.[71] However, successful flood control caused
the Chinese to lose the bottomland for vegetable growing. By
1880 only 95 out of a total of 189 Chinese truck gardeners still
lived within city limits. Eighty-three now farmed in American
Township, on the north bank of the American River, outside of
the levees.[72]

Non-Chinese gardeners were more widely dispersed throughout the city, and farmed slightly larger plots with products of a
greater value. One of the largest farms in the city had five acres
on which an Irish immigrant, T. O'Brien, grew grapes and
produced four thousand gallons of wine in 1860.[73] Although the
majority of Sacramento's truck gardeners resided in the city, it
would be erroneous to assume that the bulk of market-garden
products was grown within city limits. In 1860 Sacramento
County ranked first among California counties in the cash value
of its market-garden produce. Of a total of $139,214 reported in
the manuscript agriculture census, the bulk came from the three
townships constituting the Sacramento Delta. Sutter Township

produced $22,627 worth of market-garden crops, Franklin Township $54,070, and Georgiana Township $38,361.[74] So by far the largest value of fresh produce came from diversified farms outside of the city, which were operated by white farmers who owned or rented eighty or more acres. No Chinese truck gardeners or farmers were found in these Delta townships in the early years, but they discovered the region and its fertile soil afterward, for about half the Chinese truck gardeners in the county plus twenty-six Chinese farmers had located themselves in the Delta by 1870.

The majority of the Chinese truck gardeners farmed in partnerships, in contrast to their white counterparts who tended to farm alone. The most prevalent partnership grouping in Sacramento City consisted of six members. In comparison, Marysville's Chinese gardeners most commonly farmed in groups of two or three partners. Even though partnerships meant that the profits had to be divided into several shares, truck gardening apparently provided a relatively stable and economically satisfactory life. One indication of this is that of the fifty-nine Chinese wives living with their husbands in Sacramento County in 1860, eight were married to truck gardeners. Only laundrymen had a higher incidence of wives present.[75] (This finding challenges the common belief that only Chinese merchants were rich enough to bring wives to America with them.) As sedentary occupations that benefited from unpaid family labor, truck gardening and laundering allowed a small number of Chinese women to live in America with their husbands.

Truck Gardening in San Francisco

Compared with the mining regions and the major towns in the Sacramento Valley, San Francisco was not a particularly important center of Chinese truck gardening until the 1870s. The 1860 manuscript census of population for the city and county listed only 31 Chinese truck gardeners, and the 1870 published census reported only 23 Chinese out of a total of 220 truck gardeners in San Francisco. But the latter figures are misleading because they referred only to the city. A tally of the manuscript schedules for

the city and county together showed that there were in fact 123
Chinese, 355 European-born, 27 American, and 5 other truck
gardeners there. By 1880 the number of Chinese had declined to
35, among 280 European-born, 33 American, and 3 others.
Twenty years later only 42 gardeners, of whom 23 were Chinese,
were left in San Francisco. It appears, then, that the 1870s were
the peak years of Chinese truck gardening in the metropolis by
the Golden Gate. An account by the Polish journalist Henryk
Sienkiewicz, who traveled in California from 1876 to 1878 (and
later achieved fame as the author of *Quo Vadis?* and winner of the
Nobel Prize in literature), gives some impression of how the
Chinese truck gardeners were perceived. Since Sienkiewicz's
description is the fullest contemporary account of Chinese truck
gardeners available, it is quoted at length:

> Let us now look at the kind of work the Chinese perform
> in California. A single word describes it accurately: every-
> thing. A significant proportion of them has turned to agri-
> culture. The whole of San Francisco is situated on arid
> dunes and sandy hills, and yet whoever goes to the outskirts
> of the city will perceive at the ends of unfinished streets, on
> the hills, valleys, and slopes, on the roadsides, in fact, every-
> where, small vegetable gardens encircling the city with one
> belt of greenness. The ant-like labor of the Chinese has
> transformed the sterile sand into the most fertile black
> earth. How and when this was accomplished they alone can
> tell, but suffice it to say that all the fruits and vegetables,
> raspberries and strawberries, under the care of the Chinese
> gardeners grow to a fabulous size. I have seen strawberries
> as large as small pears, heads of cabbage four times the size
> of European heads, and pumpkins the size of our wash tubs.
>   The Chinese hut stands in the center of the garden. At
> every hour of the day you will observe the long pigtailed
> yellow gardeners now digging, now spreading manure upon
> the soil, now watering the vegetables. In the interests of
> one's own appetite it is sometimes better not to see the lat-
> ter, for I have myself had the experience of observing
> Chinamen pouring a liquid created from human excrement
> diluted with water between the leaves of heads of cabbage
> still unfolded. Yet the whole of San Francisco lives on the

fruits and vegetables bought from the Chinese. Every morning you see their loaded wagons headed toward the markets
in the center of town stopping in front of private homes. It
may even be said that in all of California this branch of
industry has passed exclusively into the hands of the
Chinese.[76]

Sienkiewicz was not entirely correct, for though the Chinese
were very important in truck gardening, they did not control it
exclusively.

FOUR

Green Gold

 H O W important was truck gardening as an occupation for the Chinese in California? What effect, if any, did truck gardeners have on their fellow immigrants and on the larger community? These questions may be answered by examining the relative dominance of Chinese among truck gardeners of all ethnicities in different regions of the state, the percentage of gardeners in the Chinese population, the social role that truck gardeners played, and how well truck gardening provided a livelihood to the Chinese in nineteenth-century California.

The Numerical Importance of Chinese Truck Gardeners

As is shown in table 8, in 1860 the counties with the largest populations in the state—San Francisco, Sacramento, Tuolumne, Calaveras, and El Dorado—also had the largest number of gardeners. San Francisco, with a population of 56,802, had 247 truck gardeners, of whom only 31 were Chinese. Sacramento ranked second with 24,142 persons, of whom 215 were truck gardeners, with 125 Chinese among them. El Dorado ranked third with 20,562 persons and 93 truck gardeners, 30 of them Chinese. Nevada County, which ranked fourth with a population of 16,446, had only 38 truck gardeners, 16 of whom were Chinese. Calaveras ranked fifth with 16,299 persons and 117 truck gardeners, 51 of whom were Chinese, while Tuolumne ranked sixth with a population of 16,229, and 161 truck garden-

ers, of whom 42 were Chinese. The close coincidence between
the ranking of the counties in total population and in the number
of gardeners strongly suggests that in this period vegetables were
grown primarily for local consumption.

The importance of the Chinese in truck gardening can be
clearly seen both in absolute numbers and in the percentage of
Chinese in the total gardening population. The proportion of
Chinese among truck gardeners ranged from a low of 12.6 percent in San Francisco to a high of 75.3 in Sacramento City. In
the southern mining region, the percentage ranged from 26.1 in
Tuolumne County to 43.6 in Calaveras County. In the northern
mining region, the absolute number of Chinese gardeners was
smaller, but the relative percentage was larger, ranging from
35.0 percent in Sierra County to 52.2 percent in Placer County.
The greater dominance of the Chinese in the northern mining region can best be explained by the timing of the entry of
Chinese into this region. In San Francisco and the southern
mining region, where there was already a large American and
foreign white population by the time the Chinese came, Chinese
found it difficult to get started in truck gardening because white
gardeners had already established themselves in those localities.
Those who did grow vegetables managed to find a market primarily among their fellow Chinese. In the northern mining
region, on the other hand, Chinese arrived at about the same
time as the white population and found it easier to capture the
Chinese and part of the non-Chinese market as well through the
quality of their produce and their willingness to peddle their
wares from door to door.

Truck gardening was an occupation that drew mainly non-
native speakers of English. Foreign-born immigrants were able
to compete in the occupation because success in it did not depend on skills in English communication, and relatively little
capital was required to get started in the business. In certain
counties, European immigrants were more numerous than
Chinese, but if they are separated into discrete groups according
to national origins, no single one outnumbered the Chinese.
European immigrant gardeners were most prominent in San
Francisco, where they constituted 78.5 percent of the total,
with the largest number coming from France, Italy, Switzerland,
and the states that eventually unified into the German nation.

Table 8 *Chinese and Non-Chinese Truck Gardeners Listed in the 1860 Manuscript Census of Population by County and Nativity, Selected Counties*

	Total Number	Born in China		Born in Europe		Born in U.S. and Canada		Born Elsewhere	
		Number	Percent	Number	Percent	Number	Percent	Number	Percent
SOUTHERN MINING COUNTIES									
El Dorado	93	30	32.3	22	23.7	39	41.9	0	0
Amador	30	8	26.7	12	40.0	9	30.0	1	3.3
Calaveras	117	51	43.6	39	33.3	14	12.0	13	11.1
Tuolumne	161	42	26.1	68	42.2	47	29.2	4	2.5
Mariposa	22	2	9.1	18	81.8	2	9.1	0	0
NORTHERN MINING COUNTIES									
Plumas	12	0	0	7	58.3	5	41.7	0	0
Butte	47	21	44.7	7	14.9	19	40.4	0	0
Sierra	83	29	34.9	23	27.7	31	37.4	0	0
Yuba (except Marysville)	31	15	48.4	6	19.4	4	12.9	6	19.4
Nevada	38	16	42.1	9	23.7	13	34.2	0	0
Placer	69	36	52.2	18	26.1	15	21.7	0	0

SACRAMENTO VALLEY									
Marysville City (Yuba County)	33	19	57.6	8	24.2	0	0	6	18.2
Sacramento City	146	110	75.3	27	18.5	8	5.5	1	0.7
Rest of Sacramento Co.	69	15	21.7	28	40.6	26	37.7	0	0
SAN FRANCISCO	247	31	12.6	194	78.5	21	8.5	1	0.4

SOURCE: Tallied and computed from U.S. National Archives, Record Group 29, "Census of U.S. Population" (manuscript), 1860.

NOTE: The above counties contained the majority of both the Chinese and the non-Chinese population in California in 1860. No attempt was made to search for truck gardeners in the rest of the counties in the state.

In this and the following tables, the cities of Marysville and Sacramento have been shown separately in order to highlight conditions there. Marysville is part of the Sacramento Valley, but the rest of Yuba County is more properly considered part of the northern mining region.

Though large-scale Italian immigration into the United States did not begin until the 1880s, Italian gardeners, along with Italian miners, were nonetheless prominent in 1860. Gardeners of English and Scottish origins were few in number. The only other foreign-born truck gardeners of importance were Mexicans and Chilenos. There were few American-born gardeners, and only in El Dorado, Tuolumne, and Sierra counties were they of any significance.

The Chinese had a greater tendency than non-Chinese to take up truck gardening as an occupation. A comparison of table 8 with table 2 in chapter 2 indicates that in 1860 the percentage of Chinese among truck gardeners was higher than the percentage of Chinese within the general population. Although only 4.8 percent of San Francisco's population were Chinese, 12.6 percent of the truck gardeners were. In Sacramento County, where only 7.2 percent of the people were Chinese, 58 percent of the truck gardeners had been born in China. In the city of Sacramento, 75 percent of the gardeners were Chinese. In the southern mining region, approximately 20 percent of the total population were Chinese, but Chinese gardeners made up about one-third of all gardeners. In the northern mining region, about one-sixth of the population but over 40 percent of the gardeners were Chinese.

The relative importance of the Chinese among all gardeners did not mean, however, that a large percentage of the Chinese population worked as truck gardeners. In 1860 truck gardening provided a living for as few as 0.6 percent of the Chinese in El Dorado County and as many as 7.2 percent of those in Sacramento City. In most counties, only a little over 1 percent of the Chinese population earned a living growing truck crops, the desire of the majority still being to mine for gold. More important, in a market economy, where total income depended on unit price times the quantity of a commodity sold, an increase in the supply did not necessarily result in an increase in total revenue, because a greater supply was likely to drive the unit price downward. This was especially true of perishable fresh produce. If more individuals had become truck gardeners and increased the total volume of fresh vegetables available in the market, and if demand did not simultaneously increase, the result would have been a drop in the unit price of vegetables. In contrast to manufacturing, where economies of scale could be realized with a greater input of capital, vegetable growing was labor-intensive, and the unit cost of production remained fairly constant regardless of the amount of capital input or the scale of total output. A decrease in selling price without a concomitant reduction in the unit cost might have caused the profit margin to fall to such a low level that truck gardening would no longer be economically viable. General market conditions, therefore, regulated the total number of truck gardeners, including Chinese, who could survive at any given time.

By 1870, as table 9 shows, the number of gardeners of all nativities had declined drastically in the southern mining region, owing to the exodus of miners from this area. Chinese gardeners either disappeared altogether from certain localities, such as Tuolumne and Amador counties, or they became proportionately more important, as in Calaveras and El Dorado counties. The contradictory fates of Chinese gardeners in these counties cannot be explained by differential rates of decline in the Chinese population. There is insufficient information to account for why Chinese truck gardeners were able to maintain themselves in two of the counties but not in the others.

A similar pattern of differential development was apparent in the northern mining region, but here the development is more explicable. Both Chinese and non-Chinese gardeners had virtually disappeared in remote Plumas and Sierra counties, but they had increased in Butte, Nevada, and Placer counties, where Chinese now made up approximately two-thirds of all the truck gardeners. Demographic and economic changes were responsible for these developments. In the 1860–70 decade, the population of Plumas and Sierra counties, consisting mainly of miners, fell greatly, so the market for fresh produce shrank. In Butte and Placer counties the total population also fell, but the decline was compensated by a population increase in Nevada County— made possible by the boom in quartz mining—which brought a general prosperity to Nevada City and Grass Valley and their vicinities and sustained the market for fresh produce, a luxury consumption item.

The most dramatic increases in the number of market gardeners were in the Sacramento Valley and the San Francisco Bay Area. Again, changes in the centers of concentration of gardeners followed the general demographic shift. With the exception of

Table 9 Chinese and Non-Chinese Truck Gardeners Listed in the 1870 Manuscript Census of Population by County and Nativity, Selected Counties

	Total Number	Born in China		Born in Europe		Born in U.S. and Canada		Born Elsewhere	
		Number	Percent	Number	Percent	Number	Percent	Number	Percent
SOUTHERN MINING COUNTIES									
El Dorado	36	17	47.2	11	30.6	8	22.2	0	0
Calaveras	61	30	49.2	23	37.7	4	6.6	4	6.6
NORTHERN MINING COUNTIES									
Butte	68	41	60.3	16	23.5	11	16.2	0	0
Yuba (except Marysville)	13	6	46.2	5	38.5	2	15.4	0	0
Nevada	82	52	63.4	25	30.5	5	6.1	0	0
Placer	88	63	71.6	11	12.5	14	15.9	0	0
SACRAMENTO VALLEY									
Shasta	29	15	51.7	8	27.6	6	20.7	0	0
Tehama	6	5	83.3	0	0	1	16.7	0	0
Marysville City (Yuba Co.)	67	60	89.6	5	7.5	2	3.0	0	0
Sacramento City	80	35	43.8	30	37.5	14	17.5	1	1.3
Rest of Sacramento Co.	51	37	72.5	10	19.6	4	7.8	0	0
Yolo	30	23	76.7	3	10.0	4	13.3	0	0

SAN FRANCISCO BAY AREA

Alameda	204	118	57.8	69	33.8	16	7.8	1	0.5
San Francisco	510	123	24.1	355	69.6	27	5.3	5	1.0
Santa Clara (gardeners)	246	104	42.3	110	44.7	26	10.6	6	2.4
Santa Clara (berry cultivators)	256	256	100.0	0	0	0	0	0	0
LOS ANGELES	26	19	73.1	7	26.9	0	0	0	0

SOURCE: Tallied and computed from U.S. National Archives, Record Group 29, "Census of U.S. Population" (manuscript), 1870.
NOTE: Counties with only a handful of Chinese truck gardeners have not been included in this table.

Sacramento City, which lost a large fraction of its Chinese truck gardening population—a loss explained by the entry of Chinese into larger-scale farming in the Sacramento Delta—Chinese now formed the majority of gardeners in these areas, with the Bay Area experiencing the largest increase in absolute numbers.

The increase in the number of Chinese gardeners not only kept pace with but surpassed the increase in the overall Chinese population. In 1860 gardeners made up more than 2 percent of the Chinese population only in Sacramento and Tuolumne counties. By 1870, in five of the mining, four of the Sacramento Valley, and two of the Bay Area counties, as well as in Los Angeles, truck gardeners constituted 2 or more percent of the total Chinese population, which itself had increased by 50 percent in the preceding decade. Among gardeners of other nativities, the distribution of European immigrants remained about the same as a decade earlier, with Italians gaining prominence, especially in the Bay Area. Irish immigrants had also entered the field, and Latin American gardeners had virtually disappeared. The number of American gardeners had declined so much that they exceeded a dozen persons only in the Bay Area counties.

The decade between 1870 and 1880 saw a number of contradictory developments. As table 10 shows, Chinese truck gardeners had become truly dominant in those mining counties where they remained and in the Sacramento Valley. Their relative percentage ranged from 63.4 percent in Solano County to 100 percent in Sutter County. In Shasta, Tehama, Sutter, and Solano counties, there were hardly any gardeners in 1870, so the Chinese gardeners found there in 1880 were pioneers. Gardeners formed one-quarter and one-sixth of the total Chinese population in Tehama and Sutter counties, respectively, so it is most likely that truck gardening had drawn the Chinese there in the first place. In these areas, a small number of Europeans remained, but Americans and Latin Americans had departed almost completely from the scene.

An opposite development had taken place in the San Francisco Bay Area in the 1870–80 decade. Here the number of Chinese truck gardeners had plummeted from the peak reached ten years earlier; they now formed only about 10 percent of the gardeners in Alameda, San Francisco, and Santa Clara counties. Americans had also become unimportant, whereas European immigrants had become paramount. Persons of Italian, German, French, Irish, and English birth remained most numerous, and the Portuguese became noticeable. The most likely explanation for the precipitous decline of Chinese truck gardeners in the Bay Area is the general anti-Chinese atmosphere in San Francisco. In the late 1870s the Workingmen's Party of California, under the leadership of the demagogic Denis Kearney, staged many demonstrations against the Chinese, campaigning under the single slogan "The Chinese Must Go!" Chinese in San Francisco found it increasingly difficult to survive in nonservile businesses that depended on white patronage, and vegetable gardening was one occupation in which they lost out to European immigrants.

The most significant development between 1870 and 1880 was the growth of Chinese truck gardening in Los Angeles. Fully 208 of a total of 234 market gardeners in Los Angeles—that is, 89 percent—were Chinese, though the latter were only 3.5 percent of the county's total population. Gardening provided a living for 17.8 percent of the county's Chinese population. Since there had been only 26 gardeners, all nationalities combined, in Los Angeles ten years earlier, it is obvious that the Chinese had introduced truck gardening into this area also. As elsewhere, the early entry of the Chinese enabled them to dominate the market.

Since the manuscript schedules of the 1890 census were destroyed in a fire, no information on developments between 1880 and 1890 is available. By the end of the nineteenth century, as table 11 shows, Chinese truck gardeners still formed very large proportions of all gardeners in some of the counties in the Sierra Nevada foothills and the Sacramento Valley. They made a slight comeback in the San Francisco Bay Area, declined in importance in Los Angeles, and made a strong showing in Fresno County. European-born gardeners continued to control the market in the Bay Area and had captured the Los Angeles market by 1900, where, interestingly, American-born gardeners had also entered in force. The only other important development was the entry of a number of Japanese immigrants into the field. It should be pointed out, however, that there is some uncertainty with regard to the number of Japanese "gardeners" because a number of them—especially those in Alameda County, where they were most numerous—were in the nursery business or tended residential gardens and were not vegetable growers as such.

Table 10 *Chinese and Non-Chinese Truck Gardeners Listed in the 1880 Manuscript Census of Population by County and Nativity, Selected Counties*

	Total Number	Born in China		Born in Europe		Born in U.S. and Canada		Born Elsewhere	
		Number	Percent	Number	Percent	Number	Percent	Number	Percent
MINING COUNTIES									
Butte	94	82	87.2	7	7.4	5	5.3	0	0
Yuba (except Marysville)	47	37	78.7	6	12.8	4	8.5	0	0
Nevada	128	93	72.7	23	18.0	12	9.4	0	0
Placer	91	75	82.4	8	8.8	8	8.8	0	0
El Dorado	20	13	65.0	4	20.0	2	10.0	1	5.0
SACRAMENTO VALLEY									
Shasta	69	49	71.0	4	5.8	16	23.2	0	0
Tehama	186	183	98.4	3	1.6	0	0	0	0
Marysville City (Yuba Co.)	51	50	98.1	1	1.9	0	0	0	0
Sutter	43	43	100.0	0	0	0	0	0	0
Sacramento City	127	89	70.1	30	23.6	8	6.3	0	0
Rest of Sacramento Co.	117	95	81.2	17	14.5	5	4.3	0	0
Yolo	64	46	71.9	16	25.0	2	3.1	0	0
Solano	41	26	63.4	13	31.7	2	4.9	0	0

SAN FRANCISCO BAY AREA								
Alameda	301	21	7.0	237	78.7	42	14.0	1	0.3
San Francisco	351	35	10.0	280	79.8	33	9.4	3	0.9
Santa Clara	110	18	16.4	78	70.9	13	11.8	1	0.9
SAN JOAQUIN VALLEY								
San Joaquin	193	86	44.6	97	50.3	10	5.2	0	0
Tulare	34	33	97.1	0	0	1	2.9	0	0
LOS ANGELES	234	208	88.9	18	7.7	6	2.6	2	0.9

SOURCE: Tallied and computed from U.S. National Archives, Record Group 29, "Census of U.S. Population" (manuscript), 1880.
NOTE: Counties with only a handful of Chinese truck gardeners have not been included in this table.

Table 11   *Chinese and Non-Chinese Truck Gardeners Listed in the 1900 Manuscript Census of Population by County and Nativity, Selected Counties*

	Total Number	Born in China[a]		Born in Europe		Born in U.S. and Canada		Born Elsewhere[b]	
		Number	Percent	Number	Percent	Number	Percent	Number	Percent
MINING COUNTIES									
Butte	36	31	86.1	3	8.3	1	2.8	1	2.8
Yuba (except Marysville)	18	13	72.2	1	5.6	2	11.1	2	11.1
Nevada	15	11	73.3	2	13.3	2	13.3	0	0
Placer	11	8	72.7	2	18.2	1	9.1	0	0
El Dorado	1	0	0	1	100.0	0	0	0	0
SACRAMENTO VALLEY									
Shasta	43	31	72.1	4	9.3	8	18.6	0	0
Tehama	17	7	41.2	1	5.9	2	11.8	7	41.2
Marysville City (Yuba Co.)	74	70	94.6	3	4.1	1	1.4	0	0
Sutter	53	50	94.3	1	1.9	2	3.8	0	0
Sacramento City	94	29	30.9	21	22.3	38	40.4	6	6.4
Rest of Sacramento Co.	101	29	28.7	62	61.4	6	5.9	4	4.0
Yolo	12	7	58.3	5	41.7	0	0	0	0
Solano	35	2	5.7	32	91.4	1	2.9	0	0

SAN FRANCISCO BAY AREA									
Alameda	280	54	19.3	161	57.5	31	11.1	34	12.1
San Francisco	42	23	54.8	15	35.7	3	7.1	1	2.4
Santa Clara	229	41	17.9	127	55.5	54	23.6	7	3.1
SAN JOAQUIN VALLEY									
San Joaquin	52	5	9.6	35	67.3	10	19.2	2	3.8
Fresno	87	71	81.6	10	11.5	6	6.9	0	0
Tulare	8	6	75.0	2	25.0	0	0	0	0
Kern	12	8	66.7	3	25.0	1	8.3	0	0
COASTAL COUNTIES									
Monterey	39	20	51.3	11	28.2	8	20.5	0	0
Los Angeles	452	95	21.0	181	40.0	170	37.6	6	1.3

SOURCE: Tallied and computed from U.S. National Archives, Record Group 29, "Census of U.S. Population" (manuscript), 1900.

[a] A small number of the Chinese gardeners were born in California.

[b] Most of those born "elsewhere" were of Japanese ancestry.

The Social Impact of Truck Gardening

In the days prior to the widespread use of refrigeration, the success of growing and marketing fresh produce, unlike that of less perishable staples, depended on the growers' ability to meet the changing needs of a nonfarming population: they must respond to almost daily fluctuations in the market for fresh produce. The high perishability of market garden crops makes it difficult for their growers and distributors to maintain any market control. Two of the available regulatory mechanisms are spacing out their harvests by growing crops that ripen at different times of the year, and distributing their produce in more than one market.

Chinese truck gardeners managed to prosper because they did grow a sufficiently large variety of crops and tapped into several markets. While the first crops they cultivated were almost certainly Chinese vegetables sold to fellow Chinese, they also learned at an early date to cultivate vegetables that white customers would buy. Supplying both a Chinese and a non-Chinese population increased the variety of the crops they could grow and enabled them to space out their harvests to provide a more even flow of income over a longer period. In addition, itinerant peddling increased the geographic range of their markets and gave them two sources of income—one from production, the other from distribution. Later, selling their produce through San Francisco commission merchants opened up yet larger markets.

Combining production with distribution helped the first Chinese truck gardeners to make a go of their business. Itinerant peddling was a practice they had brought over from China and had quickly adapted to California conditions. In a provocative study of marketing networks in traditional China, G. William Skinner has argued that there were two, not one, hierarchical systems of social organization there—one administrative, the other economic. He hypothesized that the "standard marketing community," comprising an average of eighteen villages, with a modal population of some 7,000—rather than a single village— set the boundaries of the peasants' social world. Markets in traditional China were not held every day, so that marketing activities were periodic rather than continuous. Not all goods were exchanged or services performed on the actual marketing days themselves; rather, marketing days allowed peasants and towns-people to make contracts, and the delivery of goods and services could just as easily be carried out on the "cold days" between marketing days. Itinerant traders helped to maintain the economic ties that bound together the marketing communities.

Skinner called producers who peddled their own wares "mobile firms." Peasants who grew crops for sale, and artisans who traveled with their wares and "workshops" on their backs purveying services of all kinds, were trying to make their means of livelihood economically viable: "The total amount of demand encompassed by the marketing areas of any single rural market is insufficient to provide a profit level which enables the entrepreneur to survive. By repositioning himself at periodic intervals, the entrepreneur can tap the demand of several marketing areas and thereby attain the survival threshhold.... By concentrating demand on certain specific days, marketing periodicity enables such entrepreneurs to combine sales with production in an optimally efficient manner."[1]

If Skinner was correct, some Chinese emigrants to California must have brought with them extensive experience in buying and selling goods and services to outsiders. It is not surprising, then, that when some of them began to grow truck crops they would market these themselves, becoming mobile firms that combined production with trading, retaining the essential goals of periodic marketing and itinerant peddling without adhering strictly to their finer aspects. A reporter for the *Newark Courier* in New Jersey who visited California in 1869 noted:

> Every morning you meet Chinamen by the dozen, in the principal streets, with long poles on their shoulders, balancing at either end an enormous basket filled with vegetables, designed for sale in localities remote from the market. These baskets sometimes contain two or three bushels each, but the Chinaman, being an expert in balancing, carries his load, under which any person of ordinary strength would sink, with perfect ease—moving with a swinging, rollicking gait, indicative of neither weariness or exhaustion. It is said that these small dealers in fruit and vegetables drive a thriving business with people of moderate means, who are not able to purchase in other than small quantities.[2]

Such "thriving business" soon attracted the attention of white persons. In the June 5, 1886, issue of the *Wheatland Graphic*, the editor asked why there was "no white man in Wheatland who could run a wagon with fruit and vegetables between said town and Grass Valley." He had noticed that there were

> three or four chinese [*sic*] running wagons on this road, and as I understand, are clearing from $100 to $150 per month. The people in Spenceville, and no doubt on the whole road, are obliged to buy from the chinese [*sic*], but are willing to patronize a white man if there were only one on the road. I think it would be a good investment for a man with a capital of $250 to $300. He could make a good living and a little money besides.[3]

In addition to increasing their own income, itinerant peddlers played important social roles. They became one of the first groups of Chinese to establish ongoing, face-to-face relations with whites, when they brought fresh produce to the latter's doors. Given their frequent interaction with whites, it was possible that gardeners served as intermediaries between Chinese and whites even in matters not related to the growth and sale of vegetables and fruit.

Truck gardeners also served as a link between the urban and rural Chinese populations. (Though the boundary between the two groups was a fluid one, given the large number of Chinese who led a migratory existence in nineteenth-century California, it is possible to think of separate urban and rural populations.) The vegetable peddlers who traveled routinely between town and country carried information regularly from one locality to another. Producers in rural or suburban areas garnered information about the living habits of people in the towns, while urban Chinese learned about farming opportunities in the countryside.

Finally, having learned to plant crops under California conditions, and having caught glimpses of the kinds of farming that were possible, some truck gardeners began to grow other crops, and others eventually operated diversified farms that produced potatoes, onions, beans, cereal, and tree fruit. Truck gardening thus served yet another funneling function by providing an avenue for Chinese to enter different and larger economic undertakings.

Chinese vegetable peddler with a one-horse wagon, Sutter County, California. (Photograph courtesy of the California Historical Society, San Francisco)

Chinese Strawberry-Growers in the San Francisco Bay Area

Besides vegetables, Chinese truck gardeners also grew strawberries. They first did so around the southern end of San Francisco Bay in the stretch of land between the town of Alviso and
the city of San Jose; later, the center of strawberry growing
moved a little farther west to the area between San Jose and the
town of Santa Clara. In 1860 there were only 22 Chinese listed in
Santa Clara County, even though San Jose was a transit point
for gold miners who traveled to the Southern Mines on foot, but
soon the area's suitability for strawberry cultivation attracted
many Chinese. The first lease of a strawberry farm by a Chinese
was recorded in 1868, so it can be assumed that some had begun
to work as strawberry cultivators and pickers a few years prior to
that date. By the 1870s the cultivation of strawberries, raspber-
ries, blackberries, and gooseberries had become one of the most
important means of livelihood for Chinese residents in Santa
Clara County. In 1870, 256, or 16.8 percent, of the 1,525 Chinese
listed by census takers in the county were strawberry growers,
and in 1880, 361 persons, or 13.4 percent, of the Chinese population of 2,695 were.

Actually, a far greater number of persons were involved in
strawberry cultivation than the census indicated, because the
number of extra workers hired during picking season was about
six times the number needed during other times. The Chinese in
fact monopolized strawberry cultivation and picking in Santa
Clara County: the 1860, 1870, and 1880 manuscript population
census listed no white berry-cultivators at all.

Several local landowners were responsible for introducing
strawberry cultivation to the Chinese. Between the mid-1860s
and the turn of the century, over two dozen landowners leased
parts of their farms to Chinese for planting, cultivating, and harvesting strawberries and other berries. Given the multiple leases
made by Isaac Bird, William Erkson, and William Boots with
Chinese tenants, these men appear to have been instrumental in
bringing Chinese into the area.

Isaac Bird was an early American settler in Santa Clara County who was already a prominent local citizen by the late 1850s, as
his name first appeared in county archival records in his capacity
as trustee of the First Baptist Church in 1858. By 1862 he had

acquired two tracts—one was part of the Spanish-Mexican land grant, the Rancho Yerba Buena y Socayrin confirmed by the federal government to Antonio Chabolla, and the other was a tract purchased from Antonio Sunol, who owned part of the Rancho de los Coches.[4]

In the fall of 1868 he signed three separate leases with Chinese tenants. In August he leased thirteen acres to Ah Sing, [Ah] Tom, and [Ah] Jim for a six-year period. One acre of this land had already been set out in strawberries, while four acres were meadows. The three tenants were instructed to plant eight acres to strawberries and four acres to blackberries. The lease was on a sharecropping basis, with Bird and the tenants each to receive 50 percent of the net profit. The cost of shipping-boxes and freight would be deducted from the proceeds before they were divided. The tenants did not need capital of any sort. Bird provided them with a plow, a harrow, and all other tools, as well as the horses and harnesses and a wagon to haul the berries to the railroad depot. He also gave them lumber and nails for building themselves a house. Since berry cultivation required irrigation, Bird saw to it that the tenants would dig ditches and build small reservoirs; he also provided them with lumber for building troughs. The tenants were to plow the land and to perform all the necessary labor. Unlike later leases, no mention was made in this initial lease of how the berries were to be marketed.[5]

The following month Bird found another set of tenants, Ah Chi, Ah Lang, [Ah] Ang, and Sam Long Charley. Sam Long Charley was to become a prominent Chinese farmer in the area: over the coming years, his name would appear repeatedly in many county documents as he leased small berry patches and larger farms. Bird's second lease with Chinese tenants turned over to them twenty acres in two parcels, one of which was to be leased for six years, the other for five. The tenants were to plant sixteen acres to strawberries while the four acres which adjoined the Guadalupe River were to be set out in blackberries. A house stood on one of the parcels, and the tenants were given permission to live in it. The terms of the lease were the same as the first one, with the additional provision that the berries were to be shipped and sold under the landlord's name through San Francisco commission merchants.[6]

In October 1868 Bird made a contract with his third set of

Chinese tenants when he leased sixteen acres adjoining the Guadalupe River to Boe Chi, Ah Poy, Ah Tong, and Ge Lam for five years on a sharecropping basis and under the same terms as the first two leases.[7] In all these contracts, the terms made it clear that Bird was to supervise every phase of the cultivation, harvesting, packing, shipping, and marketing. The close supervision called for was understandable, since at this point the Chinese had not yet acquired a reputation for expertise in strawberry cultivation.

The terms of the leases signed later on between Chinese tenants and other white landowners for the most part followed the pattern set by Bird. Compared with leases signed by Chinese tenant farmers in other counties, these berry cultivation leases were unusual in two ways. Only in this area did sharecropping predominate; in other counties, Chinese tenants who paid fixed cash rents were more prevalent than sharecroppers. It was also unusual for the landlords to provide *everything* except the labor needed for cultivation; elsewhere, the Chinese tenants often contributed some tools or draft animals of their own.

Others who leased land to Chinese for strawberry cultivation were William Erkson and William Boots, who, together with James Fogarty, William Welsh, and James Lick, plus four heirs of Domingo Alviso, were tenants in common of the southern half of the Rancho Rincon de los Esteros, which had been confirmed to the children of Domingo Alviso. This land lay between Coyote Creek and the Guadalupe River and had been parceled out so that the heirs received tracts ranging from 100 to 483 acres.

William Erkson leased sixteen acres in September 1868 to Ah Yung, Ah Sang, Ah King, and Ah Hoy for eight years. He agreed to plow and harrow the land to prepare it for cultivation, and to haul the berries to the depot without charging the tenants for either service. He provided his tenants with a dwelling house, draft animals, and tools; he also advanced them $300 worth of provisions, which they were to repay from the proceeds of the first crop. Since Erkson's tract adjoined neither Coyote Creek nor the Guadalupe River, he agreed to dig a well for the tenants should there be insufficient water for irrigating the crops. Even though Erkson provided more services himself, he was still willing to divide the crop half and half with his tenants.[8] In subsequent years, Erkson continued to lease land to Chinese tenants.[9]

William Boots, Erkson's fellow inheritor, was a prominent citizen of Santa Clara County, being a trustee of the town of Alviso. His share of the Rancho de los Esteros measured 193 acres and adjoined Coyote Creek.[10] Boots signed his first lease with Chinese tenants in March 1869, when he leased 8 acres to Ah Chun, Ah Sah, Ah Tou, and Ah Lou for a six-year period for 50 percent of the crops. The tenants were to plant asparagus, blackberries, and currants, which were to be marketed in their landlord's name through San Francisco commission merchants. Boots provided all the needed farming implements and agreed to haul the produce to the depot himself. However, he charged the tenants wages equivalent to half a day's labor for each trip he made.[11] In the next few years, Boots signed four other leases with Chinese tenants for parcels ranging from 16 to 25 acres.[12] In all of them, he had the tenants plant one or two other crops in addition to strawberries, probably as an insurance against fluctuating prices in the strawberry market.

Charles Wade, a neighbor of Erkson and Boots, came into possession of his share of the Rancho Rincon de los Esteros when he married Estafana Alviso. Perhaps because he was a judge and was therefore busy with his professional duties, he had leased his farm to a white tenant, A. L. Peebles, in November 1868. Peebles then sublet ten acres of Wade's tract one month later to Chinese strawberry-growers. Like Erkson, Peebles also agreed to haul the fruit to the depot for free and to provide all the lug chests, as well as all the provisions required by the tenants and their hired hands until the latter had the means to buy their own food after the first crop had been sold.[13]

Strawberry cultivation on a sharecropping basis by Chinese spread in Santa Clara County as a result of the initial example set by these pioneer landowners and their Chinese tenants. A number of the landlords, by virtue of their head start, soon acquired a reputation as the most prominent strawberry growers in the area, even though the bulk of the work was in fact done by Chinese. In 1877 William Boots and Charles Wade, together with a Mrs. Shields, were named in the *Pacific Rural Press* as three of the largest strawberry growers in the Alviso–San Jose area.[14]

There is some evidence that landlords competed with each other for Chinese tenants. As time passed, landlords seemed to offer their tenants more incentives. Landowner A. Malavos tried harder to please his tenants by providing them with the kind of food they preferred. In the lease he signed, he promised to bring them rice, Chinese cooking oil, and meat from the butcher, in contrast to some of his competitors who merely allowed their tenants to buy provisions on credit from the Farmers' Union, which most likely did not stock any Chinese items.[15] Other landlords, like William Erkson who set the example, allowed their tenants to have one acre rent free to grow vegetables for their own use.[16] In a few cases, during the first year of berry cultivation, instead of dividing the crop half and half, the tenants were to receive two-thirds.[17] These small but telling concessions show that the Chinese possessed considerable bargaining power, and that the relationship they established with their landlords was by no means a completely lopsided one.

For the most part, the Santa Clara landowners were trying to get additional income from those sections of their land which were not yet in cultivation, and since the income per acre from strawberry cultivation was high, they were quite willing to make whatever small concessions were necessary to attract Chinese tenants to work for them. Moreover, some of them were willing to perform part of the work themselves. In that sense, their relationship was more or less a business partnership, with the Chinese providing the labor and the expertise for cultivation, and the landlords providing the land, the farming implements, the plants, the teams and wagons, irrigation water, some of the labor, the connections with San Francisco commission merchants, and even the provisions needed by their tenants for survival.

That the landowners and their tenants lived and worked in close proximity is corroborated by clauses in some of the leases. A landlord named W. Oliver seemed concerned that his collaboration with Chinese tenants not violate prevailing mores: his lease with Ah Tack, Ah Foon, and Ah Ping signed in 1878 specified that the tenants were not to cultivate the berry patches on Sundays. Rather, during the harvest season, should it be necessary to do so, the tenants were to "put on more men on Fridays and Mondays" in order to prevent the berries from becoming overripe on the vines.[18] Another landlord, A. Malavos, in signing a lease with Ah Wing in 1878, specified that the tenant shall not employ "quarrelsome or lazy or incompetent Chinamen."[19]

In a lease signed in 1890, Rosalie L. Younger made it clear that "no immoral institutions or any Chinatown or undue congregation of Chinamen" were to be allowed on her ranch; moreover, her tenants could not have "pig pens, Chinese women, opium dens or gambling" on the premises.[20]

Some time during the 1870s, labor contractors began to organize teams of strawberry cultivators and pickers. In 1870 Chinese strawberry-cultivators lived in households that averaged five members, and no sign of large-scale labor contracting was yet apparent. By 1880, however, seventeen Chinese foremen were listed in the manuscript population census in Santa Clara Township, Santa Clara County, as heads of households—six of them had strawberry cultivators as household members, while the rest had farm laborers living with them. Those listed in the census schedules as "China bosses" or foremen in charge of strawberry pickers had an average of six individuals under them; those in charge of farm laborers had an average of ten persons under them. These labor contractors ranged in age from twenty-seven to sixty, and almost all of them could speak some English, as is indicated in the census schedules. In addition to the households organized by labor contractors, there were thirty-one other households of berry cultivators and pickers in the area.

County records show that the size of berry patches leased to Chinese tenants in the 1870s and 1880s remained about the same as those leased in the late 1860s. The most likely reason that strawberry growing remained on a small scale was the extraordinary labor-intensity of the activity. An article in the *Pacific Rural Press* in 1877 indicated that in general, 1 Chinese cultivator could care for 2 acres of vines except during the picking season, when 3 Chinese were needed to tend and pick each acre.[21] Thus, a strawberry farm of 15 acres would require 45 men during picking season. County archival records indicate that the total number of acres leased by Chinese berry-growers fluctuated considerably during the 1870–1890 decades. On the average, between 150 and 200 acres were under cultivation (or at least were recorded as being under cultivation). That meant at least 500 to 600 pickers were needed each harvest season. The *Pacific Rural Press* had estimated in 1877 that there were "10,000 Chinamen" engaged in berry cultivation in Santa Clara County.[22] Such a large figure was improbable, since it represented over 40 percent of the total

Chinese population of San Francisco and Santa Clara counties combined at this time. The actual number involved was perhaps about 1,000.

To even out the large differential in the demand for labor between the picking season and other times, the landlords and their tenants resorted to intercropping. Onions were the most common crop interplanted with strawberries. In strawberry fields, the earth was piled up in ridges two to two-and-a-half feet apart. The berry vines were planted on the sloping sides of the ridges, while the onions were planted on the ridges themselves. Runners from the strawberry vines were carefully pruned, except when they were allowed to grow for the express purpose of providing new plants. Irrigation water flowed in the furrows between the ridges, and both berry and onion plants benefited simultaneously.[23] Where onions were intercropped, the landlords provided all the seed onions the first year, but thereafter they and their tenants each provided half.[24] In those cases where the landlords did not desire to grow onions, the Chinese tenants were given permission to plant whatever vegetables they wished on the ridges so long as these did not interfere with the growth and productivity of the vines.

The strawberries were packed in boxes of eight to ten pounds each. The boxes were placed like drawers in a chest, one on top of another, and when they were thus loaded with care, little injury was done to the fruit during transportation to market. The berries were shipped either by railroad or by boat from the Alviso wharf to San Francisco, which served as an entrepôt port for the strawberries grown in the south bay. From San Francisco, the fruit was transshipped to other places, including out-of-state destinations.

The price for strawberries fluctuated from season to season. An abundant crop drove prices down to as low as four cents a pound in the late 1870s; in other years the fruit could fetch about twelve cents a pound. In the 1870s, according to the estimate of a knowledgeable landlord, the average net profit was $400 per acre, which meant that the landlord and his tenants each received $200 per acre.[25] In terms of per capita income, if a household of strawberry cultivators averaged ten persons, and they took care of an average of twenty acres—since each person could care for two acres during nonpeak season—then an income of $4,000 would be divided into ten shares to give each individual $400. If the Chinese tenants were responsible for paying the wages of the seasonal pickers, however, their annual income would be considerably less. Even so, their income was in the same range as the per capita income of vegetable growers and larger than the annual per capita income of Chinese laborers in the 1870s and 1880s.

The Chinese derived two additional advantages from strawberry cultivation: they became familiar with the commission system of marketing crops—knowledge that would stand them in good stead as they became large-scale fruit growers in the mid-1870s—and they established a reputation for themselves as expert farmers. In 1886 a reader wrote the *Pacific Rural Press* inquiring whether it was true that Chinese applied a solution of soda, or soda and ammonia, to the berry plants to produce foliage, and then one of potash to induce them to bear fruit. The editor responded that he was not sure if the method worked; however, he said, "the Chinese are expert gardeners and [have] brought their own experience [with them]," so presumably their method was worth trying.[26]

The Economics of Chinese Truck Gardening

There is very little information on the earnings of Chinese truck gardeners, but the manuscript schedules of the census of agriculture provide some limited information on their scale of operations, while the manuscript schedules of the census of population contain information about their relative wealth. Although there were hundreds of Chinese truck gardeners shown in the 1860 census of population, only fourteen of them were listed in the census of agriculture. They were located in El Dorado, Yuba, and Sacramento counties. Table 12 shows that most of them cultivated only two or three acres, and only three produced truck crops worth more than $1,000 during 1859–60. The average value per acre of the crops produced by the thirteen truck gardeners who gave figures to the census takers was $537, but

Table 12   *Chinese Truck Gardeners Listed in the 1860 Manuscript*
Census of Agriculture

County and Township	Name of Person	Number of Acres	$ Farm Value	$ Value per Acre
EL DORADO COUNTY				
Mud Springs Twp.	Chin	4	300	75
YUBA COUNTY				
Marysville Twp.	Sing Mow	3	150	50
	Ah Ming	2	200	100
	Ah Ki	2	200	100
	Ah You	2	200	100
	Ah See	3	300	100
	Ah Yow	3	300	100
Linda Twp.	Ah Ling	4	100	25
Long Bar Twp.	Ah Tong	1	100	100
	Ah Quong	1	50	50
SACRAMENTO COUNTY				
Cosumnes Twp.	Ah Sam	8	275	34
	Ah Yun	21	400	19
	Chon	11	400	36
	Ah Chee	16	560	35

SOURCE: U.S. Bureau of the Census, "Census of Productions of Agriculture" (manuscript), 1860. Figures for farm value per acre and value of products per acre were computed by the author.

NOTE: Only those farms with given acreages and products have been included in this table. Those Chinese listed in the manuscript census of agriculture who raised only livestock have been excluded. Dashes mean no information was given for those columns.

a NC = not computed, owing to the difficulty of assigning relative value to the different crops produced on the same farm.

b Po = potatoes, Pe = peas, W = wheat. (Even though Ah Sam produced no market garden crops, he grew 180 bushels of peas, so data on him are included in this table.)

this figure is inflated because three gardeners had large values for their products. If they are excluded, then among the other ten the average value of truck crops produced was $325.

In El Dorado County in 1860, Chin, the sole Chinese gardener listed in the manuscript census of agriculture, produced $1,000 worth of truck crops. In comparison, there were 31 non-Chinese gardeners, who produced a total of $7,245, with the average at $234 per farm. Chin was the third-largest truck

$ Value of Implements	Kind of Livestock	$ Value of Livestock	$ Market Gardens	$ Value Products per Acre	Other Crops
—	2 horses	100	1,000	250	—
50	6 pigs	110	1,200	400	—
50	4 piglets	12	700	350	—
50	2 piglets	6	500	250	—
50	10 pigs	110	160	80	—
30	—	—	200	67	—
50	2 piglets	14	1,500	500	—
75	1 piglet	10	500	125	—
—	2 piglets	10	400	400	—
—	8 piglets	50	300	300	—
25	—	120	—	NC[a]	40 bu. Po, 180 bu. Pe,[b] 200 lb. butter
30	—	50	200	NC	250 bu. Po
—	—	—	90	NC	100 bu. Po, 20 bu. W[b]
40	—	40	200	NC	300 bu. Po

gardener in the county, surpassed only by one who produced $1,560 and another who produced $1,100 worth of truck crops.

In Marysville, Linda, and Long Bar townships in Yuba County—the only three townships where Chinese gardeners were listed—42 non-Chinese gardeners produced $28,260 worth of truck crops, at an average value of $673, but this figure is also inflated because 7 of them produced large crops of $1,000 to $5,000. If these 7 are excluded, the other 35 non-Chinese garden-

Table 13 Chinese Truck Gardeners Listed in the 1870 Manuscript Census of Agriculture

County and Township	Name of Person or Company	Number of Acres	$ Farm Value	$ Value per Acre	$ Value of Implements
TEHAMA COUNTY					
Tehama Twp.	Mow Chung & Co.	15	300	20	200
	Ah Ling & Co.	25	500	20	250
	Georgie Wee & Co.	70	1,000	14	250
	Win Lee & Co.	30	600	20	300
	Yung How & Co.	30	600	20	300
BUTTE COUNTY					
Oregon Twp.	Hop Wo & Co.	10	1,000	100	400
	Fung Sing	7	1,000	143	600
	Yen Chu	3	500	167	—
	Hong Goon	2	200	100	—
	Ing Goon	4	600	150	100
YOLO COUNTY[a]					
Grafton Twp.	Ah Fook	10	300	30	—
Washington Twp.	Ah Puck	3	1,000	333	—
	Sung Sing	10	500	50	50
EL DORADO COUNTY					
Coloma Twp.	Toy Gee & Co.	8	500	63	—
MARIPOSA COUNTY					
	Ah Tong	5	100	20	—
SANTA CLARA COUNTY					
Alviso Twp.	Ah Kay & Co.	14	2,800	200	1,125
	Ah Lock & Co.	13	2,600	200	50
LOS ANGELES					
Los Angeles Twp.	Yang Van & Co.	15	600	40	200

SOURCE: U.S. Bureau of the Census, "Census of Productions of Agriculture" (manuscript), 1870.

NOTE: Only those farms with given acreage and products have been included. All acres given in this table are tilled, improved acres.

[a] Three Chinese listed in Merritt Township in Yolo County have been placed in the table on Chinese farmers in the Sacramento–San Joaquin Delta (table 19). Chinese farmers located in Franklin and Georgiana townships of Sacramento County are also in table 19.

[b] The difference between the total value of output and the value of the market garden crop indicates that there were other products, but these were not stated in the manuscript census.

NS = product not specified.

$ Wages Paid in 1869–70	Kind of Livestock	$ Value of Livestock	$ Market Gardens	$ Other Products	$ Value Total Products	$ Value Products per Acre
3,000	6 horses	360	9,000	—	9,000	600
3,000	8 horses	500	9,000	—	9,000	360
4,500	4 horses	240	19,500	—	19,000	279
4,500	4 horses	240	16,000	—	16,000	533
4,000	6 horses	300	15,500	—	15,500	517
1,000	4 mules	300	5,000	—	5,500	550
900	2 mules	150	4,000	—	4,000	591
—	2 horses	100	1,500	NS[b]	1,800	600
—	2 horses	100	800	NS[b]	3,000	1,500
100	3 mules	160	3,000	—	3,000	750
—	2 horses	100	600	—	600	60
—	—	—	600	—	600	200
—	—	—	600	—	600	60
—	2 horses	115	150	600	750	94
—	2 horses	100	—	—	—	—
1,700	2 horses	200	5,600	—	5,600	400
100	2 horses	200	675	—	675	52
20	2 horses	100	—	—	—	—

ers produced a total of $12,760, at an average of $365 per farm. In comparison, the 9 Chinese gardeners in Yuba County averaged $607 per farm. If Sing Mow, who produced $1,200, and Ah Yow, who produced $1,500, are not counted, then the average value for the rest was $394. The Chinese in Yuba County, therefore, produced about the same average value of truck crops as their white counterparts in 1860.

In Sacramento County there were many non-Chinese gardeners, so the Chinese will be compared only with other gardeners in Cosumnes Township where the listed Chinese resided. Twenty-one non-Chinese gardeners in this township produced $4,690, at an average of $223 per farm. Compared with them, the three listed Chinese produced smaller amounts, two at $200 and the third at $90.

Although the average value of their truck products was comparable to that of white gardeners, the total income from all crops of the Chinese gardeners was much lower than that of whites because the Chinese specialized completely in truck gardening, which meant that truck crops provided their sole income from agriculture. Non-Chinese gardeners, who grew vegetables on diversified farms, had larger total incomes because only part of their income derived from truck crops. The only Chinese listed in the 1860 manuscript agriculture census who operated diversified farms were the four persons in Cosumnes Township, Sacramento County, who grew potatoes, peas, and some wheat. One, Ah Sam, churned $200 worth of butter—an unusual product, since Chinese in those days did not eat butter.

As table 13 makes apparent, by 1870 the value of crops grown by Chinese truck gardeners had greatly increased. The 1870 manuscript agriculture census listed sixteen Chinese gardening households that gave the census takers information on the value of their truck crops, seven of whom grew more than $5,000 each. Five of the seven were in Tehama Township, Tehama County, which spans both sides of the upper Sacramento River. Their plots, which ranged from 15 to 70 acres, were considerably larger than the Chinese truck gardens found elsewhere in the state and were thus farms, not gardens. The relatively large acreage and high income of these few gardening households skewed the average value of truck crops produced upward for Chinese truck gardeners in the state as a whole.

Chinese truck gardeners in Tehama County, who started the trend of cultivating truck crops on a field rather than a garden scale, handled the increased scale of their operations by using partners. In contrast to the gardeners listed in the 1860 manuscript agriculture census who farmed singly, all five of the Tehama County Chinese truck-gardening households were organized as companies. County archival records, which recorded many leases between white landowners and Chinese *yuen* (the Chinese word for garden or estate),[27] corroborate the existence of the partnership system.

The Chinese gardeners in Tehama County monopolized vegetable production in the area. The 1870 manuscript population census listed only one non-Chinese truck gardener in Tehama Township, Tehama County, while the 1870 manuscript agriculture census listed only five non-Chinese farms that produced truck crops, which together produced only $1,200 worth, at an average of only $240 per farm, compared with the Chinese, who produced a total of $68,500, at an average of $13,700 per farm. Three of the farms operated by white farmers producing truck crops were grain farms that grew vegetables only as a supplementary crop. The Chinese, on the other hand, specialized in vegetables and fruit.

Another location where Chinese truck gardeners flourished was Paradise Ridge, Oregon Township, Butte County—a mining region with a concentration of Chinese, where in 1870 five Chinese gardeners produced a total of $14,300, or an average of $2,860 per farm. In comparison, six non-Chinese gardeners, who also specialized in truck crops, grew $19,400, or an average of $3,233 per farm. Only one produced some barley, and three grew some grapes for wine-making. Here, too, specialization went hand in hand with an increase in the scale of production.

Chinese truck gardeners were also found in Yolo County, where in Grafton Township Ah Fook produced $600 worth of truck crops, whereas seven white gardeners produced a total of $1,950 at an average of $279 per farm. In Washington township in the same county, 73 non-Chinese gardeners produced a total of $50,709, at an average of $695 per farm, whereas Ah Puck and Sung Sing each produced $600. In Santa Clara County, although we cannot be entirely sure, it is most likely that the market garden crops of Ah Kay and Company and Ah Lock and

Table 14 Chinese Truck Gardeners Listed in the 1880 Manuscript Census of Agriculture

County and Township	Name of Person or Company	Number of Acres		$ Farm Value	$ Value /Acre[b]	$ Value of Implements
		Tilled	Other[a]			
SACRAMENTO VALLEY						
Tehama County						
Lassen Twp.	Pin Huh (CR)	20	27	—	—	—
	Ah Shew (O)	200	175	15,000	64	200
	Shen Bow (CR)	30	2	—	—	—
	Onn Lop (CR)	30	10	—	—	—
	Sing Lem (CR)	50	4	8,000	157	200
	Ah Jin (CR)	15	6	—	—	—
	Ah Seow (CR)	45	6	—	—	—
	Hop Sing & Co. (CR)	200	125	15,000	67	200
Butte County						
Oregon Twp.	Lee Hip (CR)	30	0	1,000	33	100
	Lee Hop (CR)	25	0	700	28	—
	Ah Chung (CR)	47	0	1,200	26	100
Ophir Twp.	Shipp & Co. (CR)	15	0	—	—	—
	Wing On (CR)	10	0	1,000	100	250
	Ah You (CR)	3	0	300	100	50
	Ah Soon (CR)	5	0	800	160	80
	Fong (CR)	2	0	150	75	10
Colusa County						
Newville Twp.	Ah Hen (CR)	10	0	200	20	125
Yuba County						
East Bear River Twp.	Hong Cum (CR)	6	0	600	100	—
Marysville Twp.	Ah Pon (CR)	10	0	5,000	500	—
Linda Twp.	Ah Quing (CR)	6	0	500	83	—
Nevada County						
Grass Valley Twp.	Ah Wing (CR)	10	80	1,500	58	—
	Ah Sing & Co. (CR 1)	12	8	3,000	214	50
	Ah Sing & Co. (CR 2)	75	10	3,000	39	—
	Tin Loy (CR 1)	9	3	3,000	300	15
	Tin Loy (CR 2)	5	4	3,000	500	10

$ Wages Paid in 1879–80	Kind of Livestock	$ Value of Livestock	$ Market Gardens	Other Products	$ Value of Total Products	$ Product Value /Acre[c]
500	2h, 1p, 24x	150	2,000	100 bu.B, 700 bu.P, $2,200 F, 50 tons H	5,000	143
3,000	6h, 2mc	1,000	4,000	600 bu.B, 1,200 bu.P, $3,400 F	8,500	35
150	—	5	1,000	100 bu.B, 1,100 bu.P, $100 F	1,500	47
200	—	—	800	80 bu.B, 150 bu.P, $650 F	1,800	50
2,000	5h, 24x	600	4,000	400 bu.B, 650 bu.P, $550 F, 25 tons H	5,800	107
100	—	—	500	60 bu.B, 600 bu.P, $600 F	2,000	98
2,000	5h, 24ch, 24x	400	1,000	200 bu.B, 1,050 bu.P, $700 F, 12 tons H	6,000	118
3,000	6h	800	4,300	300 bu.B, 6,850 bu.P, $3,200 F, 40 tons H	9,000	38
—	10h	200	3,000	—	3,000	100
2,000	12h	250	3,000	—	3,000	120
280	5h	100	2,000	—	2,000	43
910	4h	250	3,625	—	3,625	242
800	4h	250	2,000	500 bu.B, 300 bu.P, $20 F, 855 lb. wool	2,500	250
—	—	—	1,000	120 bu.B, 50 bu.P	1,000	333
50	3h, 3p, 2ch	150	1,200	400 bu.B	1,200	240
—	—	—	200	—	200	100
200	6h, 6p	150	3,000 [sic]	3 tons H	2,180 [sic]	238
—	2h	50	1,000	150 bu.P	1,000	167
—	2h	75	500	454 bu.P	500	50
—	2h	100	900	—	900	150
—	6p, 3mc, 15x	170	700	80 bu.P, $30 F, 250 lb. grapes	700	50
768	4mc, 1c, 12ch	220	500	5 bu.P, $500 F, 150 lb. butter	1,500	107
—	—	75	—	—	—	—
—	2p, 12ch	10	500	132 bu.P, 13 bu.F	500	46
—	—	—	500	132 bu.P	500	71

Table 14 *(continued)*

County and Township	Name of Person or Company	Number of Acres		$ Farm Value	$ Value /Acre[b]	$ Value of Implements
		Tilled	Other[a]			
Sutter County						
Yuba Twp.	Ah Sing (CR)	7	0	200	29	50
	Seng Weu (CR)	5	0	150	30	16
	Ah Fong	9	0	400	44	20
Solano County						
Vallejo Twp.	Ah Sam (SC)	40	0	1,600	40	—
	Ah Tim (CR)	20	0	1,000	50	—
	Charley Tim (CR)	37	0	1,600	43	—
	Sam Lee (CR)	20	0	—	—	—
Vacaville Twp.	Ah Hing (SC)	37	15	5,000	125	200
Yolo County						
North Putah Twp.	Song Sing & Co. (CR)	54	0	320	6	70
East Grafton Twp.	Ah Low (CR)	10	0	500	50	25

SAN JOAQUIN VALLEY AND SOUTHERN INTERIOR

County and Township	Name of Person or Company	Number of Acres		$ Farm Value	$ Value /Acre[b]	$ Value of Implements
Amador County						
Township No. 2	Ah Sue & Co. (CR)	20	0	1,500	75	—
	Ah Gun & Co. (SC)	10	0	—	—	100
	Ah Chum & Co. (CR)	8	0	2,000	250	50
Merced County						
Township No. 14	Ah Wong & Co. (SC)	40	0	—	—	—
Fresno County						
3rd District	Ah Gon (CR)	20	0	1,000	50	75
	Ah You (CR)	20	0	—	—	—
Kern County						
5th District	Ah Sing (CR)	16	0	600	38	50
	Sing Lee (CR)	10	10	1,000	83	50
	Ah Ki (CR)	25	0	750	30	40
San Bernardino County						
Riverside Twp.	Ah Sing (SC)	10	0	—	—	—
San Bernardino Township	Heong	10	0	1,200	120	—

CENTRAL AND SOUTHERN COAST

County and Township	Name of Person or Company	Number of Acres		$ Farm Value	$ Value /Acre[b]	$ Value of Implements
Sonoma County						
Santa Rosa Twp.	Jim & Co. (CR)	10	0	3,000	300	110
	Thom & Co. (CR)	8	0	2,500	313	110

$ Wages Paid in 1879–80	Kind of Livestock	$ Value of Livestock	$ Market Gardens	Other Products	$ Value of Total Products	$ Product Value /Acre[c]
—	4h	200	600	100 bu.P	600	86
—	2h	30	1,000	500 bu.P	1,000	200
—	1h	50	500	200 bu.P	500	56
—	2h, 100ch	120	800	—	850	21
300	—	—	1,800	—	1,800	90
300	2h	100	2,000	—	2,200	59
250	2h	100	1,800	—	1,800	90
100	2h, 40ch	150	400	400 bu. barley, $1,200 F, 15 tons H	1,800	35
500	7h, 11p	200	3,500	500 bu.P, 10 tons H	3,850	71
170	3h, 4p	100	550	—	550	55
—	1h	50	200	$1,500 F	1,700	85
—	2h	80	200	2,500 bu.P	1,200	120
—	—	150	2,000	—	2,000	250
80	2h, 4p, 12ch	115	600	1,000 bu. corn, 20 bu.B, 80 bu.P	600	15
—	6h, 2mc, 50ch	100	1,500		1,537	77
—	2h	135	1,500	5 bu.B, 550 bu.P	1,778	89
—	3h, 24ch	154	1,300	200 bu.P	1,300	88
80	1h, 24ch	15	100	80 bu. corn, 5 bu.B, 33 bu.P	600	60
238	2h, 12ch	40	100	1,500 bu. corn	350	14
—	1mc, 1c, 6ch	100	1,300	50 lb. butter	1,300	130
500	4h	200	2,000	—	2,000	200
—	1h, 8p	75	1,500	—	1,500	150
—	1h, 2p	40	1,200	—	1,200	150

Table 14 *(continued)*

County and Township	Name of Person or Company	Number of Acres		$ Farm Value	$ Value /Acre[b]	$ Value of Implements
		Tilled	Other[a]			
Napa County						
Napa City	Ah Jim (SC)	13	5	5,000	357	100
	Ah Yen (CR)	4	0	2,500	625	50
St. Helena Twp.	How Fung & Co. (CR)	0	6	—	—	20
Alameda County						
Brooklyn Twp.	Ah Hong (CR)	65	0	36,000	554	—
	Ah Fang (CR)	5	0	—	—	—
	Hop Ching & Co. (CR)	90	0	50,000	556	200
	Ah Tom & Co. (CR)	44	0	40,000	909	200
	Ching & Co. (CR)	8	0	3,000	375	—
	Ah Look	30	0	9,000	300	—
	Ah Gim & Co. (CR)	50	0	40,000	800	200
	Ah Kei (CR)	40	0	8,000	200	—
	Ko Tong & Co. (CR)	70	0	25,000	359	200
Santa Clara County						
Gilroy Twp.	Moun Fung & Co. (SC)	35	0	2,000	57	20
Santa Clara Twp.	Tye On (CR)	18	0	4,000	222	—
San Luis Obispo County						
San Luis Obispo	Long Ne & Co. (SC)	10	0	8,000	800	50
City	Ah Sing & Co. (SC)	8	0	1,000	125	75
Los Angeles County San Antonio/						
Vernon District	Ah Chung (CR)	20	0	3,000	150	300
Wilmington Twp.	Yak Sa (CR)	50	0	3,000	60	—
San Diego County						
San Diego Twp.	Ah Bank & Ah Shoe (CR)	16	0	2,000	125	200

SOURCE: U.S. Bureau of the Census, "Census of Productions of Agriculture" (manuscript), 1880.

NOTE: Only those Chinese farmers with given values for market garden produce have been included.

CR = cash rent; O = owner-operator; SC = share crop; h = horse; p = pig; ch = chicken, mc = milk cow; c = cattle; x = other animals; B = beans and peas; P = Irish and sweet potatoes; F = fruit; H = hay.

[a] Other acres include two categories: (a) pastures, orchards, and vineyards that were combined in the census schedules, and (b) permanent meadows and woodlands.

[b] In computing $ Value/Acre, the number of "other" acres has been divided by 5 and

$ Wages Paid in 1879–80	Kind of Livestock	$ Value of Livestock	$ Market Gardens	Other Products	$ Value of Total Products	$ Product Value /Acre[c]
450	1h	50	400	$200 F	780	46
200	2h	75	1,000	—	1,000	250
—	2h	50	950	—	950	158
300	4h, 2mc	150	1,300	40 bu.P, 20 tons H	1,400	22
—	—	—	400	500 lb. Canada peas	600	120
1,000	4h	200	2,000	50 tons H	4,600	51
300		300	600	10 tons H	1,000	23
100		100	300	4 tons H	400	50
300	3h	300	600	4 tons H	800	27
—	2h	100	1,000	2,500 bu.P	1,500	30
200	2h	100	600	4 tons H	600	15
1,000	4h	150	2,000	50 tons H	2,000	29
2,000	5h, 6p, 60ch	125	1,600	480 bu. corn, 1,000 bu.P	2,100	60
1,500	1p		1,800	—	1,800	100
—	2h, 50ch	100	800	260 bu.P	1,000	100
110	1h, 12p, 100ch, 50x	75	800	230 bu.P	—	—
—	7h, 15p, 50ch	250	1,500	—	1,800	90
500	6h	175	1,200	10 tons H	1,200	26
500	4h, 24ch	200	1,500	280 bu.P	2,000	125

then added to the number of "tilled" acres to form the denominator. The manuscript schedules of the census of agriculture show that the acreage in pastures, meadows, and woodlands far exceeds that in orchards and vineyards, so "other" acres consist mainly of untilled land whose value is estimated to have been one-fifth the value of tilled land.
[c] The denominator used in computing $ Product Value/Acre is the sum of the "tilled" acres and the acres in orchards and vineyards. The latter are shown in the manuscript schedules of the census of agriculture, but due to lack of space are not listed separately in this table but are subsumed under "other" acres. Therefore, the denominator used to compute $ Product Value/Acre on those farms containing orchards and/or vineyards is not explicitly given in this table.

Company were strawberries rather than vegetables. Overall for the state, then, the Chinese dominated market gardening in certain areas of Tehama and Butte counties, while elsewhere their product value per acre was comparable to that of whites.

In the 1870–80 decade, three developments became apparent: instead of growing truck crops exclusively, Chinese gardeners had become farmers operating diversified farms; this resulted in a decrease in the product value per acre of their farms, because other crops were worth less per acre than truck crops; and differential changes in land values made economic considerations really important in terms of the location of truck gardens and farms.

As table 14 shows, there were 67 Chinese truck-gardening units (cultivated by 65 individuals or groups) listed in the manuscript agriculture census of 1880, only 19 of which specialized completely in truck crops. This is in distinct contrast to earlier decades when Chinese gardeners grew vegetables and almost nothing else. Because the Chinese truck gardeners now grew truck crops along with other crops, it is no longer possible to calculate the value per acre of their truck crops per se. While the product value per acre has been computed (see the last column), it is not possible in most cases to say which part of the total value came from vegetables, because no information on the number of acres devoted to truck crops was given in the 1880 agriculture census.

It is possible, however, to compute the product value per acre for all crops taken together. In terms of dollars—without taking changes in the price index into account—the product value per acre seemed to have decreased drastically in the 1870–80 decade. Not only that, but by this time the ratio of average product value per acre to average farm value per acre had fallen below one, which means that while the value of land and improvements on it had increased during the decade, the value of crops grown on the land had not kept pace with the increase in land values. To compensate for the lower product value per acre, truck gardeners of all ethnicities had to increase the *scale* of their operations. Census information indicates that Chinese truck gardeners indeed did that. Though the smallest truck farm listed in the 1880 manuscript agriculture census was only 4 acres located within Napa City, most farms producing truck crops were larger, the average being approximately 40 acres—a significant increase

over the size of most Chinese-operated truck farms a decade earlier. In 1880 the largest Chinese-operated truck farm, of
200 acres, was located in Lassen Township, Tehama County,
to the east of Tehama Township.

A great divergence in land values had occurred in the 1870s.
The gap between the value of farms located within urban or suburban areas and of those located in the countryside was now very
large. In Brooklyn Township, Alameda County, south of the city
of Oakland, the value per acre of Chinese truck farms ranged
from $200 to $909. In the valleys north of San Francisco Bay,
truck farms in Santa Rosa, Sonoma County, and Napa City,
Napa County, were worth $300 to $625 per acre. In Marysville,
Yuba County, truck farms were valued at $500 per acre. Even in
a small city like San Luis Obispo, farms were valued at $125
to $800 per acre. In contrast, property values in more rural
regions ranged from approximately $40 to $100. Yet, the product
value per acre was *higher* in the farms situated in rural areas. A
logical outcome of these changes in land value was that the most
rational decision truck gardeners could make was to lease land
in suburban fringes outside of city limits. Here, land values—
and consequently rent—would be lower than within city limits,
but distance from markets would not be so great that the distribution and sale of their fresh produce would be impaired.

The geographic shifts of Chinese truck gardening as a result of
population movements and changes in land values are well illustrated in maps 4, 5, and 6. Map 4 shows that in 1860 there were
many clusters of Chinese truck gardeners scattered throughout
the mining counties. In general, they were rather evenly distributed, each cluster being located less than ten miles away from
the next. All were close to the major mining camps. Sacramento
City contained the only large urban concentration. The relatively widespread distribution of Chinese truck gardeners can be
explained by three factors: the wide dispersal of the mining
population; a relatively primitive system of transportation; and
the lack of storage facilities for perishable fresh produce. Truck
gardeners therefore had to be scattered in many different localities, relatively close to their customers—the miners and the
urban dwellers—so that their fresh produce could be taken to
market without undue delay.

May 5 shows that geographic concentration had begun to take

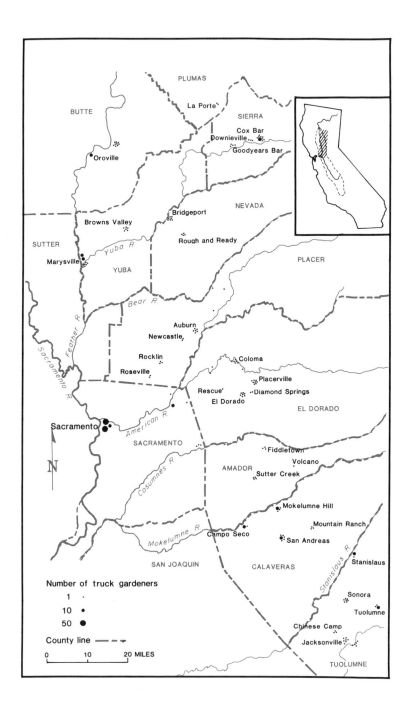

PLUMAS

La Porte

BUTTE

SIERRA

Cox Bar
Downieville
Goodyears Bar

Oroville

NEVADA

Bridgeport

Browns Valley

SUTTER

Yuba R.

Rough and Ready

PLACER

Marysville

YUBA

Bear R.

Feather R.

Auburn
Newcastle

Sacramento R.

Rocklin

Roseville

Coloma

Placerville

Rescue
El Dorado

Diamond Springs

EL DORADO

Sacramento

American R.

SACRAMENTO

Fiddletown
Volcano

Cosumnes R.

AMADOR
Sutter Creek

N

Mokelumne Hill

Mountain Ranch

Mokelumne R.

Campo Seco

San Andreas

SAN JOAQUIN

CALAVERAS

Stanislaus R.

Stanislaus

Number of truck gardeners

1   ·
10 •
50 ●

County line — — —

0   10   20 MILES

Sonora

Tuolumne

Chinese Camp

Jacksonville

TUOLUMNE

place by 1870. Instead of being scattered in clusters throughout the mining regions, Chinese truck gardeners now were congregated on the outskirts of growing towns, such as Placerville in El Dorado County, Auburn in Placer County, and Grass Valley in Nevada County. These towns had developed merchandising functions, and the cultivation and sale of fresh vegetables was one of them. Marysville and Sacramento City now shared prominence as truck gardening centers in the Sacramento Valley.

The process of geographic concentration can be seen even more clearly in map 6, which shows the distribution of Chinese truck gardeners in 1880. The small clusters of Chinese gardeners had almost disappeared from the smaller mining camps, mainly because many Chinese and other miners had left those camps. On the other hand, mining camps which had developed into towns and cities—such as Placerville, Auburn, Grass Valley, and Nevada City—now supported larger groupings of Chinese truck gardeners. Nevada County, in particular, was experiencing an economic boom as hardrock mining in the North Bloomfield and Empire mines brought great prosperity to the region. Sacramento and Marysville maintained their prominence, and fresh produce from their truck gardens was now transported to outlying areas, given improvements in the means of transportation.

It is impossible to compare the income of Chinese truck gardeners with that of Chinese engaged in other occupations, because there is little reliable information on the income of Chinese in nineteenth-century California. Available figures do not reflect regional variations or changes over time, and it is hard to decide how representative they might have been. More reliable information exists on the value of personal and real property held by Chinese, as recorded in the manuscript schedules of the 1860 and 1870 censuses of population. (Later censuses did not collect similar information.) Economists consider property holdings to be a measure of wealth—a concept distinct from

Map 4   *Northern California: Chinese Truck Gardeners, 1860*
SOURCES: *Based on data in U.S. National Archives, Record Group 29, "Census of U.S. Population" (manuscript), 1860, and on historical maps of Butte, Plumas, Sierra, Yuba, Nevada, Placer, El Dorado, Sacramento, Amador, Calaveras, and Tuolumne counties. (See list of historical maps for exact citations.)*

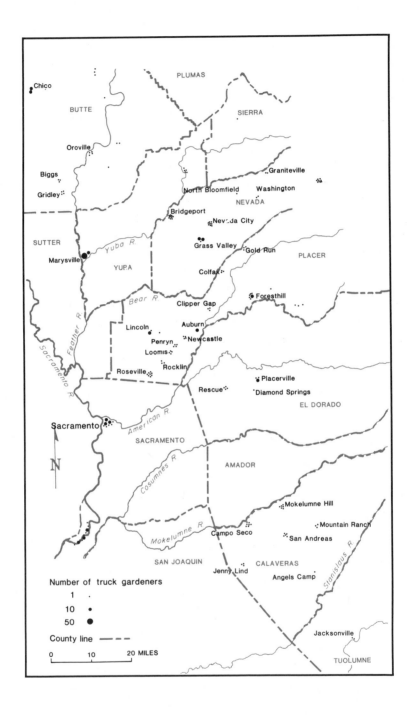

Number of truck gardeners

1   .
10   •
50   ●

County line   - - -

0   10   20 MILES

Chico

BUTTE

PLUMAS

SIERRA

Oroville

Biggs

Gridley

Graniteville

North Bloomfield   Washington

NEVADA

Bridgeport

Nevada City

SUTTER

Yuba R.

Grass Valley

Gold Run

Marysville

YUBA

Colfax

PLACER

Bear R.

Foresthill

Clipper Gap

Lincoln

Auburn

Penryn

Newcastle

Loomis

Rocklin

Roseville

Placerville

Rescue

Diamond Springs

EL DORADO

Sacramento

American R.

SACRAMENTO

Cosumnes R.

AMADOR

Mokelumne Hill

Mountain Ranch

Mokelumne R.

Campo Seco

San Andreas

SAN JOAQUIN

CALAVERAS

Jenny Lind

Angels Camp

Stanislaus R.

Jacksonville

TUOLUMNE

Feather R.

Sacramento R.

N

income—and even though only a small number of Chinese reported property holdings, the information provides a rough indication of how well truck gardeners fared vis-à-vis other Chinese.

The comparison of different groups of Chinese property owners will be confined to those in the counties of the Northern and Southern mines. For unknown reasons, no Chinese except a lone servant was listed as a property holder in the city of Sacramento in 1860; the number of property owners in San Francisco was also far smaller than expected; likewise, census takers listed almost no Chinese as owners of property in the Klamath/Trinity mining region.

Chinese property owners, for the sake of comparison, have been divided into six groups: miners, truck gardeners, merchants and grocers, physicians and druggists, laundrymen, and miscellaneous persons with property. The last category encompasses a variety of individuals, including butchers, bakers, restaurant keepers, hotel keepers, tailors and seamstresses, cooks, teamsters and packers, blacksmiths, carpenters, barbers, cigar makers, woodchoppers, laborers, servants, clerks, prostitutes, and housewives. These are lumped together simply because the number of each was too small to be meaningfully analyzed separately.

No attempt will be made to compute the proportion of property holders to the total number of persons in a particular occupational category because there is no way to ascertain how many Chinese with property gave accurate or honest information to the census takers. Given their experience with discriminatory taxes and license fees, Chinese had every reason to withhold information about their wealth and income. It is probable, therefore, that only persons with visible property holdings, such as merchants with their stocks of goods, herbalists with their apothecary, or truck gardeners with their plots of land and tools, reported their holdings with any accuracy. Miners, in particular,

Map 5   *Northern California: Chinese Truck Gardeners, 1870*
SOURCES: *Based on data in U.S. National Archives, Record Group 29, "Census of U.S. Population" (manuscript), 1870, and on historical maps of Butte, Plumas, Sierra, Yuba, Nevada, Placer, El Dorado, Sacramento, Amador, Calaveras, and Tuolumne counties. (See list of historical maps for exact citations.)*

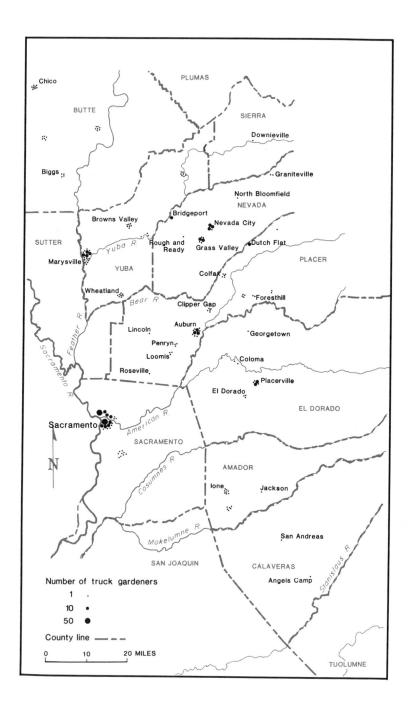

Chico

BUTTE

PLUMAS

SIERRA

Downieville

Biggs

Graniteville

North Bloomfield

NEVADA

Bridgeport

Browns Valley

Nevada City

SUTTER

Yuba R.

Rough and
Ready

Grass Valley

Dutch Flat

Marysville

YUBA

Colfax

PLACER

Wheatland

Bear R.

Foresthill

Clipper Gap

Feather R.

Lincoln

Auburn

Georgetown

Penryn

Loomis

Coloma

Sacramento R.

Roseville

Placerville

El Dorado

EL DORADO

Sacramento

American R.

N

SACRAMENTO

Cosumnes R.

AMADOR

Ione

Jackson

Mokelumne R.

San Andreas

SAN JOAQUIN

CALAVERAS

Stanislaus R.

Angels Camp

Number of truck gardeners

1 .

10 •

50 ●

County line — —

0   10   20 MILES

TUOLUMNE

probably underreported the amount of the gold dust in their possession, since that kind of property was easier to hide. Neither
has an effort been made to compare the property holdings of
Chinese and non-Chinese in each occupational grouping,
because only truck gardeners are under consideration here.

As table 15 shows, many more Chinese in the southern mining
region in 1860 seemed to own property than those in the northern mining region. However, it cannot be assumed with certainty
that this was an accurate rendering of the true situation.
Nonetheless, it is possible to make comparisons within each
occupational group. Chinese miners in the Southern Mines
seemed to hold a smaller amount of personal property but a considerably larger amount of real property than their peers in the
north. This was probably an accurate depiction because, in
general, as corroborated by county archival records, relatively
more Chinese in the southern area bought mining grounds and
agricultural land than those in the north. The average value of
personal property held by truck gardeners in the two regions was
almost the same—$178 and $172—but too few were shown with
real property to make any meaningful statements about the
latter. Gardeners in the south owned more personal property
than miners, holding an average of $178 versus $131, while those
in the north were on an even par with miners. Overall, gardeners
ranked below merchants and those in nonagricultural occupations, but they were better off than laundrymen.

Among property owners who resided in towns or mining
camps, merchants seemed most wealthy. However, it should be
remembered that much of the personal property they held was
inventory, which cannot properly be regarded as things they truly
owned, because these were stocked only temporarily for trade
and exchange. There was little difference in the amount of personal property owned by Chinese merchants in the two mining
regions. However, the average value of the real property mer-

Map 6   *Northern California: Chinese Truck Gardeners, 1880*
_____   SOURCES: *Based on data in U.S. National Archives, Record*
Group 29, "Census of U.S. Population" (manuscript), 1880,
and on historical maps of Butte, Plumas, Sierra, Yuba, Nevada,
Placer, El Dorado, Sacramento, Amador, Calaveras, and Tuo-
lumne counties. (See list of historical maps for exact citations.)

	Southern Mining Region[a]		Northern Mining Region[b]	
	Personal Property	Real Property	Personal Property	Real Property
Miners				
Number listed with property	756	325	127	2
Total amount of property	$98,769	$245,555	$22,590	$600
Average amount held	$131	$756	$178	$300
Truck Gardeners				
Number listed with property	70	9	51	1
Total amount of property	$12,425	$1,500	$8,780	$100
Average amount held	$178	$167	$172	$100
Merchants and Grocers[c]				
Number listed with property	205	64	177	18
Total amount of property	$112,075	$16,140	$101,236	$7,650
Average amount held	$547	$252	$572	$425
Physicians and Druggists				
Number listed with property	33	4	26	2
Total amount of property	$7,975	$310	$7,510	$2,000
Average amount held	$242	$ 78	$289	$1,000
Laundrymen				
Number listed with property	12	2	27	0
Total amount of property	$935	$200	$2,585	0
Average amount held	$ 78	$100	$ 96	0
Other Persons with Property				
Number listed with property	160	26	34	0
Total amount of property	$27,552	$5,030	$8,340	0
Average amount held	$172	$193	$245	0

SOURCE: Compiled and computed from U.S. National Archives, Record Group 29, "Census of U.S. Population" (manuscript), 1860.

[a] Includes Amador, Calaveras, El Dorado, Mariposa, and Tuolumne counties.

[b] Includes Butte, Nevada, Placer, Plumas, Sierra, and Yuba counties.

[c] Includes all persons listed as merchants, storekeepers, traders, and grocers; excludes peddlers, butchers, and fish vendors.

chants possessed in the southern mining region was considerably smaller than that in the north—$252 versus $425. Both figures confirm their top economic ranking within Chinese communities. The personal property held by physicians and druggists also cannot be considered all theirs, since they too stocked herbs and other medicines needed in their professional practice. In both mining regions, this group ranked next to merchants in the average value of personal property in their possession. Laundrymen were at the bottom of the economic ladder, owning very little personal or real property, while the miscellaneous other persons with property seemed about as well off as miners and gardeners.

A decade later, as is showned in table 16, the number of Chinese property owners had declined greatly, and a noticeable divergence had appeared between the prosperity of the Chinese in the southern and that of those in the northern mining region. Miners in the north now were more than five times as wealthy as those in the south in terms of personal property held, and eight times as well-to-do with regard to real property. Truck gardeners in the north held more than twice the average amount of personal property as their peers in the south, while the average value of their real property was triple. Truck gardeners in the south were better off than miners, who were not doing well at all, while those in the north were less so. Compared with property holders in the urban trades, truck gardeners in the southern region ranked below merchants and the miscellaneous group, but above physicians and laundrymen in the average value of personal property held. They also had the second highest average value of real property possessed, next only to merchants. In the northern region, gardeners held the smallest average amount of personal property, but the largest average value of real property. Such a discrepancy may be explained by the great increase in land values over the 1860–70 decade discussed earlier. The greatest change in economic status was experienced by northern laundrymen—at least, those with property seemed to be doing much better than ten years earlier.

Though truck gardeners in 1860 did not possess as much tangible property as miners, in many ways gardening as an occupation had certain nonpecuniary advantages over mining— advantages that might have been more important than the monetary ones in causing certain individuals to take up truck

Table 16 Personal and Real Property Held by Chinese in the Mining Regions, 1870

	Southern Mining Region[a]		Northern Mining Region[b]	
	Personal Property	Real Property	Personal Property	Real Property
Miners				
Number with property	126	62	481	165
Total amount of property	$25,778	$7,320	$120,815	$33,250
Average amount held	$205	$118	$251	$202
Truck Gardeners				
Number with property	8	9	24	17
Total amount of property	$1,650	$2,250	$10,550	$10,750
Average amount held	$206	$250	$440	$632
Merchants and Grocers[c]				
Number with property	95	64	132	52
Total amount of property	$87,000	$24,100	$162,750	$25,150
Average amount held	$916	$377	$1,233	$484
Physicians and Druggists				
Number with property	1	2	26	5
Total amount of property	$200	$300	$12,450	$1,500
Average amount held	$200	$150	$479	$300
Laundrymen				
Number with property	2	4	18	7
Total amount of property	$100	$350	$14,500	$2,500
Average amount held	$ 50	$ 88	$806	$357
Other Persons with Property				
Number with property	47	21	232	28
Total amount of property	$11,200	$3,050	$58,575	$7,600
Average amount held	$238	$145	$252	$271

SOURCE: Compiled and computed from U.S. National Archives, Record Group 29, "Census of U.S. Population" (manuscript), 1870.

[a] Includes Amador, Calaveras, El Dorado, Mariposa, and Tuolumne counties.

[b] Includes Butte, Nevada, Placer, Plumas, Sierra, and Yuba counties.

[c] Includes all persons listed as merchants, storekeepers, traders, and grocers; excludes peddlers, clerks, butchers, and fish vendors.

gardening as a livelihood. From the 1850s on, Chinese miners were frequently subjected to physical violence from both white miners and tax collectors. Gardeners experienced less harassment because they were not perceived to be removing the "natural wealth of the land," which many white miners felt "rightfully" belonged only to them. Also, a few acres of land provided opportunities to earn money in supplementary ways. For example, in Yuba County, the truck gardeners supplemented their income by raising pigs; in Sacramento County they grew potatoes, peas, and wheat. Finally, the income from truck gardening was more evenly distributed over the year, whereas mining was always a gamble, with a few rich strikes barely making up for the days, weeks, months, and years of hopeful but not necessarily remunerative labor.

Although the urban enterprises were obviously more lucrative, they required capital. Merchants, herbalists, restaurant or boardinghouse keepers all needed capital to set up their businesses. Moreover, gardening did not involve any complicated professional skills such as those needed by druggists, blacksmiths, or tailors. Requiring only a few simple tools and a small plot of ground, truck gardening was a relatively easy occupation to enter. In general, it was a pedestrian occupation that provided a modest but steady living. To place in perspective the relative importance of truck gardening as an occupation among the Chinese population, table 17 shows that more than 5 percent of the total Chinese population were truck gardeners in only Sacramento County in 1860; but this was true in Alameda, Los Angeles, Santa Clara, and Yolo counties in 1870; in Los Angeles, Sutter, Tehama, Tulare, and Yolo counties in 1880; and in Shasta, Sutter, Tehama, and Yuba counties in 1900.

Eventually, there were even occasional success stories among Chinese truck gardeners. In 1886, for example, an article in the *Pacific Rural Press* made truck gardening sound almost lucrative. Commenting on the Los Angeles area, the writer said:

> There is much said about the money-making in oranges, lemons and grapes in Los Angeles, but the Chinamen are supposed to be making more money out of common plebeian cabbages than any orange grower or vigneron [wine-maker] in the country. There is an enormous demand for cabbage for shipment to the East.... The supply is at all

Table 17   *Chinese Truck Gardeners as a Percentage of the Total Chinese Population by County, Selected Counties, 1860–1900*

	1860			1870		
	Number of Chinese Gardeners	Total Number of Chinese	Percent Who Are Gardeners	Number of Chinese Gardeners	Total Number of Chinese	Percent Who Are Gardeners
MINING COUNTIES						
Butte	21	2,177	0.96	41	2,082	1.97
Sierra	29	2,208	1.31	1	810	0.12
Yuba	34	1,781	1.91	66	2,337	2.82
Nevada	16	2,147	0.75	52	2,627	1.98
Placer	36	2,392	1.51	63	2,410	2.61
El Dorado	30	4,762	0.63	17	1,560	1.09
Calaveras	51	3,657	1.39	30	1,441	2.08
Tuolumne	42	1,962	2.14	2	1,524	0.13
SACRAMENTO VALLEY						
Shasta	–	(415)	–	15	574	2.61
Tehama	–	(104)	–	5	294	1.70
Sutter	–	(2)	–	–	(208)	–
Sacramento	125	1,731	7.22	72	3,195	2.25
Yolo	–	(6)	–	23	395	5.82
Solano	–	(14)	–	2	920	0.22
SAN FRANCISCO BAY AREA						
Alameda	–	(193)	–	118	1,939	6.09
San Francisco	31	2,719	1.14	123	12,022	1.02
Santa Clara	–	(22)	–	104	1,525	6.82
SAN JOAQUIN VALLEY						
San Joaquin	0	139	0	0	1,626	0
Fresno	–	(309)	–	–	(427)	–
Tulare	–	(13)	–	–	(99)	–
Kern	–	(–)	–	–	(143)	–
COASTAL COUNTIES						
Monterey	–	(6)	–	–	(230)	–
Los Angeles	–	(11)	–	19	234	8.12

SOURCE: Number of gardeners tallied from U.S. National Archives, Record Group 29, "Census of U.S. Population" (manuscript), 1860, 1870, 1880, and 1900. Total number of Chinese from U.S. Bureau of the Census, *Census of Population*, 1860, 1870, 1880, and 1900.
NOTE: A dash (–) means no tally of gardeners was made for the county.

times less than the demand.... The Los Angeles cabbage is greatly preferred by the Eastern dealers.... The price at present is $1.50 per hundred head or $30 a ton. In this climate there is no trouble to grow two crops of cabbage per year.... They are planted two feet apart which will give 10,000 to the acre. This is 25 tons to the crop or 50 tons a year off an acre of good ground. The gross results of such a

1880			1900		
Number of Chinese Gardeners	Total Number of Chinese	Percent Who Are Gardeners	Number of Chinese Gardeners	Total Number of Chinese	Percent Who Are Gardeners
82	3,793	2.16	31	712	4.35
2	1,252	0.16	–	(309)	–
87	2,146	4.05	83	719	11.54
93	3,003	3.10	11	632	1.74
75	2,190	3.42	8	1,050	0.76
13	1,484	0.88	0	206	0
2	1,037	0.19	–	(148)	–
9	805	1.12	–	(158)	–
49	1,334	0.37	31	102	30.39
183	774	23.64	7	729	0.96
43	266	16.17	50	274	18.25
184	4,892	3.76	58	3,254	1.78
46	608	7.57	7	346	2.02
26	993	2.62	2	903	0.22
21	4,386	0.48	54	2,211	2.44
35	21,745	0.16	23	13,954	0.16
18	2,695	0.67	41	1,738	2.36
86	1,997	4.31	5	1,875	0.27
7	753	0.93	71	1,775	4.00
33	324	10.19	6	370	1.62
20	702	2.85	8	906	0.88
27	372	7.3	20	857	2.33
208	1,169	17.79	95	3,209	2.96

crop is $1,500 an acre.... Even with labor [costs ... there would be] a net profit of $1,000 per acre. It is said that some of the Chinese make that much.[28]

The plebeian cabbage indeed! The Chinese had come to mine for gold; for a few of them, the nuggets they found turned out to be green.

FIVE

New World Delta

WHEN Chinese first began to cultivate the soil in California, they did so on a relatively small scale, at least by California standards. In the 1860s the size of the farms leased by Chinese ranged from 5 to 20 acres in the mining regions and the valley towns, 10 to 25 acres in Santa Clara County, 40 to 60 acres in Alameda County, and about 10 acres in San Mateo County. Vegetables and small fruit were planted on the majority of these farms. A departure from this pattern of small-scale cultivation came in the early 1870s when Chinese began to lease larger plots—many more than 100 acres—in the Sacramento–San Joaquin Delta and along the Sacramento River and its tributaries to grow other kinds of crops, including wheat, corn, barley, alfalfa, potatoes, beans, and onions. An early example of Chinese who planted field crops was Ah Chow and Company, who leased 35 acres from Charles M. Swinford in 1869 to grow wheat on land adjoining the Sacramento River in Colusa County.[1] Five years later, in the same county, Yep Ah Chung leased 200 acres on the west bank of the Sacramento River from Joseph Hamilton to plant alfalfa.[2] In 1873 Chou Ying and Wee Ying leased three plots totaling 551 acres from George D. Roberts and Joseph Roberts, Sr., in the Sacramento Delta, but it is not known what they used the land for;[3] most likely, they grew potatoes. In 1874 Ah Sung and Company leased four parcels totaling 420 acres from Parham Wall to grow castor beans on bottomland along the east bank of the "Old Bed of the Yuba River" in Yuba County.[4] Other leases similar to these followed. By the late 1870s Chinese tenant farmers could be found in many places throughout northern and central California.

In the 1880s Chinese tenants began to enter the semiarid San Joaquin Valley, where they planted vineyards and orchards along rivers and dug miles of irrigation ditches—work for which they seldom received any compensation. They also became more numerous in the valleys along the northern, central, and southern California coast, where they practiced diversified farming. Although scattered narrative sources indicated that Chinese also worked in the citrus groves of Los Angeles, Orange, Riverside, and San Bernardino counties, they did so only as seasonal workers. County archival records provide little evidence that they leased any citrus groves. Chinese tenants in southern California were primarily vegetable and celery growers. In Santa Barbara they also produced seeds for sale. In general, Chinese entry into tenant farming in a particular region coincided with the date when that region was opened up for intensive cultivation, but it would be wrong to assume that they undertook only intensive agriculture.

In this study, California's agricultural regions have been divided into the Sacramento Valley, the Sacramento–San Joaquin Delta, the San Joaquin Valley, the smaller coastal valleys, the southern California coastal plains, and the Imperial and Coachella valleys in the southern interior desert. Chinese farmed actively in all but the last region. Of the other five, Chinese tenant farmers were most numerous and their historical significance greatest in the Sacramento–San Joaquin Delta. It would add a romantic touch to our story were it possible to claim that Chinese immigrants came to farm in the Delta because it reminded them of home—the Pearl River delta of Kwangtung Province, where most of them had lived. Such a case cannot be made, however, because though the Chinese no doubt saw the deltaic marshlands as they rode steamers up the Sacramento River on their way to the mining regions, they did not enter the Delta to farm on their own initiative. They first went to work in the area when white landowners recruited them to do reclamation work. Only after they became aware of the extraordinary fertility of Delta soil did they lease land to grow crops.

What is this place called the Delta?[5]

The Delta

California's two great river systems, the Sacramento and the
San Joaquin, and their respective tributaries, drain more than
one-third of the state's area. Their floodplains make up the
great Central Valley, the northern half of which is called the
Sacramento Valley, the southern half, the San Joaquin Valley.
The Central Valley, one of the world's richest farming areas, is
an elongated plain about four hundred miles long and an average
of fifty miles across, surrounded on all sides by mountains.
The coastal ranges which form the western edge of the valley are
broken at only one spot, east of the Carquinez Straits. This single
break, flanked to the north by the Montezuma Hills and to the
south by the Mount Diablo Range, is so narrow that the waters
of the two great river systems back up for miles as they push
through this outlet to the sea. Over the centuries, a marshy
deltaic swamp had formed. Unlike most estuarine deltas, which
broaden as they approach the ocean, the Sacramento–San Joaquin Delta diminishes in width as it approaches the highlands
abutting the Carquinez Straits, its streams merging rather than
separating as they proceed seaward. The Delta, twenty-four
miles across at its widest point and forty-eight miles along its
longest axis, is thus shaped like a rough parallelogram, with its
northern, eastern, southern, and western apices marked by the
cities of Sacramento, Stockton, Tracy, and Antioch, respectively.

Tall bulrushes, known as tules, covered the marshy swamp.
Before people tamed this land, the tules grew and died each year,
decaying to become a rich organic soil known as peat. In the
islands of the central Delta, the peat layer is as thick as forty
feet. Many of the peat islands are shaped like shallow bowls
with central depressions—known as backswamps—rimmed by
natural and man-made levees, the former formed by sediment
carried by the waters of the meandering sloughs, the latter built
to protect the land from periodic floods.

Farming in the Delta has been both profitable and risky. Aside
from raging torrents that have sometimes washed out entire
islands, the Delta suffers from soil subsidence: over time, the
centers of the islands have sunk lower and lower so that today
almost three-fifths of the Delta's area is at or below sea level.
Modern roads have been built on top of the levees, and many a

tourist has had the strange experience of driving along roads that
are on the same level as the tops of fruit trees growing in the
orchards on both sides. The problem of subsidence has been
compounded by the rise in the height of riverbeds as more and
more sediment settles in the channels. The danger of floods was
most acute during the 1860s and 1870s, when hydraulic mining
prospered in California and vast amounts of "slickens" were
washed down from the foothills, clogging many rivers, raising
their beds, forming sandbars, making them unnavigable, and
flooding adjacent farmland, the mud ruining stream water for
irrigation or drinking. Ironically, reclamation itself has created
increased risks: as levees were built to enclose more and more
islands and tracts, whenever there was a large runoff the water
in the rivers—being confined to their constricted channels—had
a greater tendency to overflow the levees into the depressed fields
on the islands. After floods subsided, the existence of levees also
made it harder to drain the areas that were wet. Powerful pumps
have to be used, adding to the cost of farming. In more recent
years, salinity has also become a cause for worry as more and
more of the water from the Sacramento River and its tributaries
is transported through man-made channels to the southern part
of the state, leaving less to flow through the Delta to flush out
the salt brought in by seawater flowing inland through San
Francisco Bay during high tides.

Delta soil conditions have affected various crops differently.
Potatoes are a case in point: virgin peat soil provides extraordinarily large yields, but at the same time, peat soil with its high
organic content develops a fungus after it has been planted to
potatoes for several years.[6] So farmers who wish to grow potatoes
from one year to the next must move from field to field. Another
important Delta crop, asparagus, depletes certain nutrients in
the soil, and after twenty years or so the yields drop drastically.
But unlike land planted to other crops, fertilizers do not improve
the yield of old asparagus fields on peat land. Yet, it is uneco-
nomical to dig the aging asparagus plants out because their
rhizomes—the source of food for the asparagus shoots—grow
very large, spreading at least eight feet across in the ground.
Digging up old asparagus fields is a backbreaking job, and fields
are generally not dug up until yields have greatly declined.[7]
Various fruit trees, too, have had to be planted selectively. The

Delta's high water table is detrimental to peach, plum, apricot, and cherry trees, which, accordingly, grow well only on the high natural levees. The Bartlett pear, on the other hand, grows well despite the high water table and has been the most important tree fruit produced in the area.[8]

The Delta has also had a rich social history. Perhaps nowhere else in California has an area's social structure so closely mirrored its biological and physical ecological balance. People of many different ethnic backgrounds have worked together closely for 130 years to make the Delta into one of the state's most productive agricultural areas. The Delta is one of the few places left in rural California where third-, fourth-, and even fifth-generation descendants of some of the early settlers continue to farm the land in a complex pattern of interaction between man and nature, and between one ethnic group and another. In this historical development, Asian immigrants have played a central role. More than any other place in rural California, the Delta has provided the conditions conducive to the establishment of relatively stable and settled Asian communities. Chinese have been involved in all stages of farm-making in the Delta: reclaiming the swamps, clearing the land, breaking up the sod for cultivation, leasing part of the land to grow crops, and harvesting and packing them for marketing.

Public Land Disposal, Reclamation, and Agricultural Development

Although the Delta is usually thought of as a single entity, in fact subareas within it—containing different combinations of organic and mineral content and reclaimed at different dates— exhibit two divergent patterns of landownership and land use. Many small and medium-sized farms are found in the northern mainland tracts along both banks of the Sacramento River in Yolo, Solano, and Sacramento counties and on the natural levees of the peat islands around the periphery of the Delta. Larger farms are found in the centers of the islands of the northern Delta in Sacramento and Solano counties and on both the mainland tracts and the islands of the southern Delta in San Joaquin and Contra Costa counties. During the nineteenth century, in general, the smaller farms grew vegetables and fruit in combination with small amounts of grain, while the larger farms in the back-swamps of the peat islands produced primarily grain, potatoes, and beans. Asparagus—first grown commercially in 1892 on Bouldin Island, and for many years considered the queen crop of the Delta—was grown on farms of different sizes.

Almost all the land in the Delta was granted by the federal government to the state of California, which then disposed of it. When California became a state, with the exception of land in the Spanish-Mexican land grants ultimately confirmed by the federal government, the rest of the state became part of the public domain. In subsequent years, a number of federal enabling acts created the legal and administrative framework under which public land passed into the hands of the state of California, railroad companies, and private individuals. California gained title to part of the public domain under four major acts: the Act of September 4, 1841—enacted to allow the federal government to grant 500,000 acres each to certain states for the purpose of internal improvement—under which California received the allotted amount; the Act of September 28, 1850—known as the Arkansas Act—which granted California over 2,200,000 acres of swamp and overflowed lands, proceeds from the sale of which were to be "applied, exclusively, so far as necessary, to the purpose of reclaiming such lands by means of levees and drains"; the Act of March 3, 1853, which granted 5,500,000 acres to California for the establishment of public schools; and the Act of July 2, 1862—known as the Morrill Act—which granted the state 150,000 acres for setting up "colleges for the cultivation of agriculture and mechanical science and arts."[9] Altogether California received almost 8,500,000 acres from the federal government, which made it one of the largest recipients of federal grants from the public domain. The other two large recipients were claimants to the Spanish-Mexican land grants, who as a group received a total of 8,800,000 acres, and railroad companies, which received about 11,500,000 acres.[10] Compared with these major recipients, individual settlers had greater difficulty in acquiring land, a fact that gave rise to an outcry against "land monopoly."[11]

Approximately one-quarter of the swamp and overflowed lands granted to California lay in the Sacramento–San Joaquin

Delta. This land did not pass to the state all at once. The federal government was supposed to survey the land, and the state was then to select (or "locate") sections from surveyed areas. However, California passed laws in 1855, 1858, 1859, 1863, 1868, and 1872 to provide for the disposal of swamp and overflowed lands *prior* to the completion of the federal survey.[12] The state was thus selling land it had not yet formally received, and it created great confusion and endless legal problems in the process.

The manner in which California disposed of its swamp and overflowed lands favored the creation of large holdings. Originally, there was an acreage limitation attached to the sale of swamp lands—the federal government had set a limit of 160 acres on each individual entry, but California raised this limit to 320 acres. Purchasers who desired large tracts preferred therefore to obtain their land from the state. Moreover, many loopholes existed in California's laws. The 1855 Act was poorly worded and added to the confusion of titles arising from the federal-state dispute. Worse, the 1855 Act did not even require that land applied for as swamp and overflowed be under water or at least be wet part of the year![13] Finally, even though the main purpose of the grant was to enable the state to get such land reclaimed, persons who purchased the land with cash were not required to reclaim it at all, while those who bought the land on credit had to make only a small down payment, allowing them to hold the land for speculation at a very small cost. The 1858 law abolished credit purchases, but it was superseded by the 1859 law, which not only reinstated the credit purchase option but increased the acreage limitation to 640 acres. In addition, affidavits of the intent to settle on the land and to reclaim it were no longer required. In 1868 acreage limitations were abolished altogether, so all obstacles to the purchase of thousands of acres by single parties had now been removed.[14]

Removal of acreage limitations on swamp land sales initially led to an increase in the number of large parcels held by individual owners, but after land reclamation began, there was a tendency to subdivide some of the largest tracts. However, there was a floor below which subdivision never fell, because landowners were careful not to subdivide to a point beyond which they would lose control over a reclamation district: the more owners there were, the harder it was to get agreement on whether and when to build reclamation works. Smaller landowners tended to be more reluctant to share the cost of building levees, drains, and ditches.[15]

The earliest reclamation projects were undertaken on a small scale by private individuals using their own funds. There was incentive for undertaking such work, because after making a down payment of only 20 percent on their purchase price, landowners were credited with payment in full if they convinced a commission of three persons that they had reclaimed and cultivated their land for three years. After 1872 landowners were credited with payment in full if they spent only two dollars per acre on reclamation,[16] which allowed individual buyers to obtain hundreds and even thousands of acres at almost no cost.

The state assumed responsibility for reclamation in 1861 by creating a three-person Board of Reclamation Commissioners with authority to form reclamation districts if one-third of the landowners in an area so petitioned. One dollar per acre was to be allocated from the state's swamp land funds to pay for surveys and reclamation works. Later, reclamation districts were given taxing and bonding powers to enable them to raise more revenue, for reclamation proved to be far more expensive than anyone had envisioned. The danger of repeated floods meant there was no guarantee that an area, once reclaimed, would remain so.

The Board of Reclamation Commissioners accomplished little. Many landowners who did not stand to profit from the building of dikes, drains, and levees refused to be taxed by their neighbors who stood to benefit from such works. In 1868 the board was dissolved, and responsibility for reclamation was turned over to each county's board of supervisors. Counties could then authorize the formation of reclamation districts upon petition by one-half the landowners in an area.[17] From the 1870s onward, reclamation was undertaken primarily by two kinds of corporate entities: reclamation districts formed by (mainly local) individual landowners, and large corporations financed with outside capital formed for the express purpose of reclaiming swamp lands as an investment venture.

The reclamation of swamp and overflowed lands in the Sacramento–San Joaquin Delta took place in two stages. During the first period, which lasted from the early 1850s to the early 1880s, the work was done mainly by "wheelbarrow brigades"

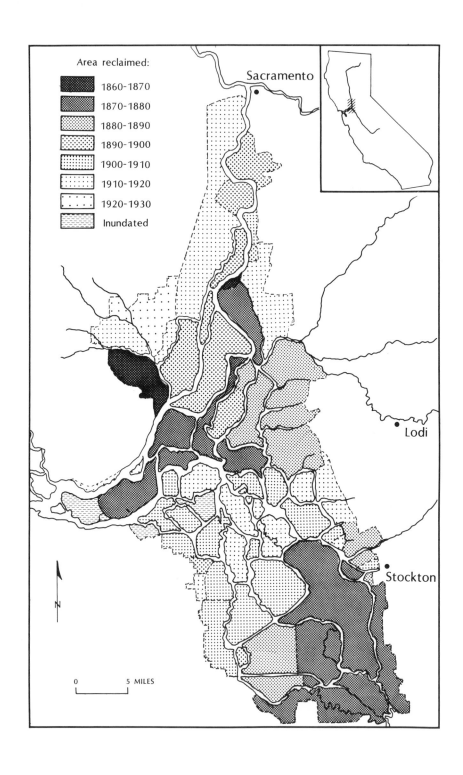

Area reclaimed:

1860-1870
1870-1880
1880-1890
1890-1900
1900-1910
1910-1920
1920-1930
Inundated

Sacramento

Lodi

Stockton

N

0   5 MILES

of Chinese laborers employed by both individual landowners and land-reclamation corporations. As map 7 shows, during these early decades the mainland tracts around the periphery of the Delta—where the soil had a lower organic content, was less swampy, and therefore was easier to drain and dike—were reclaimed. In the later period, the clamshell dredge—a barge outfitted with a long boom at the end of which were two claw-like buckets—introduced in 1879, soon replaced the Chinese wheelbarrow gangs. The giant buckets scooped earth out of the riverbeds, and the boom then swung the buckets over to the levee to dump the fill.[18] The introduction of heavy equipment made it possible to reclaim the swampier central peat islands. Almost all of this work was carried out by large land companies because reclaiming the central Delta was so costly that individuals could not afford to undertake the task.[19]

Once it had been shown that Delta land could be reclaimed, people were eager to establish farms in the area. Swamp land proved to be so fertile that some farmers could grow four crops a year. Another advantage was that the Delta was one of the few areas in the state that did not suffer from summer drought. Furthermore, with its maze of waterways, crops could be transported in bulk on barges to outside markets fairly cheaply.[20] Sacramento and Stockton each developed inland ports to handle the export of the area's products. From the late 1850s to the 1880s, Delta counties were major producers of wheat, which was loaded at the two ports and taken to San Francisco for transshipment overseas. Later, potatoes, onions, beans, various green vegetables, asparagus, and fruit were also exported by water and by rail.

Land use patterns depended on location and on the kind of crops grown.[21] Location determined the proportion of each holding that could be tilled, because in the less swampy mainland tracts and along the natural levees almost all the land on farms

Map 7   *The Sacramento–San Joaquin Delta: Reclamation Sequence,*
          1860–1930
          SOURCE: *John Thompson,* The Settlement Geography of
          the Sacramento–San Joaquin Delta, California *(1957),*
          map 16, p. 219.

could be cultivated, but in the backswamps, even though land
was held in large parcels, only a small portion of it could be
farmed. First, water had to be pumped out, the land had to dry,
and drains, sluices, dikes, floodgates, and levees had to be built
before farming was possible. The kinds of crops planted also
influenced how much of the land on a farm needed to be used.
Cereals took up a lot of land, but fruits and vegetables needed far
smaller acreages to bring the same returns. One peculiarity of
the Delta was that potatoes and beans—not normally thought of
as high-return crops—turned out to be the most profitable crops
on virgin peat soil.

In Franklin Township, Sacramento County—which may be
taken as representative of the northern mainland tracts—most
of the landowners did not own much more land than they could
cultivate. In 1860, of the 191 farms listed in the manuscript
census of agriculture, 38 (20 percent) were under 40 acres, 18
(9 percent) were between 41 and 80 acres, 102 (53 percent) were
between 81 and 320 acres, 23 (12 percent) were between 321
and 640 acres, and 10 (5 percent) were larger than 640 acres.
Regardless of size, almost all the land on these farms was under
cultivation. Between 1860 and 1870, only 11 new farms were
established, but the number of medium and large farms had
increased. Now only 18 farms were smaller than 40 acres, and
24 farms were larger than 640 acres. Farms ranging from 81 to
320 acres now numbered 118.

In 1860, 43 percent of the farms in Franklin Township were
diversified farms growing hay, cereals, beans, potatoes, deciduous fruit, grapes, and vegetables, along with raising livestock,
while 35 percent specialized in hay, grain, and livestock. In the
1860–70 decade, with the boom in wheat cultivation, the number of farms devoted only to extensive grain cultivation increased
from 66 to 101, while diversified farms decreased from 82 to 30.
By 1880 the total number of farms in Franklin Township had
fallen to 159. Surprisingly, the smaller farms under 80 acres,
most of which were diversified, not only survived but increased
in number (with 20 under 40 acres, and 21 between 40 and 80
acres), but many of the larger ones had been taken out of cultivation, as newly opened wheatland in Colusa County in the northern Sacramento Valley outstripped Delta farms in productivity.

In contrast to Franklin Township, a different picture emerged

in Georgiana and Union townships in the central Delta, where
more large farms existed from the beginning, but where fewer
acres on each farm were under cultivation. Two-thirds of the
farms were between 81 and 640 acres in 1860, but on more than
90 percent of these, less than 40 acres on each were tilled, reflecting the nascent stage of agricultural development in this part of
the Delta. Most farms in the central peat islands in 1860 were
also diversified farms. In the following two decades, the number
of medium-sized farms (81 to 320 acres) increased from 62 in
1860 to 84 in 1870 to 107 in 1880, while farms larger than 320
acres increased from 10 to 17 to 36. Though reclamation in the
central Delta was more difficult and expensive, agricultural development did take place, as is shown by the fact that the number of farms with less than 40 acres under cultivation fell from 84
in 1860 to only 7 in 1880. At the same time, the number of farms
with 81 to 320 acres under cultivation increased from 2 in 1860 to
52 in 1870 to 107 in 1880. The number of farms with more than
320 acres under cultivation rose spectacularly from 1 in 1870 to
32 in 1880.

As in the northern Delta, the central Delta also showed a trend
toward specializing in cereal production in the 1860s. Between
1860 and 1870, the number of diversified farms producing at
least one item from each of the three combinations—hay, cereal,
and livestock; potatoes and beans; and deciduous fruit, grapes,
and vegetables—decreased from 69 to 25, while farms engaged
only in hay-cereal-livestock production increased from 1 to 47.
The following decade, however, saw a reversal of trends. Diversified farms flourished once more as cereal cultivation became
less important, but the most significant development in these
central Delta peat islands was a tripling of the number of farms
specializing in potato and bean cultivation.

Specialization and diversification have always existed alongside each other in Delta farming. In the 1860s many farms were
diversified, each growing a variety of crops. When Delta farmers
began to specialize, they did so in wheat and barley (staples not
normally thought of as specialty crops), followed by beans and
potatoes (again not specialty crops). Only at the turn of the century did asparagus, beets, tomatoes, celery, and various fruits
begin to be grown as specialty crops on a large scale. (Though
fruit had been grown on the natural levees along the Sacramento

Chinese laborers grading a levee, Sacramento Delta, California. (Photograph courtesy of the Sacramento River Delta Historical Society, Locke, California)

River between Clarksburg and Freeport since gold rush days, the field crops have outranked them in importance.) Today, most Delta farms specialize in one or two crops, but over the area as a whole there is great diversification.

Chinese tenant farmers and farm laborers were intimately involved in every phase of the Delta's land reclamation and farm-making. In fact, it can be argued that without them the Delta would have taken decades longer to develop into one of the richest agricultural areas in the world.

The Chinese and Swamp Land Reclamation

There is insufficient historical evidence to show exactly which locations in the Delta had been reclaimed with Chinese labor. It is possible only to state in general that Chinese were employed to build levees in those areas which were reclaimed before the mid-1880s. A document published by the Tide Land Reclamation Company stated that Chinese were hired to build the first levees on the mainland tracts in Rio Vista Township in Solano County, on Twitchell Island, on Brannan Island, and on part of Roberts Island.[22] There is also documentary evidence that Thomas H. Williams hired as many as a thousand Chinese workers to reclaim Union Island, which originally comprised the present Union Island, the Fabian Tract, Victoria Island, and the upper half of Woodward Island.[23]

In addition, a 1913 map of roads and steamboat landings in the Delta provides indirect evidence on where the Chinese might have been employed to build levees.[24] Until recent decades, the major means of transportation in the Delta was by boat, so steamboat landings existed in close proximity to each other throughout the Delta. Such landings were constructed on the levees and were usually named after their owners or the persons most frequently identified as using them. Map 8 shows the location of steamboat landings which were named after Chinese and Japanese and which in all likelihood were built and used primarily by them. It is also probable that Chinese had built the levees at those locations with Chinese-named steamboat landings. Many of the names, such as Hop Sing, Quong Lee, Quong Goon, Hop Goon, and Tai On, were names of companies which had leased large tracts for farming and apparently had their own steamboat landings. Chinese-named landings were shown on the west side of Roberts Island and on the northern tip of Union Island, and it is known for certain that Chinese had been used to reclaim those islands. Other names, such as China Landing or Jap Camp, were less specific, and those landings might not have been owned by any particular firm or individual, though they obviously must have been used extensively by Chinese and Japanese.

There may have been some inaccuracies in the placement of the landings. The largest cluster of Chinese-named steamboat

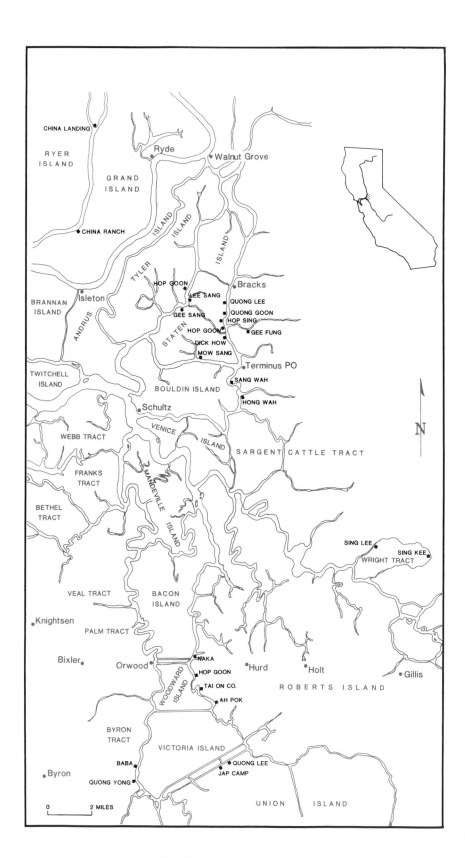

landings shown on the map was on Staten Island. County
archival records contain only a few documents showing that
Chinese leased land on that island. However, Chinese had leased
almost the entire Brack Tract on the shore across from Staten
Island year after year in the 1890s. The firm Hop Sing, for which
one of the landings on Staten Island was named, for example,
had leased the east 100 acres of Lot 7 of the Brack Tract from
Murphy and Frankenheimer in 1896.[25] It is possible that the
cartographer had placed the landings on the wrong side of the
river! Alternately, it could mean that some leases with Chinese
were not recorded.

Swamp land reclamation was an extremely unpleasant
occupation. Working in waist-deep muck under the most in-
salubrious conditions, many a reclamation laborer probably
caught chills or pneumonia. Malaria was also prevalent in the
swamps.[26] It is not known how many Chinese laborers died
while working in levee construction. Aside from accidents and
disease, Chinese were also exposed to danger during heavy
floods, when they were ordered to pile sandbags around the
inside base of the levees to strengthen them. During the 1878
flood, dozens of Chinese laborers burdened with sandbags and
splattered with mud were seen scrambling up and down the
levees shouting at passing steamers to pick them up, but it is not
known whether the steamers did so. The steamers had been sent
by landowners to rescue their cattle, which were probably
considered more precious than the Chinese laborers.[27]

Three landowners—Reuben Kercheval, P. J. van Löben Sels,
and George D. Roberts—will serve as examples of men who employed Chinese to build levees in reclamation work. Kercheval
is a good example of a pioneer settler who undertook land reclamation on his own initiative and at his own expense. He wanted
to protect his 320-acre farm and the farms of his neighbors from
floods. Van Löben Sels was associated with a savings and loan

Map 8   *The Sacramento–San Joaquin Delta: Steamboat Landings
Named after Chinese and Japanese, 1913*
SOURCE: Map of Sacramento and San Joaquin Rivers,
Showing All Landings to Sacramento and Stockton and
Roads Leading to Them, *1913. (Map in California State
Archives.)*

association and therefore had financial backing, but much of the reclamation he undertook was also for the protection of his own land. He eventually owned over 5,000 acres, part of which he farmed himself and part of which he leased out to tenants. George D. Roberts, on the other hand, undertook reclamation purely as a speculative venture. He bought swamp land at very low prices from the state, reclaimed it, and then quickly resold it for a good profit. In testimony before the Joint Special Committee to Investigate Chinese Immigration of the United States Congress in 1877, he boasted that he once owned a quarter million acres of swamp land.[28]

Reuben Kercheval was probably the first person to have used Chinese laborers for swamp land reclamation. Kercheval's personal history is similar to that of many of the early landowners in the area. Born in Ohio in 1820, he was taken by his parents to Illinois at the age of ten and lived there until he was twenty-nine years old. He came to California in 1849 and briefly mined for gold, but he soon realized that a more steady income could be made by provisioning other miners. His uncle, Armstead Runyon, who had also come from Illinois, had already settled near a spot upon which the town of Courtland was to be built, some twenty miles downstream from Sacramento City along the Sacramento River. Kercheval decided to join his uncle as a farmer and settled on a tract at the northern tip of Grand Island. There is no record that he ever leased any of his 320 acres to white or Chinese tenants.

Kercheval was something of a visionary. Even before he experienced his first flood, he became convinced of the necessity to build levees. He tried to involve his neighbors in this enterprise, but no one was interested except Sam Brannan, the Mormon merchant who promoted the idea of building levees to protect Sacramento City itself. Even after the flood of 1852, Kercheval failed to persuade his neighbors to join him in his enterprise, so he decided to proceed alone. The first crew of reclamation laborers he hired consisted of Chinese, California Indians, and Hawaiians—then colloquially called Kanakas. (John Sutter had first brought Hawaiians to California to labor on his farm.) Kercheval had twelve miles of levees, which measured thirteen feet at the base and three feet across at the crest, built solely at his own expense. The reclamation laborers were paid 13.5 cents for every cubic yard of earth they dug and piled up. Over the next quarter century, Kercheval built bigger and stronger levees each time after a flood washed them away. In 1876, for example, he employed 200 Chinese and 40 Hawaiians to build levees a hundred feet wide at the base, five feet wide at the crest, and twelve feet high.[29] During this period when he employed Chinese laborers, Kercheval served in the State Assembly (1873–74 and 1877–78), but legislative records do not show what stand, if any, he took on the question of Chinese immigration and the use of Chinese labor.

An account written in longhand on school notebook paper by Pietre Justus van Löben Sels (popularly known as P. J.), who settled in the Delta in the mid-1870s, provides a glimpse of how the Chinese wheelbarrow gangs worked. Van Löben Sels was a Dutch lawyer who had come to the United States for a visit in 1876. While in San Francisco, he met the daughter of the president of the San Francisco Savings Union, fell in love with her, and married her. At that time the San Francisco Savings Union owned several thousand acres in the Delta. Van Löben Sels's father-in-law, assuming that a Dutchman must know something about dikes and levees, put him in charge of reclaiming the land.[30]

Like Kercheval, P. J. also decided to hire Chinese levee-builders. In his journal, P. J. drew a picture of how levees were built. Most of the delta islands and tracts already had low natural levees. Planks two feet wide were placed between a "borrow pit" and the natural levee, and these boards formed a gently upward sloping plank-way, supported at short intervals by pilings sunk into the mucky ground. Chinese laborers dug dirt from the borrow pit, shoveled it into wheelbarrows, and slowly wheeled the barrows up the plank-way, dumping the fill on top of the natural levee. P. J. discovered that pushing these wheelbarrows up the slope was arduous work: attempting to push such a wheelbarrow up the plank-way himself, he found he could not even lift it. There was also a constant danger of slipping because the dripping mud made the planks wet and very slippery. Many of the laborers fell, and some doubtless were injured, but P. J. claimed that none were ever killed. P. J. expressed marvel at the patience and strength of "those sturdy little Orientals."[31]

P. J. engaged in a friendly rivalry with his Chinese workers.

For his part, he sometimes tried to lure them to sign contracts by false promises. Two kinds of fill were used—peat and clay—and P. J. knew that the Chinese preferred to work with peat because it was much lighter than clay. Even though the Chinese were paid only 7.5 cents for each cubic yard of peat dug and filled, and 13.25 cents for each cubic yard of clay, they still preferred peat. Landowners like P. J., however, preferred clay because it made a more tightly packed fill. Peat had two contradictory tendencies: when it got soaked with water, the weight of the water would pull the peat blocks down, causing the levee to sink as much as three to six feet a year; on the other hand, since water itself has buoyancy, it was impossible to pack the peat blocks tightly. Clay did not possess such peculiarities. To entice potential Chinese contractors and their laborers to work for him, van Löben Sels promised them they could use peat, but after the work started he would try to cajole them into using clay.[32]

The Chinese also used some tricks to get a better deal. Since the wages of the reclamation laborers were calculated according to the amount of earth dug and piled up, it was crucial that an accurate way be found to measure the volume so moved. Van Löben Sels and his neighbor, O. R. Runyon (son of Armstead Runyon and cousin of Reuben Kercheval), measured the size of the hole in the borrow pit every four or five days, but owing to undulations in the earth, it was difficult to figure out the exact amount of dirt that had been dug out. To leave a record of their work, the Chinese left a column or a pyramid of dirt in the middle of each pit to serve as a yardstick to show the depth of the hole. Van Löben Sels noticed that they always tried to leave this pyramid at what had been the highest point on the ground before the pit was dug. Occasionally, he and Runyon discovered that "a pyramid sometimes had undergone an 'operation' " in the night, having been sliced across horizontally and a layer of dirt inserted in between! Van Löben Sels and Runyon fined their Chinese workers for such practices. He claimed that when found out, the Chinese—"like little naughty but good children"—would accept his judgment and never made any fuss. P. J. did not seem to look askance at such "cheating," because as he wrote in his journal, "I have often been cheated by white men, but never by a Chinaman." Often, when there was a disagreement over the total amount of wages to be paid, the Chinese would take out an abacus and make their calculations in less than ten minutes—a feat that impressed P. J. enormously, since it took him and O. R. Runyon at least four or five hours to arrive at their figures.[33]

Van Löben Sels obtained his levee builders through Chinese labor contractors. There were two levels of contracting. A head contractor would initially bargain for the job, but this person seldom stayed around to supervise the actual work. Subcontractors acted as foremen and worked alongside their men. Both the head contractor and the subcontractors deducted commissions from the laborers' pay. P. J.'s account indicated that the laborers did not fully trust the head contractor. The laborers were supposed to be paid once a week. At first, van Löben Sels paid the head contractor in cash, but later on, at the urging of the laborers themselves, he began to issue the head contractor checks. The laborers told P. J. that sometimes they were not paid by the head contractor, who claimed that the money in the payroll had been "waylaid by highbinders!" Ming Sing Gue, the subcontractor who supervised the levee builders on the van Löben Sels land, always wanted P. J. to tell him when he had paid the head contractor, so that he and his laborers could "take care of the rest" to insure that the head contractor did not "escape" with their wages.[34]

The laborers on the van Löben Sels land lived in "miserable tents which they constructed by themselves" and hired their own cooks. They worked long days, rising with the 5:00 A.M. bell and working until 6:00 P.M. Their first task in the morning was to clean the horses; at 5:40 A.M. they had breakfast; and at 6:00 A.M. they began work. They took a break at 11:30 A.M., ate lunch at noon, and resumed work at 1:00 P.M. Between 6:00 P.M., when they stopped work, and 6:40 P.M., when they ate dinner, they cleaned themselves and put their tools in order.[35] Horses, when used to work in reclamation projects, were outfitted with special "tule shoes" which the Chinese had made.[36] These were large ski-like woven mats of tule attached to the horses' hoofs to prevent them from sinking too deeply into the mud. Sometimes accidents occurred, when a poor horse was sucked under and could not be extricated. It is not known if the Chinese themselves also wore tule shoes.

P. J. perceived himself to be a friend of the Chinese, and

seemed proud that they called him "Bossee Selsee." His ran
his farm somewhat like a plantation. In 1898, for example, he
erected a store and rented it for $10 a month to Chan Tin San, a
prominent Chinese merchant and tenant farmer who played an
active part in establishing the town of Locke, a town built solely
by the Chinese in the Sacramento Delta. P. J. got the Chinese
laborers on his ranch to agree to buy things at Tin San's store.
Every three months, he advanced money to his laborers (up to
half the amount of their total wages) so they could buy goods
there. In return, Tin San gave P. J. a 5 percent commission on all
the business done at the store.[37] Besides this store, van Löben
Sels in later years also leased large tracts to both Chinese and
Japanese tenant farmers.[38]

The best-known landowner who used Chinese labor extensively to reclaim the thousands and thousands of acres he owned was
George D. Roberts, the president of the Tide Land Reclamation
Company which had been established by San Francisco and
Oakland capitalists in 1869 "to buy, sell, cultivate, dyke, ditch,
drain, reclaim and otherwise improve tide and other lands in
the State of California."[39] The corporation was capitalized at
$12,000,000 issuing 120,000 shares at $100 each. Its principal
place of business was San Francisco, and its first seven trustees
were Roberts himself, Archibald Peachy, Lloyd Tevis, Edward
B. Dorsey, Solomon Heydenfeldt, Layfayette Maynard, and
John S. Hagar.[40] At least one of these men—Lloyd Tevis—
besides Roberts also dealt extensively in California real estate
during this period.

Roberts first obtained swamp land tracts in his own name or
through what were known as "dummy entrants"—men he paid
to make entries at the land office in their own names. They would
then sell or assign title to their parcels to him. After the acreage
limitation on the purchase of swamp lands was removed, it was
no longer necessary for Roberts to use such dummies, but he
continued to buy tracts from individuals. In most cases, after
acquiring the land in his own name, Roberts transferred title to
the Tide Land Reclamation Company, which relieved him of
personal liability in case his venture failed. Between 1868 and
1872, San Joaquin County deed records show that Roberts
made twenty-one separate purchases in his own name, six of
them directly from the state of California. In addition, the

Houseboats that some Chinese workers lived in while building levees in the Sacramento Delta, California. (Photograph courtesy of the Sacramento River Delta Historical Society, Locke, California)

company bought twelve tracts under its name.[41] In Sacramento County, between 1871 and 1877, Roberts received title to eleven different tracts.[42] In Yolo County, Roberts bought 17,459 acres.[43] Roberts reported to the congressional committee that the land he acquired had cost him "very little.... We paid nominally a dollar an acre to the State, and when it is reclaimed that dollar is credited to us and we get it back." Buying from individuals, Roberts paid "from two to three dollars an acre," plus a dollar to the state.[44]

Although Roberts claimed to have owned a quarter million acres—120,000 of which were held in the name of the Tide Land Reclamation Company[45]—it is doubtful whether he owned all of

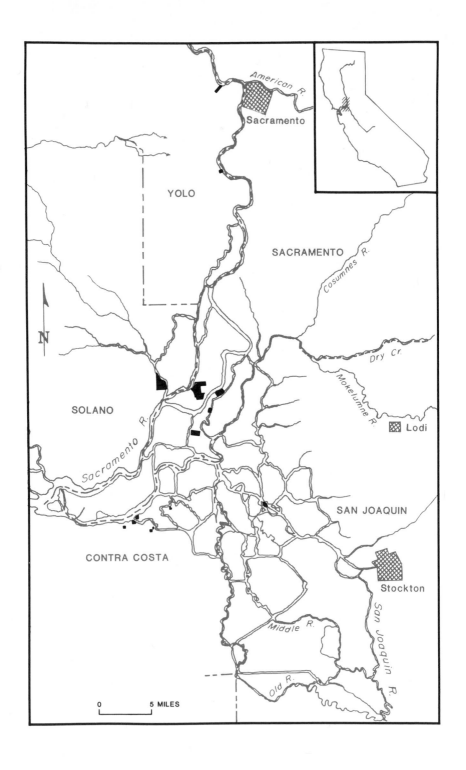

this amount at the same time, because the entire Delta—even
according to the maximum definition of its boundaries—encompasses only half a million acres. By 1878 Roberts had lost all this
land except for Twitchell Island, which he kept, partly because
he may have been sentimentally attached to it, since it
was the site of his first reclamation efforts in 1869,[46] and partly
because it was unsalable, having been ruined by the floods of
1875 and 1878.[47] In 1875 Roberts had already turned over all
his holdings on Roberts Island to J. P. Whitney,[48] and a few
years later he and the Tide Land Reclamation Company
transferred some 6,000 acres on Grand Island, Union Island,
and in the Yolo Basin to Thomas O. Williams and David
Bixler—themselves large land speculators and landowners—to
whom Roberts owed $60,000.[49] These figures indicate that
Roberts's partially reclaimed land was assessed at $10 per acre
for the purpose of debt settlement. In his testimony, he stated
that in general, after his land was reclaimed, it fetched between
$20 and $100 per acre.[50]

   Roberts used Chinese reclamation laborers exclusively in his
speculative enterprise. He estimated that he and his company
had partially reclaimed 30,000 to 40,000 acres using Chinese
workers, and in 1876 he had "three or four thousand employed,
mostly under contract."[51] He reported that "in building the
docks we contract by the yard, so much a yard. We go to some of
the Chinese merchants or business men, and tell them we want
to give a contract for a certain nuumber [*sic*] of miles of levee.
They will contract, then, sometimes in large and sometimes in
small bodies of land. Sometimes the contracts are for five, six,
seven, or eight hundred or a thousand [cubic] yards, and sometimes less, with one individual, as the case may be. We pay so

Map 9   *The Sacramento–San Joaquin Delta: Farms Leased by Chinese
Tenants, 1870–1879*
   SOURCES: *Contra Costa County, "Leases" and "Chattel Mortgages"; Sacramento County, "Leases," "Chattel Mortgages,"
and "Crop Mortgages"; San Joaquin County, "Book G of Miscellaneous" (contains leases) and "Book I of Miscellaneous"
(contains crop and chattel mortgages); Solano County, "Leases"
and "Chattel Mortgages"; and Yolo County, "Leases."*

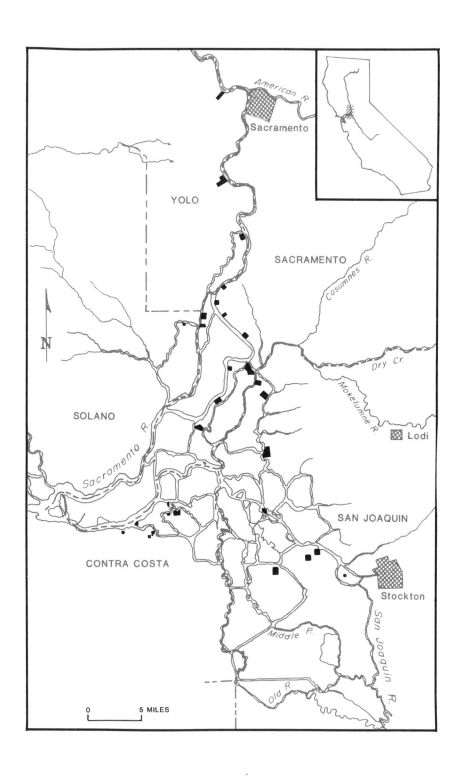

much a yard, and measure the work after it is done, and they receive their pay."[52] Roberts testified that Chinese labor cost $6 to $7 for each acre of land reclaimed, but other sources gave a figure of $12 per acre.[53] He paid them ten to fifteen cents per cubic yard, which gave each laborer a wage equal to about $1 a day. If the laborers were hired by the month, as they sometimes were, the going wage was $27 a month. He said the work done by his Chinese laborers was "satisfactory." In particular, he praised the "executive ability" of his contractors.

As van Löben Sels had noted, Roberts also found that sometimes the head contractors "defrauded" their men by not paying them. But he insisted that the Chinese workers were "entirely independent of the bosses. When the bosses do not pay them they come to me. If the boss does not pay them any wages, they tie him up and call on us. That has been the case in several instances. I find that each man has his account, and he holds the boss responsible."[54] The laborers knew how much was due them because they had a clear idea of how much work had been done. Roberts said that the measurements made by the Chinese "hold out with those of our engineers. . . . Our engineer measures the work, the Chinamen measure it, and we seldom have any disagreement. They are very accurate in their measurements."[55]

According to Roberts, the Chinese labor contractors were merchants whose chief source of income was not the commission they charged the men for finding them employment but the profits they made by provisioning the latter. "The per cent they make is simply the profit on selling rice. It is very seldom that they make anything at all on the contract per acre; but they always stipulate that they shall have the privilege of supplying the Chinamen and they make the profit in their stores. It is the storekeepers that do the contracting."[56]

Like the levee builders on the van Löben Sels land, Roberts's workers also lived together in large groups and provided their own board. "They form little communities among themselves, forty or fifty or a hundred, and they are jointly interested in the

Map 10   *The Sacramento–San Joaquin Delta: Farms Leased by Chinese*
_____   *Tenants, 1880–1889*
         SOURCES: *See sources for map 9.*

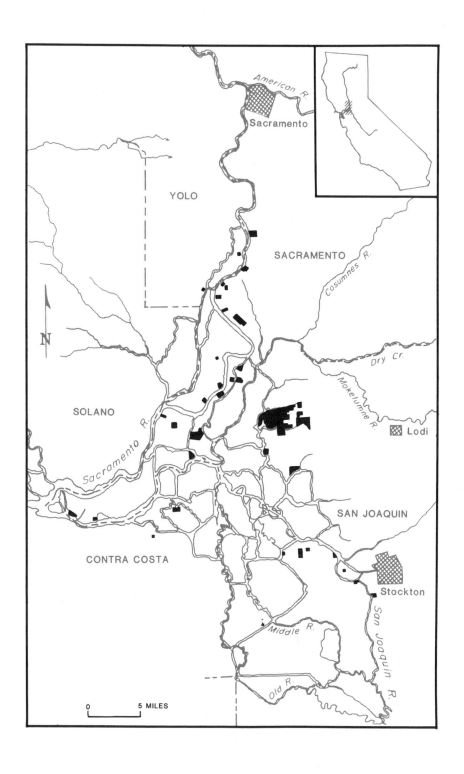

YOLO

American R.

Sacramento

SACRAMENTO

Cosumnes R.

Dry Cr.

SOLANO

Mokelumne R.

Lodi

Sacramento R.

SAN JOAQUIN

CONTRA COSTA

Stockton

Middle R.

San Joaquin R.

Old R.

N

0   5 MILES

contract. We could not get white men to do that. They would not
be harmonious and agree among themselves, but the Chinese
form little families among themselves, do their own cooking, live
in little camps together, and the work is staked off for them
separately."[57]

The profit Roberts made out of the toil and sweat of his
Chinese laborers can be estimated. He purchased swamp land
from the state for $1 to $4 per acre and spent $6 to $12 per acre
reclaiming it. He sold the reclaimed or partially reclaimed land
for $20 to $100 per acre, so he could have made a minimum of $4
($20 minus $4 minus $12) and a maximum of $93 ($100 minus
$1 minus $6) per acre. The amount of profit depended on the
original purchase price, the amount of fill required to build the
levees, and the location of the land. If the Chinese had reclaimed
40,000 acres for him, and if his average profit was $25 per acre,
he could have made a total of $1,000,000 by using their labor. If
the average profit was $50 per acre, the total gain might have
been $2,000,000. But he also experienced many failures because
floods sometimes completely washed out the levees built. The
fact that he went bankrupt indicates that he did not make such
profits, or that the profits were negated by losses, or that he
mismanaged whatever amount he had made. None of the work
would have been done, however, had Chinese laborers not been
available. When asked whether he could have obtained white
men to do the work, he said, "I do not think we could get the
white men to do the work. It is a class of work that white men do
not like.... Very few of them come here to do cheap labor.... We
could not afford to pay three or four dollars a day to white men to
do our work."[58]

Roberts not only used Chinese to reclaim his swamp land, he
also leased some parcels to Chinese tenants,[59] and also employed
Chinese farm laborers. He found the latter to be the "best field
men that we have.... Better than the Swede, and the Swede is
the best worker we have had." He liked the Chinese field hands
because he could depend on them longer: "if you can get them

Map 11   *The Sacramento–San Joaquin Delta: Farms Leased by Chinese
Tenants, 1890–1899*
SOURCES: *See sources for map 9.*

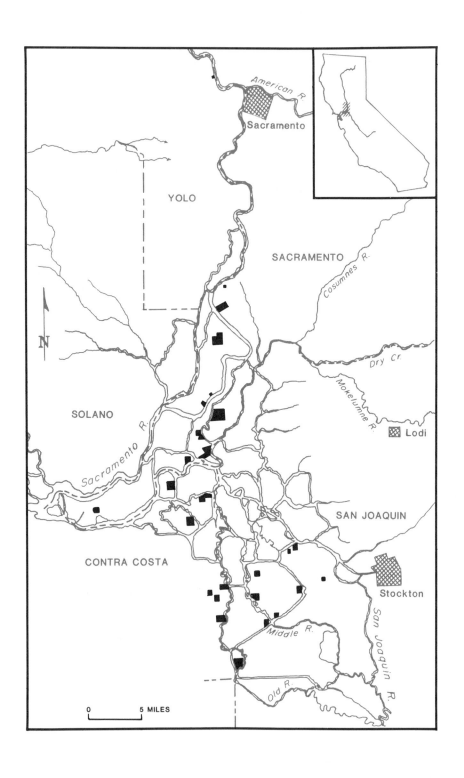

YOLO

SACRAMENTO

American R.

Sacramento

Cosumnes R.

Dry Cr.

Mokelumne R.

N

SOLANO

Lodi

Sacramento R.

SAN JOAQUIN

CONTRA COSTA

Stockton

Middle R.

San Joaquin R.

Old R.

0   5 MILES

this year you can get them next year and the year after, if you treat them well and pay them. They become attached to your place and they stay with you."[60]

The linkage between the use of Chinese labor in reclamation and Chinese entry into tenant farming in the Delta can be further demonstrated in maps 9 through 14. The maps show cumulatively, by decade, the precise location of the land leased by Chinese tenants. County archival records almost always stated the "legal description" of the land being leased. Sometimes such information was given according to United States rectangular survey coordinates, sometimes in terms of Swamp Land Survey parcel numbers, and sometimes as lot numbers in reclamation districts. In all cases, the landowner's name was given. By using county plat maps still available for the period under study, it was possible to determine the exact location, size, and shape of the land the Chinese had leased. (Those plots that could not be so located were excluded from the maps, which therefore under-represent the extent of Chinese tenancy in the Delta. Failure of the parties involved to record some leases also led to under-representation.)

Comparing this set of maps with map 7, which shows the reclamation sequence in the Delta, it can be seen that Chinese tenants generally entered an area in the same decade it was re-claimed. Map 9 shows that in the 1870s Chinese tenants farmed on the Rio Vista mainland tracts of Solano County, in the southern backswamps of Grand Island and Andrus Island of Sacramento County, and on the mainland tracts near Antioch in Township Number Five in Contra Costa County. These areas had indeed been reclaimed by the 1870s. Map 10 shows that in the 1880s Chinese tenants dispersed themselves more widely, moving into the mainland Terminous Tract, the northern tip of Tyler and Staten islands, the northern part of Roberts Island, and the Lower Jones Tract—all areas in the process of being reclaimed. In the 1890s, as map 11 reveals, in addition to areas leased in the preceding decades, Chinese tenants had begun to

Map 12   *The Sacramento–San Joaquin Delta: Farms Leased by Chinese Tenants, 1900–1909*
SOURCES: *See sources for map 9.*

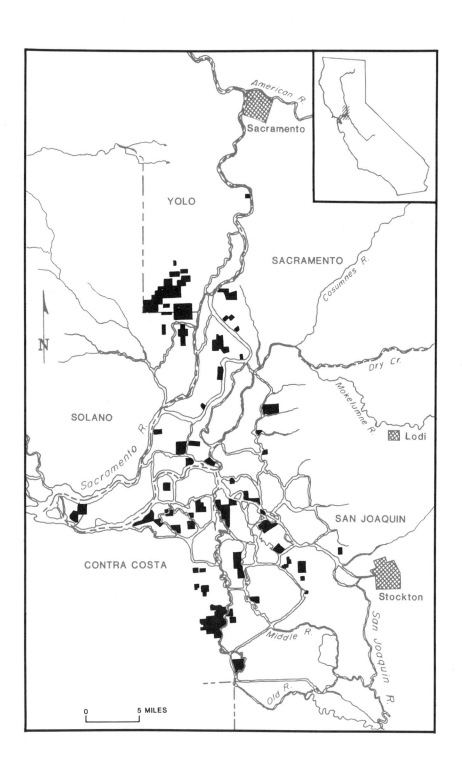

farm on Brannan, Twitchell, and Sherman islands in Sacramento County and the Bethel Tract in Contra Costa County, and they monopolized the Brack Tract in San Joaquin County. Map 12 shows Chinese tenants farming on the Bradford, Franks, Bethel, Veal, Palm, and Byron Tracts in Contra Costa County, which were reclaimed in the 1890s and early 1900s, though not with Chinese labor. By this time the clamshell dredge had replaced Chinese wheelbarrow crews, and reclamation no longer served as an entry point for land-leasing by the Chinese. Chinese tenants were also found in San Joaquin County in the Upper Jones Tract, the Lower Division of Roberts Island, and on Victoria and Coney islands. In the 1910–20 decade, as map 13 indicates, in addition to previously reclaimed and cultivated areas, Chinese moved in force into the central Delta, where they now farmed on Mandeville, Venice, Bacon, and Woodward islands, and leased large acreages in the area around the Yolo Bypass— all of which were reclaimed in these years by large corporations, such as the Holland Land Company and California Delta Farms. During this decade, California agriculture experienced a boom, owing to the high demand for foodstuffs during World War I, and Chinese tenant farmers in the Delta partook of the prosperity. Finally, as map 14 indicates, Chinese tenancy was still considerable in the 1920s and was concentrated in the Yolo Basin, where they grew asparagus for the Libby, McNeill and Libby Company. The great expansion of asparagus cultivation gave the aging Chinese the incentive to remain in Delta farming long after they left the fields and orchards elsewhere.[61]

An examination of their relationship with landlords, different patterns of tenancy, and the economics of Chinese farming in the Delta will reveal the acumen that Chinese farmers possessed. Forming a symbiotic relationship with landowners, Chinese tenant farmers and farm laborers helped to set the pattern for racial and class relations in agricultural California.

Map 13   *The Sacramento–San Joaquin Delta: Farms Leased by Chinese Tenants, 1910–1919*
SOURCES: *See sources for map 9.*

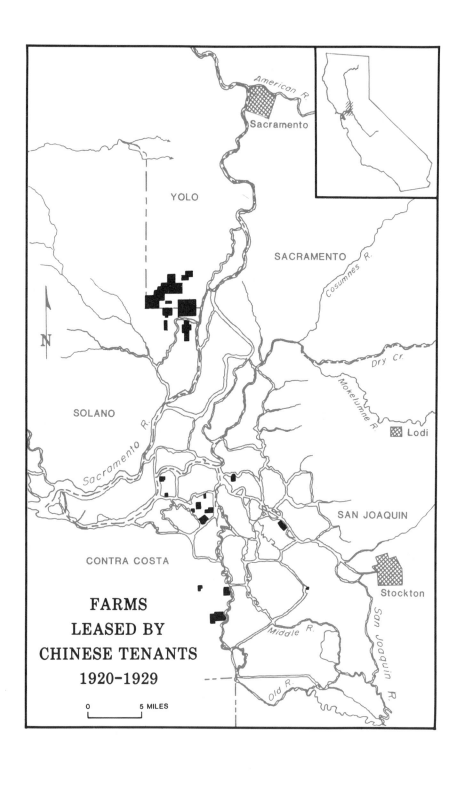

American R.

Sacramento

YOLO

SACRAMENTO

Cosumnes R.

N

Dry Cr.

SOLANO

Mokelumne R.

Sacramento R.

Lodi

SAN JOAQUIN

CONTRA COSTA

Stockton

FARMS
LEASED BY
CHINESE TENANTS
1920–1929

San Joaquin R.

Middle R.

Old R.

0   5 MILES

Map 14   *The Sacramento–San Joaquin Delta: Farms Leased by Chinese*
Tenants, 1920–1929
SOURCES: *See sources for map 9.*

SIX

Potato Kings

 IN those areas of the world which require reclamation
before the land becomes arable, many other steps have
to be taken before actual cultivation can take place. In
the Sacramento–San Joaquin Delta, after an area had
been drained and levees built to protect it, the peat
sod had to be prepared for cultivation. Breaking such
sod was very hard work, probably more difficult than prairie-
breaking, which had challenged many a pioneer settler in the
midwest in the 1820s and 1830s.[1] First the tules had to be removed. Burning proved to be the cheapest and quickest method,
but was not safe, because of the inability to control its spread. In
the dry season, given the high organic content of peat, even the soil
itself would burn. To burn the tules, one man would dig holes in
the peat turf while another followed behind, dropped clumps of
straw into the holes, and ignited them. Many landowners hired
Chinese to perform this dangerous task. After the tule stands were
removed, the land had to be plowed. It was difficult to plow virgin
peat with teams because the ground was too soft for horses to walk
on. Moreover, the sod, tangled with the woody roots of tules,
could not be cut with ordinary plows, and a special tule-cutter
with sharp steel knives attached had to be used.[2] The ground had
to be plowed at least three times before it could be sowed.[3]

Given the problems of using machines and draft animals,
Delta landowners had to rely heavily on human labor to trans-
form raw land into producing farms. It is not surprising, therefore, that having witnessed the tenacity of Chinese reclamation
workers, landowners would look to Chinese farm laborers to
bring their land under cultivation. Instead of hiring Chinese

workers directly, however, landowners found it more convenient
to lease either entire tracts or portions of tracts to Chinese
tenants, letting the latter worry about securing the requisite
labor to do all the necessary "set-up" work.[4] Chinese entered
into what Gregory A. Stiverson, describing tenancy in a different
context—the Chesapeake Bay region of Maryland in the eighteenth century—has called developmental leasing. According to
Stiverson, "the principal object of developmental leasing was to
encourage tenants to develop previously uncultivated land into
working farms and plantations so that the rental or sale value
of the tract would be increased. Ideally, developmental leasing
benefited both landlord and tenant. The tenant received a tract
of land for a low annual rent, and the landlord could charge
much higher rates when the first lease expired, counterbalancing
the years of small income from the developmental lease."[5]

It is difficult to say whether the first Chinese tenants in the
Delta indeed paid low rents. Whether rent is high or low on a
particular piece of property depends on what others in the general area are paying for similar land, the cost of bringing it under
cultivation, its productivity, the kind of crops that can be grown
on it, its proximity to markets, and a host of other factors. It is
not possible to compare the rent the Chinese paid with that paid
by others, because many Chinese tenants either paid cash rent or
made arrangements to pay partly in cash and partly with crops,
whereas almost all white tenants worked under sharecropping
arrangements. In the latter, without knowing how much of
which crop different rented farms produced, the prices of these
crops in particular years, and the actual number of acres planted
within given tracts, it is impossible to determine the value of the
crop shares that tenants turned over to the landlords. Thus, no
comparison can be made between the rents paid under cash
tenancy and through sharecropping. Even if such a comparison
were possible, given the fact that virtually no tenants of other
nationalities were found in the Delta in the 1860s and 1870s, the
extent to which landowners might have exploited Chinese cannot be calculated. It is even difficult to evaluate exactly how
much the rents paid by Chinese tenants increased in real terms
over the years. Even though archival records gave precise figures
for the rents paid, the dollar amounts are not particularly meaningful unless they can be adjusted according to changes in a price

index, but there is insufficient historical information to construct one. It is possible, therefore, only to describe the developmental work Chinese tenants and laborers did and to show how their efforts contributed to agricultural development in the area.

Chinese tenant farmers in the Delta operated different kinds of farms, depending on the location of the land they rented, the terms of their leases, and the date when they leased such land. Broadly speaking, three kinds of developmental leasing were undertaken by Chinese in the area. In the late 1860s and early 1870s, some tenants leased from absentee landlords entire tracts of virgin land in the backswamps of the northern peat islands, drained them, built levees and ditches, broke up the sod, plowed the fields, and planted whatever crops they wished. A second pattern was shown by other Chinese, who, starting in the late 1870s, began to lease the uncultivated portions of land owned by resident landowners who already had farms on sections of their holdings. Most of these holdings were on the natural levees of tracts and islands around the periphery of the northern Delta, although sections of some tracts on Roberts and Union islands in the south had also been brought under cultivation. The terms of the leases signed by these Chinese tenants seem to indicate that the landowners had decided to rent to Chinese for the express purpose of bringing unimproved portions of their farms under cultivation. The Chinese were asked to plant new orchards or hop fields or whatever else the landowners desired. Later still, beginning in the 1890s, a third pattern of developmental leasing appeared in the central Delta on the peat islands and tracts of San Joaquin and Contra Costa counties, few of which had natural levees. Here Chinese tenants leased large tracts from land corporations or individual landowners who had spent veritable fortunes reclaiming these areas with dredges and modern equipment. When landlords did not specify what crops to grow, Chinese tenant farmers usually planted these newly reclaimed tracts in potatoes, beans, onions, and sometimes asparagus.

The changing crop pattern of Chinese tenant farmers may be seen in table 18. (It should be noted that since a substantial portion of the leases gave no information on crops, the percentage for each crop should be higher, although it is impossible to say how much the percentage should be increased for which crop.) In the early period (1860–79), 23 percent of the farms

Table 18   The Sacramento–San Joaquin Delta: Crops Grown on Farms
Leased by Chinese Tenants, 1860–1920

Crops	1860–1879	1880–1889	1890–1899	1900–1909	1910–1920
Berries	0	0	1%	2%	0
Vegetables	23%	11%	18%	11%	5%
Asparagus	0	0	3%	6%	3%
Onions	5%	4%	5%	6%	5%
Potatoes	5%	10%	12%	18%	19%
Beans	5%	7%	8%	9%	11%
Alfalfa	0	4%	1%	1%	0
Grain & Hay	16%	6%	7%	6%	10%
Corn	0	0	0	0	4%
Beets	0	1%	0	0	0
Deciduous Fruit and Nuts	25%	28%	14%	5%	4%
Grapes	5%	1%	0	1%	0
Nursery Products	0	2%	0	0	0
Dairy and Livestock	2%	5%	1%	7%	8%
Pigs	0	1%	2%	0	0
Poultry	0	1%	0	0	0
Crop Not Stated	14%	15%	26%	27%	31%
Hops	0	4%	2%	0	0
TOTAL	100%	100%	100%	99%	100%

SOURCE: Compiled and computed from Contra Costa County, "Chattel Mortgages,"
Books 1–34, 1852–1924, "Leases," Books 1–12, 1862–1924; Sacramento County, "Chattel Mortgages," Books K, 27–210, 1860–1925, "Crop Mortgages," Books 66–172, 1890–
1923, "Leases," Books A–S, 1853–1923, "Personal Property Mortgages," Books 67–218;
San Joaquin County, "Book G of Miscellaneous" (contains Leases), Books G1–G49,
1850–1921, "Book I of Mortgages" (contains Crop and Chattel Mortgages), Book I1–J73,
1850–1921; Solano County, "Chattel Mortgages," Books 1–39, 1849–1920, "Leases,"
Books 1–15, 1853–1919; and Yolo County, "Leases," Books A–H, 1854–1922.

operated by Chinese tenants produced vegetables, and 25 percent raised fruit. Perhaps surprisingly, grain and hay were grown
on 16 percent of the farms. Potatoes, onions, and beans were
each produced on only 5 percent of the farms. In the 1880s only
11 percent of the Chinese tenant farmers still grew vegetables,
while 10 percent now cultivated potatoes. The percentage of
those who tended orchards had increased slightly from 25 percent to 28 percent. In the 1890s the percentage of those growing
potatoes continued to increase, but the percentage growing fruit
decreased from 28 percent to only 14 percent. By the 1900s, 18
percent grew potatoes while only 5 percent still produced fruit.

This is a reflection of the geographic shift of Chinese tenant farming from the northern Delta, which had a narrow but productive "fruit belt" on the natural levees along the banks of the Sacramento River, to the peat islands of the central and southern Delta, which were unsuitable for fruit, since most of the land there was below sea level and had no natural levees. (Over the centuries, natural levees had built up along the banks of the Sacramento River as sediment settled around the groves of willows that thrived by the water's edge, but the San Joaquin River, carrying a much smaller volume of water, brought less sediment and did not build up similar levees.)

The Early Phase of Chinese Tenancy

The first lease recorded between Chinese tenants and white landlords for Delta land in Sacramento County appeared in that county's official "Book of Leases" in 1873, but other sources indicate that Chinese had begun to farm there several years earlier. An article in the November 11, 1869, issue of the *Sacramento Bee* reported the presence of two white women married to "Chinamen" in a "colony" on land owned by J. V. Simmons near Courtland. The Chinese were "cultivating ground on a cooperative plan" and were "very successful." One of the women, a schoolteacher from Baltimore, had come to teach in San Francisco and had later gone to Salinas, where she met a Chinese and married him because she "desired to elevate him and his race." She and her husband came to live in this first reported Chinese "colony" of tenant farmers in the Delta, though they did not farm, manufacturing clothes for a living instead. No information was given about the second white woman, but the reporter commented that "the Chinese do not take kindly to the introduction of the white women into their settlement, and think their countrymen had done a wrong thing."[6]

A 1911 report to Congress dated the entry of Chinese "temporary laborers" into Sacramento to 1863 or 1864 and stated that "by 1870 the Chinese had also begun to lease ranches." According to the agents who did the investigation for the congressional committee, "the reasons assigned by the landowners for leasing

to the members of this race were that, the Chinese controlling the
supply of labor and this being organized under 'bosses,' it was
much easier for the tenant than for the owner to secure the necessary 'help,' and that the leasing was in other ways a more convenient arrangement for the rancher than farming on his own
account ... being profitable and involving less bother."[7] Leasing
was less bother because the Chinese tenants not only assumed all
managerial responsibility but also turned over greatly improved
land to their landlords at the end of the lease period.

The first recorded Chinese lease was between Chou Ying
and Wee Ying as tenants and George D. Roberts and Joseph
Roberts, Sr., as landlords. The farming ventures of Chou Ying
and Wee Ying will be used to illustrate the first pattern of developmental leasing undertaken by Chinese in the Delta. The
551 acres leased in the fall of 1873 were actually located in three
separate tracts on Grand Island. Although the lease did not say
so explicitly, it is likely that these tracts were virgin land, since
no rent was charged for one parcel during the first year. The
lease period was to be five years, and rent was payable both in a
share of the crops and in cash. A 52-acre tract was to be rent-free
during the first year, and would cost $8 per acre from the second
through the fifth year. For the other two tracts the landlords
were to receive one-quarter of the crops, to be delivered free of
charge to a shipping point of the landlords' choosing. If the
tenants neglected the land so that it became uncultivable—
which could happen easily because weeds grew abundantly in
the moist peat soil—they would have to pay a cash rent of $10
per acre. If the tenants chose to grow Chinese vegetables, then
the landlords could charge $10 per acre in lieu of one-quarter
of the crops. Mustard and kale—considered vegetables by the
Chinese, but weeds by their American landlords—were not to be
grown. The tenants were to provide all the labor for cultivation
and harvesting, and the landlords would pay all the taxes and be
responsible for levee maintenance.[8]

Chou Ying and Wee Ying apparently found Delta farming
profitable, but perhaps finding it too arduous to cultivate virgin
peat soil, the following year they leased a tract of improved land
designated as Swamp Land Survey 566. This tract belonging to
William Gwynn was located on the west side of Grand Island
and in the 1870 Sacramento County Map Book was shown to

contain 360 acres. There were already a house, a barn, and an outhouse, which the tenants were permitted to use. The tenants had the right to erect new structures, but these would become the property of the landlord at the expiration of the lease in six years. Rent was 25 percent of the crops. The tenants could plant any crop they wished, but Gwynn reserved the right to plant fruit trees on this tract.[9]

Another indication that Chou Ying and Wee Ying found Delta farming attractive was that they bought an undivided half-interest in another tract owned by William Gwynn and California's "Cattle King," Henry Miller.[10] This tract was astonishingly large: 1,375 acres. Agreeing to pay $30 per acre, Chou Ying and Wee Ying made a first down payment of $3,000 and six weeks later came up with another $20,110 in cash.[11] Since half of 1,375 acres was 687.5 acres, at $30 per acre the total purchase price was $20,625. The extra $2,485 they paid must have been six weeks' worth of interest on the unpaid balance. Where they got this capital is not known. In 1882 Wee Ying and the executor of the will of the now deceased Chou Ying sold 723 acres to Wie Se Tuck, a Chinese resident of San Francisco. The Chinese-owned tract now measured 723 acres as the result of a court decision in 1878 when the 6th Judicial District Court, in the case of *C. W. Clark v. Wee Ying et al.*, had awarded that number of acres to the defendants.[12] Wie Se Tuck, the new purchaser, paid $2,170 down, but the archival records do not indicate what the total purchase price was. In 1891 Wie Se Tuck sold the land to Wee Pack, who made a down payment of $2,500; it was not stated how much more the buyer owed.[13] Finally, in 1894 Wee Pack sold the parcel to James Hogg, who made a down payment of only $10, but again it is not known what the total selling price was.[14]

Other leases signed by Chinese in the 1870s for Delta land had more stringent terms. Ah Sing, who leased 167 acres on Andrus Island in 1875 for three years from Charles C. Perkins, a native of Connecticut, had to give the latter 50 percent of the fruit produced and 40 percent of the other crops. Perkins expected a larger share of the crops than either Roberts or Gwynn because he provided horses, tools, and several houses for his tenants' use.[15] Apparently, Perkins had leased his entire improved acreage to Ah Sing, because the 1880 manuscript census of agriculture

The first Chinatown in Isleton on the bank of Jackson Slough, the Sacramento Delta, California. (Photograph courtesy of the Sacramento River Delta Historical Society, Locke, California)

showed that he owned only 170 acres of which 1 acre was pasture and 2 acres were woodlands.

In 1876 Moy Jim Mun, Wong Hong, and Ah Yak of San Francisco leased 196 acres on Andrus Island from A. J. Donnelly, also of San Francisco, at $10 per acre for five years. The tenants had to erect a fence around the property. The landlord promised to reimburse them for the expense at an "appraised value," but he would not pay for any other improvements without prior consent. The main task facing the tenants was to transplant 3,000 eucalyptus trees "as directed."[16] It was not stated what the tenants could grow on the land, but whatever they intended to use it for, they must have considered it worth a cash rental of $1,960 a year plus the labor of transplanting 3,000 trees.

Ah Too, who leased Swamp Land Survey 363 from Josiah Pool in 1879, also gave 50 percent of his crops to his landlord. Pool, born in Illinois in 1830, had fought in the Mexican-American War and had come to mine for gold in California in 1852. He remained in El Dorado County until 1855, when he came to Sacramento County and settled on Andrus Island, eventually becoming Isleton's postmaster.[17] Five acres of Pool's land had already been planted in fruit, and Pool excluded the orchard and his own residence from the lease.[18] The kind of arrangement between Ah Too and Pool became quite common in later years, as many landowners continued to live in their houses and to cultivate a section of their farms while leasing part of their land to Chinese.

Chinese tenant farmers had learned to accept cash advances from commission merchants by the 1870s. Ah Ing of Sacramento County, farming on the ranch of C. F. Barnhisel on Grand Island in 1879, borrowed $180 from Howe and Hall, commission merchants in San Francisco, pledging his crop of fruit and vegetables as collateral. The interest rate was 1.25 percent a month.[19] In the same year, the same commission merchants advanced $1,500 at the same interest rate to Ah Wah and Company, farming on the lands of L. Winter along Georgiana Slough, accepting "all crops" and the tenants' horses, tools, and farm implements as collateral.[20] Ah Wah and his partners must have owned the horses and implements; otherwise, Howe and Hall would not have accepted them as surety.

In the 1880s a number of developments became apparent. The size of the tracts leased by Chinese tenants decreased, but more of the land was already under cultivation; rents increased along with land values as a result of greater improvements on the land; and more of the landlords as well as the Chinese tenants gave a Delta location as their address—in contrast to the 1870s, when several of the landlords and some of the tenants were listed as San Francisco residents. Most important, the main feature of the second pattern of developmental leasing undertaken by Chinese became more clearly defined: instead of leasing entire tracts to the Chinese, landlords now tended to rent out only portions of their holdings to get them cleared, cultivated, and planted with

perennial crops—crops that would continue to produce after the
Chinese tenants departed. Some of the leases signed in the 1880s
show the evolving relations between whites and Chinese and give
an idea of what kind of persons leased land to Chinese tenants.

A good example of the developmental role Chinese tenants
played is shown in the lease signed in 1882 between Koon Yek
and Company and A. S. Olsen, an immigrant from Norway, for
three years for a half-share of the crops. The Chinese leased 120
acres on Tyler Island, which represented about half of the tilled
acreage on Olsen's farm. The 1880 manuscript agriculture census listed Olsen himself as a tenant, who was leasing a farm with
235 acres of tilled land and 40 acres of pasture. Olsen's farm was
valued at $11,750, and he produced $4,000 worth of crops, consisting of 150 tons of hay grown on 40 acres, 6,000 bushels of
Irish potatoes grown on 40 acres, 1,800 bushels of sweet potatoes
grown on 9 acres, 5,500 pounds of hops grown on 4 acres, and
700 bushels of beans (acreage not stated). In addition, the farm
contained 44 acres of young fruit trees. In 1879–80, Olsen paid
$300 in wages to hired hands[21]—not a large amount compared
with other farmers in the area. The 120 acres Olsen leased to
Koon Yek and Company consisted of 60 acres of land intended
for beans and potatoes and 60 acres for hops. Fifteen acres of the
latter were already in hops, and the tenants were asked to plant
the other 45 acres to hops also. Olsen apparently decided to lease
to Chinese tenants in order to expand his hop acreage. The
planting, cultivation, harvesting, and curing of hops were all
labor-intensive activities, but hops frequently brought a larger
return per acre than other commercial crops of that period, so
farmers found it desirable to develop hopyards where soil and
climatic conditions permitted.[22]

Expanding Olsen's hop acreage involved a lot of work for
the Chinese tenants, who had to construct from scratch all the
paraphernalia for hop cultivation, including support poles and
a drying house. Although the farm was on Tyler Island, the
tenants were told they had to go to Ryer Island in Solano County
(two islands to the west of Tyler Island) to cut 100,000 hop poles
and transport them to Olsen's ranch. The tenants, however, did
not have to load and unload the poles onto barges themselves,
but the lease did not state who would do the loading. Olsen advanced money to Koon Yek and Company to hire laborers to cut

the poles and to build a twenty-by-forty-foot drying house with redwood boards that he provided. The hops were to be shipped on a joint account through mutually-agreed-upon commission merchants. To enable his tenants to farm his land, Olsen provided them with six horses, a farm wagon, various plows, harrows, hoes, and hop bars, as well as half of the seed potatoes and beans.[23]

Levi Painter was another Delta farmer who leased part of his land to Chinese tenants so they could improve it for him. Painter was born in Indiana in 1833, moved to Missouri with his parents in 1842, and started out for California in 1849. He did not get to California, however, but stopped in Nebraska for two years to trade with Native Americans. He returned to Missouri, where he farmed and worked in the mines until 1853, when he decided once again to set out for California. In 1854 he settled in Sacramento, and the following year, he acquired a 123-acre farm one mile from Courtland. In 1880 his farm was valued at $7,000, producing crops and dairy products worth $2,500. He had four acres in orchard, with the 500 fruit trees producing 1,100 bushels of fruit worth $700.[24] When Painter leased 95 acres to Tim Gan, Ah Gue, and Coon See in 1886 for a four-year period, he required them to plant an additional 1,000 fruit trees per year on his land. It was common practice in those days to plant about 100 trees per acre, so 4,000 trees would have converted 40 acres of Painter's farm to orchard. Not only were the tenants not paid for planting the trees, but they had to pay Painter $2,100 a year in rent, which came to $22 per acre. For this price, they could use the land between the saplings to grow hay and vegetables.[25]

Even though the terms sound exploitative, the arrangement must have been satisfactory to both Painter and the Chinese, because when the initial lease expired, it was renewed for seven years. Painter reduced the rent per acre to only $12.65 a year for the first three years, $14.75 a year for the fourth and fifth years, and $16.85 a year for the sixth and seventh years. The reason for the rent reduction was that as the saplings grew the Chinese could no longer use as much land between the rows of trees to grow crops for sale, but when the trees began to bear fruit, the sale of which provided the tenants with some income, Painter was able to increase the rent again bit by bit. During the second lease, the Chinese tenants were not asked to plant any additional fruit trees, but they probably tended the 4,000 they had planted previously.[26]

The conditions described in the sample leases above characterized the northern Delta. In the central Delta islands a quite different situation existed. There the Chinese specialized in growing potatoes on their own account. A number of them became some of the most prosperous Chinese in California. One, Chin Lung, became known as the Chinese "Potato King."[27] His exploits best illustrate the third form of developmental leasing that Chinese undertook in the Delta.

Potato Kings

Unlike the northern Delta, where many of the landlords who leased land to Chinese were farmers owning medium-sized farms, almost all the landlords who dealt with Chinese tenants in the central and southern Delta were holders of very large tracts. Chinese tenants had begun to farm on Roberts and Union islands in the southern Delta as early as the 1860s, but it was not until the 1890s that they entered the central part of the Delta in large numbers.[28] A firm named Murphy and Frankenheimer was responsible for bringing many Chinese there, where they began their long history of tenancy by leasing parcels of the Brack Tract. In 1895 Jacob Brack, owner of the Brack Tract, leased part of his land to J. E. Murphy; a year later, Murphy and Frankenheimer began subleasing parcels to Chinese. No information could be found on J. E. Murphy, but the Frankenheimer part of the firm's name referred to several sons of Bernhard Frankenheimer, an immigrant from Bavaria, born in 1826, who came to the United States in 1844 to join his older brother, a clothing manufacturer, worked all over the Mississippi Valley, and came to California in 1850 to work for the firm of Sutro and Frankenheimer. (Adolph Sutro, who eventually became a millionaire mining engineer, was Frankenheimer's cousin on his mother's side.) Frankenheimer began to deal in real estate in the 1880s.[29]

Jacob Brack, owner of the first tract in the central Delta leased by Chinese, was an immigrant from Switzerland, born in 1825.

He had also arrived in the United States in 1844, landing in New Orleans. From there, he traveled up the Mississippi River to Galena, Illinois, where he worked in lead mines. He joined the California gold rush in 1849, but discovered that he did not care for mining, so he worked for one of the Sargent brothers, owners of thousands of acres in San Joaquin and Monterey counties and elsewhere in California. In 1850 he bought a small tract near Woodbridge and raised sheep and cattle on the tule land. He then went into partnership with the Sargent brothers and bought 1,500 acres, which he soon sold at a great profit, with which he bought 10,000 acres for $50,000 on credit, paying only $200 down. Within two years he sold half of this tract, making enough money to pay off his debt. Brack was quite an entrepreneur: to transport his products to market, he bought a boat and built a short railroad; to process his products, he built a winery and a brewery.[30]

Murphy and Frankenheimer subdivided the Brack Tract into lots of approximately 100 acres each and leased them to Chinese tenants on a mixed sharecropping and cash rental basis. The tenants had to turn over half of their crops as rent and pay $8 per acre in addition. Almost all of Brack's tenants were Chinese companies rather than individuals; in the three-year period from 1896 to 1898, Murphy and Frankenheimer signed 36 leases with Chinese—33 of them companies—on behalf of Jacob Brack.[31] Even though he was already sixty-one years old, Brack continued to farm the remaining portion of his land himself.

In 1899 Brack's lease arrangements with Chinese tenants ended suddenly. A 1903 document reveals that the Brack Tract had suffered extensive flood damage in the preceding period. The document stated that the levees had been "greatly injured," the buildings were "entirely destroyed," and the land was "impossible to use." Brack turned over some 2,000 acres to Henry Brack, L. H. Frankenheimer, and Samuel Frankenheimer on a ten-year lease, charging no rent for the first two years in return for getting his levees and drainage ditches repaired. If the repairs cost over $8,000, the landlord and lessees would share the additional cost. Meanwhile, Brack conveyed most of his land to his children and retired to live in Lodi in 1906. At the time of his retirement, he was the second largest tax payer in San Joaquin County.[32]

After the turn of the century, most of the landlords of Chinese

tenants in the San Joaquin Delta were corporations, including
the English-Wallace Company, the Middle River Navigation
and Canal Company, the Roberts Island Improvement Company, the Venice Island Land Company, the Victoria Island
Company, the Western Company, and the Rindge Land and
Navigation Company.[33] The Middle River Navigation and
Canal Company may be taken as representative of the corporations that dealt with Asian tenants.

The company was part of an extensive corporate network controlled by capitalists from Los Angeles, who financed much of
the reclamation of the central peat islands in the San Joaquin
Delta during the first two decades of the twentieth century.
George Cochran, Stanley McClung, and Lee A. Phillips were the
three main architects of this venture. These men, together with
a small number of others—many of whom were lawyers by
training—formed the center of an interlocking directorate con-
necting many corporations operating in different parts of the
state. They served on the boards of directors of insurance companies, land companies, and water and canal companies.

George Cochran was listed in the 1900 Los Angeles City
Directory as an attorney in the law firm of Cochran and
Williams. In 1906 he had become president of the Associated
Mutual Trust and Investment Company, and in 1907 he
was listed as president of the Pacific Mutual Life Insurance
Company.[34] He served on the board of directors of the Middle
River Farming Company and the Rindge Land and Navigation
Company, held stock in the Empire Navigation Company, served
as president of the Holland Land and Water Company, and was
the largest shareholder in the Middle River Navigation and
Canal Company.[35] Each of these corporations that Cochran was
associated with leased many tracts in the Delta to Asian tenants,
particularly to Japanese.

Stanley McClung was the largest stockholder and a member of
the board of directors of the Rindge Land and Navigation Company. He also served on the board of directors of the Holland
Land and Water Company, the Middle River Navigation and
Canal Company, and the Empire Navigation Company.[36] He
was related by marriage to George Cochran. Lee A. Phillips,
also a lawyer, was the president of the Empire Navigation
Company. He was most actively involved in the dealings of these

various corporations in Delta land, serving as secretary of one corporation after another, and visiting the Delta regularly to oversee the affairs of these corporations, frequently arriving on his own yacht.[37] He was a close friend of George Shima [Ushijima Kenji], the Japanese "Potato King," and acted as the executor of the latter's will after Shima died.

The Middle River Navigation and Canal Company was incorporated in 1902 to purchase and operate water works, water transportation vehicles, land, warehouses, and farming equipment. It was capitalized at $1,500,000. Frederick Rindge, a resident of Santa Monica, who owned the Rindge Tract in the Delta, served as its president, and Lee A. Phillips was secretary.[38] Between 1902 and 1906 it acquired almost 10,000 acres of Delta land when George Cochran, Stanley McClung, Lee A. Phillips, A. J. Wallace, John Caperton, and the Middle River Farming Company turned over their individual landholdings to the new corporation.[39] The transaction between the Middle River Navigation and Canal Company and the Middle River Farming Company was facilitated by Lee A. Phillips, who happened to be secretary of both corporations. The transfer of title came about because the Middle River Farming Company, incorporated in 1901, owed $34,373 to the California and Nevada Land and Investment Company; in exchange for payment of this debt, the Middle River Navigation and Canal Company acquired 3,800 acres of land.[40]

The Middle River Navigation and Canal Company signed its first lease with a Chinese tenant in September 1905 when it leased 154 acres to the Tai On Company for growing asparagus, beans, potatoes, and fruit. The Tai On Company already had a store, which sold Chinese groceries and meat, located on the land of this corporation. The lease period was ten years; rent started at $9 per acre and gradually went up to $18 per acre during the tenth year. The Tai On Company continued to operate its store, which it rented for $120 a year, even after it went into tenant farming.[41]

Other Chinese companies leased tracts ranging from 160 acres to 500 acres. Though the scale of their operations was large— indeed gargantuan, if measured by the standards of rural China —none surpassed an individual named Chin Lung. The story of his activities forms an important chapter in the history of Chinese American entrepreneurship.

Digging up asparagus nursery roots for planting in commercial fields, Brannan Island, the Sacramento Delta, California, 1920. (Photograph courtesy of the Sacramento River Delta Historical Society, Locke, California)

Chin Lung was the fourth son of a Chinese farmer in Namshan village in a region of Heungshan (Chungshan) district called Taumoon. Born in about 1864, he emigrated to California when he was eighteen or nineteen years old to join his third brother, who had preceded him to America.[42] At the time Chin Lung arrived, his brother was working in the Sing Kee rice-importing firm, owned by someone also surnamed Chin, a fellow villager from Namshan. Chin Lung's first job at Sing Kee was sacking rice. He studied English in an evening class conducted by the Chinese Baptist Church. Soon, his knowledge of English provided him with an entry into tenant farming in the Delta. Chin

Lung first went to the Delta when he was asked to accompany a
group of his fellow villagers who had been recruited by Delta
landowners to become tenants. According to one of Chin Lung's
sons, his father knew little about agriculture.[43] What enabled
him eventually to flourish in that field was his organizational
skills: large-scale tenant farming in the Delta required an ability
to negotiate leases, to recruit labor, to house and provision
hundreds of men, and to set up a distribution network to sell
the crops produced. And all of these assumed the working
knowledge of English that Chin Lung possessed.

Chin Lung first signed a lease in his own name in 1898 when
he leased 200 acres on Andrus Island from Ernest A. Denicke on
a sharecropping basis. Ninety-five acres of the tract were already
in asparagus, and the landlord was to receive 40 percent of the
asparagus produced. On 65 acres of the asparagus field—
probably in the section where the plants were young—Chin
Lung was allowed to plant vegetables between the rows. He had
to give his landlord 45 percent of the vegetables harvested. Chin
Lung also was asked to plant vegetables on another 40 acres; but
from the latter crop the landlord expected only 25 percent of the
proceeds. Somewhat ironically, in view of his later reputation as
the Chinese "Potato King," Chin Lung was prohibited from
planting potatoes[44]—it was commonly believed that if potatoes
were grown close to asparagus, the latter's yield would be
affected.

In 1900 Chin Lung moved to the San Joaquin County part of
the Delta and leased 1,125 acres on the Sargent and Barnhart
Reclamation Tract near Stockton for $7,000 a year. The land
belonged to Ross C. Sargent, pioneer settler, land reclaimer,
and large landowner in San Joaquin County, and to Elizabeth
Barnhart, wife of Henry Barnhart, another pioneer settler and
large landowner.[45] Over the next three years, Chin Lung first extended the lease at an annual rent of $8,000 and then at $8,500.[46]
One-half of Chin Lung's crops was held in mortgage each year
until the full rent had been paid. For his operating expenses,
Chin Lung borrowed money from John M. Perry, a grain
merchant in Stockton. In 1904, for example, Perry advanced
him $2,500 in cash, with a promise of another $10,000 plus any
merchandise he desired, at an interest of 1 percent a month. As
collateral, he mortgaged the potatoes, beans, asparagus, onions,

hay, and grain he grew on four separate tracts.[47] Some of Chin
Lung's other creditors were M. D. Eaton, W. D. Buckley, and
W. D. Wallace, landowners and real estate agents in Stockton.[48]
In later years he had sufficient resources of his own and did not
require credit.[49]

Chin Lung frequently leased land from several landowners
simultaneously. When he still had the lease on the Sargent and
Barnhart tract, he also leased 100 acres on the Lower Division
of Roberts Island from Ellen Ryan for ten years at a graduated
rent, which increased from $5.50 per acre per year to $10.00 per
acre per year,[50] as well as 366 acres on the Upper Jones Tract
from the Middle River Farming Company.[51] In 1904 he leased
1,000 acres of the Palms Tract in Contra Costa County from the
Rindge Land and Navigation Company.[52] Mrs. Myra Wright
also became one of Chin Lung's landlords, leasing 628 acres to
him in 1905, and in 1906 an additional 292 acres of the Wright
Tract and 821 acres constituting the entire Elmwood Tract.[53]
During the 1910s Chin Lung farmed in other parts of the Delta,
leasing land on Venice Island, the Henning Tract, and Mildred
Island in San Joaquin County, Sherman Island in Sacramento
County, and the Byron Tract in Contra Costa County.[54] Some of
these tracts were contiguous, and others were as much as twenty
miles apart. Chin Lung must have traveled back and forth,
acting as manager of a vast business undertaking.

Each year, Chin Lung employed about 500 Chinese to help
him cultivate, harvest, and sack the potatoes, beans, onions, and
asparagus he grew. Whenever he needed temporary workers, he
could easily pick them up in Stockton's Chinatown, transporting
them to his camps in horse carts. Unlike his contemporary
George Shima, the Japanese "Potato King," who employed
laborers of many ethnic backgrounds, Chin Lung used Chinese
workers exclusively. The men were housed in bunks or tents
erected on the high ground on the levees. Like other Delta farmers, Chin Lung divided his large acreages into camps, ranging
from 100 to 500 acres in size. Each camp functioned as an
autonomous unit, with its own housing, cooking facilities, horses
and implements, laborers, and camp foremen. During the off-
season, the only persons who remained in the camps were the
cooks, the foremen, and the grooms for the horses. These men
were Chin Lung's only permanent employees, some working for

him for many years. They were paid between $400 and $600 a
year, with the teamsters receiving the highest pay. The
temporary workers were paid $1 a day plus board in the early
1900s. Wages rose to $2 a day in the 1920s. According to his
sons, Chin Lung gave some of his men the option of working for
wages or becoming his partners in his farming enterprise.[55]

Chin Lung owned over seventy horses and grew hay himself to
feed them. He owned two barges, which plied the San Joaquin
and Sacramento rivers, bringing provisions from San Francisco
and Stockton and taking his produce to San Francisco for sale.
According to the recollections of his sons and one of his nephews,
he would sell part of his potato crop on the steamboat landings in
the Delta to commission merchants who came to purchase on the
spot, and would ship part of the crop to San Francisco for sale,
docking at Pier 5.[56]

In 1912, the year before the 1913 Alien Land Act was passed
in California, Chin Lung bought 1,100 acres several miles northwest of Stockton. He named this tract the Sing Kee Tract—a
name that has persisted to this day. The tract had its own steamboat landing. It is perhaps the only tract of land in California
today bearing a Chinese name. Chin Lung's use of the name is
itself interesting: the owner of Sing Kee, the rice-importing firm
where Chin Lung worked when he first arrived in San Francisco,
was growing old, so he sold the store to Chin Lung, who then
began using its name as a personal alias. County archival
records contain many documents that indicate in parenthesis
that Sing Kee was also known as Chin Lung. After Chin Lung
acquired the store, it branched from rice importation into
other lines of business.

Chin Lung also invested in land elsewhere. After the enactment of the 1913 Alien Land Law in California, which prohibited "aliens ineligible to citizenship" (Asian immigrants) from
purchasing land, Chin Lung went to Oregon and bought 2,000
acres near Klamath Falls, on which he grew Oregon Gem potatoes. He continued to farm simultaneously there and in the Delta
until 1924, when his leases ran out and could not be renewed
because of the provisions of the 1920 Alien Land Law. Thereafter, he farmed only in Oregon for about ten years, returning to
China in 1933 when he was in his late sixties. He had something
to return to because, after Oregon also passed an Alien Land Act
in 1923, he had bought land in his native district in China.[57]

According to one of his sons, in his best year Chin Lung earned a profit of $90,000. But he did not live ostentatiously, a car being one of his few indulgences. He enjoyed driving home from the bank, holding bags of gold coins in his lap. He was much more willing to put his earnings into capital equipment, becoming the first Chinese to buy a Holt Caterpillar tractor, and paying a Hawaiian Chinese, Charlie Chin, $5 a day to drive it.[58] He invested his money several other ways. In addition to a branch of the Sing Kee firm in Sacramento, he established another import-export firm, as well as a baggage-manufacturing company, called the Shanghai Trunk Company, on Stockton Street in San Francisco. He also invested in stocks in the Wing On Department Store in Shanghai, one of the largest modern commercial establishments in China.[59]

During the farming season, Chin Lung lived in Stockton, but in the off-season he returned to San Francisco, where his family lived. He had returned to China to marry one or two years after his initial arrival and had brought his wife to the United States, setting up house for her in San Francisco. But she did not enjoy living in America and returned home. He brought her back one more time for a second prolonged visit, but he was unable to persuade her to remain. After several years, she went back to live in China permanently. Four of Chin Lung's children were born in San Francisco. When his wife returned to China, she took the children with her, but all the sons came to live and work in America when they became adults. One of his sons helped him farm on the Byron Tract; one managed the trunk company; another farmed with him in Oregon, but finding that not to his liking joined the merchant marines and later opened a restaurant; the fourth became a chrysanthemum grower.[60]

In addition to his own family, Chin Lung was responsible both directly and indirectly for bringing many people from Heungshan district to farm in the Delta. As a matter of fact, Chinese tenant farming in the San Joaquin Delta (but not the Sacramento Delta) was dominated by persons surnamed Chin. One, Chin Bow, was Chin Lung's nephew; others were related to him in more distant ways. Lease and mortgage records in Sacramento, Contra Costa, and San Joaquin counties indicate that there were at least thirty tenants surnamed Chin who farmed in the Delta during the 1900s through the 1920s. Members of the Chin clan formed the Chinese American Farms corporation in

1919 to run their enterprises more effectively. The corporation was empowered to transact real estate, acquire and develop water works, purchase and operate docks, vessels, trams and railroads, warehouses, and farm equipment, to make contracts, issue capital stocks and bonds, and do whatever else was necessary to carry on its business. Capitalized at $1,000,000, it had eleven members on its first board of directors: Chin Yen, Chin Gow, Chan [Chin] C. Wing (whose last name was Chan, not Wing), Lee Yum, Lee Cherk, Chan [Chin] Din Hoy, B. S. Yuen, Chin Get, Chin Hing, Lee Yuen, and Chin Dan Chung. All were residents of San Francisco except Chin Hing of Oakland, Lee Yuen of Fresno, and Chin Dan Chung of Stockton. Within three months of incorporation, the board of directors was increased to seventeen members. The corporation was involuntarily dissolved in 1932, at which time only four of the original board members were still serving on the now fifteen-member body.[61] At about this same time, the Potato King, entrepreneur and patriarch, also wound up his business in America and returned to China to retire, leaving many descendants, relatives, and fellow villagers behind in the New World Delta they had helped to develop.

The Economics of Potato Cultivation

Chinese specialization in potato cultivation was already apparent long before great entrepreneurs like Chin Lung entered the Delta. The 1870 manuscript agriculture census listed fourteen Chinese farmers in the Delta, all of whom grew Irish potatoes or sweet potatoes or both. As table 19 shows, during this early period the Chinese tenants grew more sweet potatoes than the Irish variety. The largest growers were Hop Lee, who produced 4,000 bushels, and Ah Yet, who produced 8,000 bushels. Many of the Chinese listed also grew beans and market garden crops. The value of their product per acre ranged from $20 to $180—a range considerably lower than that of truck gardeners, who were averaging $400 or more per acre in the same period. By 1880, as table 20 shows, the Chinese farmers in the Delta had stopped growing sweet potatoes, but were producing instead large quantities of Irish potatoes. One-third of

the Chinese farmers listed in Delta townships in the 1880
manuscript agriculture census produced more than 15,000
bushels of Irish potatoes each. Ah Yet (who may or may not be
the same person of that name listed in the 1870 census) was
again the largest grower, producing 40,000 bushels on 250 acres
of tilled land.

Why did Chinese farmers in the Delta specialize in the cultivation of the Irish potato when it was not a traditional Chinese
crop? Since the immigrants from Kwangtung province ate sweet
potatoes but not Irish potatoes as a staple, it might have been
expected that if they grew potatoes at all, it would be the sweet
variety.

To answer this question, it is necessary to examine the economics of various Delta crops in the nineteenth century. Using information from the manuscript schedules of the 1860, 1870, and
1880 censuses of agriculture (the schedules for censuses taken after
1880 have not yet been released for public perusal), farm value
per acre and product value per acre will be correlated with different crop patterns.[62] Farm value refers to the value of the land
and improvements upon it; product value refers to the dollar
value of all the crops grown on a farm. Franklin and Georgiana
townships in Sacramento County and Union Township in San
Joaquin County will again be taken as representative of the
Delta. In addition, an analysis of Sutter Township—located
north of Franklin Township and immediately south of the City of
Sacramento—will also be made to determine whether proximity
to a growing urban area had any effect on farm value and product
value per acre.

Farms devoted to the extensive cultivation of grain and hay
and the raising of livestock were found to have the lowest farm
value per acre as well as the lowest product value per acre.
Diversified farms had considerably higher per acre farm and
product values, but it is necessary to ask what kind of crops *added*
the most value to the basic grain-hay-livestock farm. The value
added by potatoes and beans, on the one hand, and the value
added by fruit and vegetables, on the other, were compared. The
1860 data were not useful for this purpose, because few farms
grew potatoes at this date. The 1870 and 1880 data revealed that
farms producing vegetables and fruit in addition to hay, grain,
and livestock were not worth much more than farms growing

Table 19   *Chinese Farmers in the Sacramento–San Joaquin Delta Listed in the 1870 Manuscript Census of Agriculture*

County and Township	Name of Person or Company	Imprvd Acres	Unimp. Acres	$ Farm Value	$ Value per Acre	$ Value of Imple- ments	$ Wages Paid in 1869–70
SACRAMENTO COUNTY							
Franklin	Ah Furt	12	40	2,000	100	–	500
Township	Ah Fee	30	100	3,000	60	–	–
	Ah Keng	30	–	1,000	33	–	–
Georgiana	Ah Gum	40	200	2,000	25	–	–
Township	Ah Yan	50	–	4,000	80	–	–
	Ah Yet	40	300	5,000	50	100	–
	Ling Gen	25	–	500	20	–	–
	Ah Ling	30	170	2,000	31	–	–
	Ah John	25	90	2,000	47	–	–
	Ah Yetta	30	225	2,500	33	–	–
	Hop Lee	60	260	3,000	27	–	–
	Ah Yute	20	140	2,000	42	–	–
YOLO COUNTY							
Merritt	Hop Hung & Co.	4	–	6,000	1,500	50	–
Township	Lee Poo	–	200	1,500	38	–	–
	Lee Sing	–	–	–	–	–	–

SOURCE: U.S. Bureau of the Census, "Census of Productions of Agriculture" (manuscript), 1870.

NOTES: In computing farm value per acre and product value per acre the number of unimproved acres was divided by 5 and then added to the number of improved acres to form the denominator.
ᵃ h= horse, p = pig.

only the latter extensive crops, but farms growing potatoes and beans in addition to hay, grain, and livestock increased both the per acre farm value and the product value substantially. That is to say, potatoes and beans, normally thought of as crops to be grown on marginal land, added more value to farmland than fruit and vegetables, commonly considered to be crops of higher value. This peculiarity may be explained by the extraordinary fertility of the peat soil in the reclaimed backswamps, which produced high yields of beans and potatoes without fertilizers, but which was not particularly good for most fruit and vegetables. The latter crops thrived only on the natural levees of the northern Delta. Besides increasing the value of the land planted to

Kind of Livestock	$ Value of Livestock	$ Market Gardens	Irish Potatoes	Sweet Potatoes	Beans	$ Other Prod.	$ Total Products	$ Value Products per Acre
–	–	–	100 bu.	500 bu.	20 bu.	–	1,200	60
2 horses	500 [sic]	–	60 bu.	–	–	–	1,800	36
3 horses	150	–	200 bu.	800 bu.	250 bu.	–	2,000	67
2 pigs	30	1,000	–	1,400 bu.	80 bu.	–	3,000	38
2h, 10p[a]	200	1,000	300 bu.	1,500 bu.	150 bu.	–	9,000	180
–	–	500	1,600 bu.	8,000 bu.	70 bu.	–	2,500	25
2 oxen	200	400	20 bu.	2,000 bu.	–	–	1,000	40
2 horses	80	100	–	2,000 bu.	80 bu.	–	1,500	23
–	–	100	100 bu.	1,000 bu.	100 bu.	–	1,000	23
2h, 6p[a]	150	100	100 bu.	1,500 bu.	100 bu.	–	1,500	20
4 horses	200	500	–	4,000 bu.	400 bu.	–	5,000	45
1 mule, 2 cattle	250	–	400 bu.	1,200 bu.	200 bu.	–	1,500	31
–	–	–	–	–	–	700	700	175
–	–	800	–	–	–	–	800	20
–	–	–	–	–	–	–	1,000	–

them, potatoes and beans apparently were also worth more on the market (as measured by the product value per acre) than comparable acreages of fruit and vegetables in that period.

Another way to look at this question is to ask whether farms that did *not* grow any hay and grain or raise livestock at all had an even higher value than those combining hay and grain with potatoes and beans. It turned out that they did. Where land was abundant, farmers could afford to sow huge tracts to wheat, but when land values began to rise, it made sense to switch to crops that gave a higher return per acre. On virgin peat soil, potatoes and beans proved to be the best choice. Elsewhere in California, fruits, berries, vegetables, hops, and other specialty crops were more suitable.

Data for Sutter Township show that proximity to Sacramento City definitely had an effect on farm values. Regardless of land use patterns, farms there were worth more. However, these farms did not have a higher product value per acre, since the fertility of the soil has no relation to the land's proximity to urban areas.

Table 20 *Chinese Farmers in the Sacramento–San Joaquin Delta Listed in the 1880 Manuscript Census of Agriculture*

County and Township	Name of Person or Company	Number of Acres Tilled	Number of Acres Other	$ Farm Value	$ Value /Acre[a]	$ Value of Implements	$ Wages Paid in 1879–80	Kind of Livestock
SACRAMENTO COUNTY								
Franklin Twp.	Lum Gui (SC)	15	80	1,000	32	–	250	2 horses
	Ah Mow (SC)	15	0	3,000	200	50	–	–
Georgiana Twp.	Ah Wing & Co. (CR)	100	0	3,000	30	75	–	–
	Et Yung & Co. (CR)	150	0	3,750	25	100	–	2 horses
	Lee Kim (CR)	108	0	4,000	37	100	–	3h, 3mc, 2p +
	Ah Him & Co. (CR)	100	4	3,000	30	100	150	4h, 8p, 10ch
	Ah Fye (CR)	15	0	600	40	–	150	–
	Ah Chaw & Co. (SC)	160	0	–	–	100	200	5h, 1mc, 10ch
	Quong Lee & Co. (SC)	99	1	3,000	30	100	100	6h, 1mc, 5p
	Ah Kim & Co. (SC)	60	0	2,400	40	40	84	2h, 1p, 6ch
SAN JOAQUIN COUNTY								
Union Twp.	Sam Kee & Co. (SC)	275	15	12,720	46	200	1,300	8h, 1p, 24ch
	Ah Hum & Co. (SC)	250	0	10,000	40	300	1,000	8 horses
	Ah Lock & Co. (SC)	100	0	4,000	40	200	500	4 horses
	Kee You & Co. (SC)	200	0	8,000	40	300	2,000	6 horses
	Ah Gee (SC)	150	0	6,000	40	200	1,500	6 horses
	Ah Moon (SC)	50	0	2,000	40	50	200	2 horses
	Ah Chu (SC)	200	0	8,000	40	200	2,000	6 horses

Ah Hong (SC)	100	0	4,000	40	100	1,500	4 horses
Ah Pook (SC)	110	0	4,000	36	100	1,000	4 horses
Ah Yet (SC)	250	0	10,000	40	150	2,000	4 horses
Ah Sung Ling (SC)	150	0	6,000	40	100	1,500	4 horses
War Ling (SC)	100	0	4,000	40	100	1,000	4 horses
Ah Coey (SC)	100	0	4,000	40	100	1,000	4 horses

CONTRA COSTA COUNTY

Township #5

Long Quong (SC)	300	0	12,000	40	500	200	12h, 1mc, 10p +
Ah Jim (SC)	120	0	–	–	25	–	4 horses
Ah Bing (CR)	65	0	2,000	31	100	–	4 horses
Ah Hoon (CR)	10	0	300	30	–	–	–
Ah Mang (CR)	15	0	450	30	–	–	–
Ah Gow (CR)	50	0	2,000	40	100	700	6 horses
Ah Jim (SC)	40	0	1,500	38	75	–	3 horses
Ah Ting & Ah Leck (CR)	100	3,000	4,000	6	150	–	6h, 20p, 50ch

YOLO COUNTY

Merritt Twp.

Ah Yow (CR)	6	0	400	67	–	–	–
Ah Joe (CR)	60	0	2,000	33	50	150	1h, 4p, 9ch
Ah Sam (CR)	20	110	1,300	31	20	100	3h, 30p

(*Table 20, continued*)

Value of Livestock	Irish Potatoes			Sweet Potatoes			Beans		Fruit		$ Total Value of Products	$ Product Value/Acre[b]
	Acres	Bushels	Yield	Acres	Bushels	Yield	Bu.	Acres	Bu.	$		
100	15	1,500	100 b/A	—	—	—	—	—	—	—	800	53
—	—	—	—	—	—	—	—	—	—	—	—	—
140	11	1,500	136 b/A	—	—	—	300	—	—	—	800	8
300	(farm was under water during 1879–80)						—	—	—	—	—	—
100	—	—	—	—	—	—	—	—	—	—	1,000	9
300	5	400	80 b/A	20	1,600	80 b/A	—	4	1,000	500	1,000	10
—	—	—	—	—	—	—	—	—	—	—	—	—
200	10	600	60 b/A	5	300	60 b/A	100	24	6,600	3,000	4,000	25
300	1	100	100 b/A	5	200	40 b/A	—	—	—	—	800	8
100	—	—	—	—	—	—	—	—	—	—	300	5
275	100	1,400	14 b/A	(+ 100 acres barley)			500	—	(+ 100 acres hay)		800	3
600	150	30,000	200 b/A	—	—	—	1,000	—	—	—	3,000	12
500	75	15,000	200 b/A	—	—	—	100	—	—	—	1,000	10
700	150	30,000	200 b/A	—	—	—	1,500	—	—	—	2,000	10
600	100	20,000	200 b/A	—	—	—	1,000	—	—	—	2,000	13
100	10	2,000	200 b/A	—	—	—	400	—	—	—	500	10
400	150	25,000	167 b/A	—	—	—	2,000	—	—	—	3,000	15

300	100	18,000	180 b/A	–	–	–	–	–	–	1,800	18
300	100	18,000	180 b/A	–	–	–	–	–	–	1,800	16
300	200	40,000	200 b/A	–	–	–	2,000	–	–	4,200	17
300	100	20,000	200 b/A	–	–	–	2,000	–	–	2,700	18
150	75	15,000	200 b/A	–	–	–	1,000	–	–	1,500	15
200	100	18,000	180 b/A	–	–	–	–	–	–	1,800	18
–	–	–	–	–	–	–	–	–	–	–	–
100	–	–	–	–	–	–	–	–	–	–	–
100	65	6,000	92 b/A	–	–	–	–	–	–	1,500	23
–	10	2,000	200 b/A	–	–	–	–	–	–	300	30
–	15	2,500	167 b/A	–	–	–	–	–	–	1,250	83
210	50	6,000	120 b/A	–	–	–	–	–	–	1,500	30
140	–	–	–	–	–	–	–	–	–	–	–
100	–	–	–	–	–	–	–	–	–	5,000	50
–	4	400	100 b/A	–	–	–	–	–	–	50	8
64	11	600	55 b/A	.1	110	110 b/A	400	–	–	400	7
90	5	500	100 b/A	7	1,120	160 b/A	160	–	–	700	35

SOURCE: U.S. Bureau of the Census, "Census of Productions of Agriculture" (manuscript), 1880.

NOTE: SC = share crop; CR = cash rent; + = plus other animals; h = horse; p = pig; ch = chicken; mc = milk cow; b/A = bushels per acre.

a In computing farm value per acre, the number of "other" acres was divided by 5 and then added to the number of "tilled" acres to form the denominator.

b See notes a, b, and c in table 14 for a full explanation on how the figures for $ Product Value/Acre are computed.

Chinese farmers winnowing beans, California. (Photograph from the Pierce Collection, courtesy of The Huntington Library, San Marino)

Having discovered that beans and potatoes added the greatest value to both farm value and product value, it is now easy to understand why Chinese tenant farmers chose to specialize in those two crops, particularly potatoes. Though Chinese traditionally have valued land for noneconomic reasons—peasants, merchants, artisans, all coveting the status of landowners— Chinese immigrants in America had to be more rational (in the classical economic definition of rationality) and treat land primarily as a factor of production. Since they had to pay rent as tenant farmers, they *had* to specialize in those crops which brought the highest returns. Rent was a drain on their total income, which meant that if they wished to survive or prosper as farmers, they had to grow those crops which brought a sufficiently large income so that even after a surplus had been drained off, there was enough left for a living. In short, they had to be more adept at capitalist calculations than white owner-operators.

Furthermore, once Chinese immigrants discovered the advantages of a certain economic activity, they worked to dominate it. This was true of Delta potato-farming as it was of truck gardening. In 1870 approximately 40 percent of the farmers in the three Delta townships who specialized in potato and vegetable production were Chinese; by 1880 two-thirds of the bean and potato specialists there were Chinese.

Comparing the yield per acre and the scale of operations of white and Chinese potato-growers in 1880, it was found that in all parts of the Delta the majority of both white and Chinese potato-growers produced yields ranging from 100 bushels to 299 bushels per acre, indicating that the productivity of the two groups of farmers was comparable. In terms of the scale of operations, however, the Chinese had a greater tendency to engage in the large-scale cultivation of potatoes. A majority of the European immigrant and American potato-growers had less than 20 acres of their farms in potatoes, whereas the Chinese potato-growers were divided into two groups—a small group who planted less than 20 acres, and a larger group who planted more than 50 acres. The majority of Chinese potato-growers devoted 100 to 199 acres to potatoes.

Despite high yields and good prices, potato cultivation in the Delta had drawbacks that may help to explain why Chinese, more than whites, specialized in it. All the problems were associated with the peculiarities of peat soil. Virgin peat was capable of producing from 300 to an astonishing 800 bushels of potatoes per acre, but the yield started to fall precipitously after two or three years, owing to the growth of a fungus. According to agronomists sent in 1908 by the Bureau of Plant Industry of the United States Department of Agriculture to investigate the causes of the rapid decline in the yield of potato fields in the San Joaquin Delta, moist, warm peat soil was conducive to the development of *Fusarium oxysporum* Schlecht, commonly known as potato wilt. It is a disease that is related to the wilt diseases of cotton, watermelon, cowpea, and tomato. Leaves of potato plants afflicted with the disease turn a lighter green color, rolling and curling around the margins. Soon, the tops wither, causing the plants to ripen prematurely. The roots of the diseased plants become brittle, and the wilt fungus appears as a white or pink mold. In the early stages, evidence of the disease is hard to detect, but in time the potatoes themselves become rough and scabby. If the potatoes are stored in a cold place, nothing much happens, because the fungus requires warmth to develop, but when they are stored in a warm place, dry rot soon appears in the tubers. Sometimes bacteria cause a secondary soft rot.[63]

Delta farmers thought that the disease had something to do with the depletion of particular nutrients in peat soil, but the agronomists assured them such was not the case. They confirmed that Delta soil was extremely fertile and that nutrients were not quickly depleted. But the potato disease was difficult to control, since it originated and spread underground. The best remedy seemed to be crop rotation. Barley and beans planted after potatoes were especially effective in reducing the severity and spread of the fungus. Potatoes could be planted on a piece of ground again—with good results—after it had been planted to the other two crops for several years. Consequently, potato farmers had to move from one field to another, which meant that tenants could specialize in continual potato cultivation more easily than farm owners. Chinese who did not own land, therefore, found it easier to become potato and bean specialists.

Peat soil caused other problems. Initially clumpy, after several years of cultivation it becomes extremely fine, so fine that it gets into everything—into houses and inside clothes, impossible to wash off. Tule farmers—that is, those who cultivated the backswamps in the center of islands—never lived there, escaping each evening to houses and shacks built on higher ground or even set on barges moored in the rivers. Most whites would rather not work "behind the levees." Aside from Asian immigrants, only Italians and Portuguese became tule farmers. These groups were willing to till the land behind the levees because they had come from poor countries and were eager to "make it" in America. They put up with physical inconveniences in exchange for substantial economic rewards. Landowners welcomed them because the rich peat soil would have yielded nothing without people to cultivate it. Less desperate old-stock Americans and German and British immigrants who also farmed in the Delta lived and worked almost entirely on the more pleasant natural levees. Thus the land and its peculiarities profoundly influenced the social ecology of a community where race, ethnicity, and class were all aligned according to the contours created by nature.

SEVEN

So Many Fecund Valleys

THE CHINESE in California did not play the same pioneering role in the cultivation of fruit and other specialty crops as they had done in truck gardening and Delta potato-growing. With the exception of the Vina district in Lassen Township, Tehama County, where they planted and tended the first large orchards, Chinese entered most of the fruit-growing areas after the first orchards had been planted. In these districts—which differed considerably in their historical development—the pattern of Chinese involvement has also shown substantial regional variation.

Climate and soil, human settlement patterns, and man-made calamities have all influenced the location of fruit-growing districts in California. Climate and soil are important because fruit trees grow well only in light, well-drained soil. Sandy loam is best for orchards, whereas clay is detrimental because it retains too much water, causing the roots of fruit trees to rot. Temperature is also important because late frosts kill buds, but some species do require a certain period of cold weather in the winter to produce good fruit. For example, cling peaches require about 1,500 hours of temperatures below 45 degrees from September to February to allow the trees a sufficient rest period before they bloom in the spring. During the blooming and fruiting seasons, the peach trees need a good balance of rather high temperatures (but not excessive heat), abundant sunlight, and low humidity. Summer rains are harmful because high humidity enables brown rot fungus—one of the major diseases affecting cling peaches— to thrive.[1] The availability of water is also crucial, since in California irrigation is necessary for most species of fruit. Therefore, the distance from streams and the quality of the irrigation water also help to determine the location of orchards.

In the early years, proximity to consumers was another important factor influencing the location of orchards because perishable fresh fruit had to be readily marketed. For that reason, where climate and soil permitted, orchards sprang up close to the earliest settled areas in the state. As is shown in table 21, in 1860 four of the five top-ranking fruit-producing counties were in or near the Sierra Nevada foothills where tens of thousands of miners had earlier congregated. Yuba County ranked first in the value of fruit produced in 1859–60, but this fact is misleading because a single orchard—owned by G. G. Briggs—produced over 60 percent of the total value of fruit grown in the county. If Briggs's orchard is not counted, then Coloma Township along the American River in El Dorado County, and the fruit belt in the northern Sacramento Delta, produced the most fruit. Coloma Township had forty farms with orchards, twenty-four of which each produced more than $1,000 worth of fruit, which was consumed by people in the southern mining region. The three townships in the Sacramento Delta had ninety-one farms that cultivated fruit, eight of which produced more than $1,000 each. The fruit grown here was sold in Sacramento city or shipped to the northern mining region.

In California, perhaps more than elsewhere in the world, manmade calamities played a critical role in determining the location of orchards because mining debris destroyed many areas that started out as, or could have become, thriving fruit-growing districts. Debris was a threat beyond the farmers' control. Though farmers had organized to oppose hydraulic mining, they did not win their battle until the late 1880s when a court injunction prohibited miners from using powerful jets of water to wash down entire mountains in order to get at the gold ore deposited there.[2] But before that victory was achieved, many orchards had already been "completely destroyed by debris," as was noted in the manuscript schedules of the 1870 and 1880 censuses of agriculture.

A glance at table 21 will indicate that in 1870, while Sacramento and Yuba counties—whose riverine fruit-growing belts were already suffering from the cumulation of silt—

Table 21   The Ten Top-ranking Counties in Terms of the Value of Orchard Products, California, 1860, 1870, 1880, and 1900

Rank	1860		1870		1880		1900	
	County	$ Value	County	$ Value	County	$ Value	County	$ Value
1	Yuba	142,490	Sacramento	165,385	Santa Clara	228,923	Santa Clara	2,994,927
2	El Dorado	84,815	Sonoma	151,804	Alameda	210,745	Fresno	1,104,294
3	Sacramento	70,360	Yuba	127,110	Sacramento	179,028	Sonoma	927,831
4	Los Angeles	57,290	Solano	125,630	Sonoma	168,767	Solano	914,222
5	Tuolumne	54,980	Santa Clara[a]	98,437	Los Angeles	122,920	Tulare	849,490
6	Napa	30,215	Alameda	91,826	Solano	91,880	Sacramento	751,619
7	Santa Clara	30,095	Los Angeles	74,090	El Dorado	82,648	Placer	535,307
8	Sonoma	29,131	El Dorado	60,857	Napa	82,460	Kings	522,188
9	Alameda	28,530	Yolo	47,479	Placer	64,697	Los Angeles	502,472
10	Solano	26,785	Amador	43,350	San Bernardino	56,012	Alameda	477,403

SOURCE: U.S. Bureau of the Census, Eighth, Ninth, Tenth, and Twelfth *Census of the United States: Statistics of Agriculture*, 1860, 1870, 1880, and 1900. (The data for 1890 are not given in terms of dollars, but are in bushels for each fruit, so they are not comparable with the data for other census years and have not been included here.)

NOTE: The above figures, which come from the published census, in many cases do not match the figures derived from adding up the figures for individual farms listed. Discrepancies are less than 5 percent in most cases. I have used the published figures whenever the discrepancies do not affect the rank order of the county. The only exception is Santa Clara County for 1870, for which I have used the sum of the individual farms from the manuscript schedules. The published figure is $69,746, which would have placed Santa Clara County in seventh place instead of fifth.

[a] The published figure is $69,746, which would have placed Santa Clara County in seventh place instead of fifth.

Table 22 Chinese Fruit Growers Listed in the 1880 Manuscript Census of Agriculture by County

County and Township	Name of Person or Company	Apples		
		Acres	Trees	Bushels
SACRAMENTO VALLEY				
Tehama County				
Lassen Twp.	Pin Han	6	300	1,200
	Ah Shew	10	500	1,500
	Shon Bon	1	20	80
	Onn Top	2	120	360
	Sing Lem	3	150	350
	Ah Jim	8	400	1,600
	Ah Smow	1	50	150
	Hop Sing & Co.	4	200	700
Sutter County				
Nicolaus Twp.	Ah Sing	½	15	30
Soland County				
Vacaville Twp.	Ah Hing	–	–	–
Sacramento County				
Georgiana Twp.	Lee Kim	2	100	800
	Ah Chaw & Co.	2	300	600
SIERRA NEVADA FOOTHILLS				
Nevada County				
Grass Valley Twp.	Ah Ning	2	100	30
	Ah Sing	2	300	300
	Tin Loy	1	50	–
SAN FRANCISCO BAY AREA				
Santa Clara County				
San Jose Twp.	Ah Jim	3	1,200	1,200
San Mateo County				
5th Twp.	Sun Lee	½	25	50
Napa County				
Napa City	Ah Jim	4	300	400
SAN JOAQUIN VALLEY				
Kern County				
5th District	Sing Lee	2	100	200

SOURCE: U.S. Bureau of the Census, "Census of Productions of Agriculture" (manuscript), 1880.

NOTE: Other products grown on the above farms have not been included in this table.

	Other Fruit		Total $ Value of Fruit	Average $ Value/Acre	Average $ Price/Bushel	% Total Prod. Value in Fruit
Acres	Trees	Bushels				
9	900	3,000	2,200	147	0.52	44
30	3,000	10,000	3,400	85	0.29	40
1	50	100	100	50	0.55	6
4	400	1,300	650	108	0.39	36
4	350	1,200	550	79	0.35	9
2	200	450	600	60	0.29	30
5	500	1,200	700	115	0.51	11
30	3,000	10,000	3,200	94	0.29	35
–	–	–	20	40	0.67	1
15	1,600	–	1,200	80	–	67
2	200	200	500	125	0.50	50
20	3,000	6,000	3,000	136	0.45	75
1	25	–	30	10	0.10	4
⅛	50	4	500	24	1.64	33
2	150	13	–	–	–	–
–	–	–	500	167	0.41	100
–	–	–	25	50	0.50	2
–	–	–	200	50	0.50	26
2	150	250	400	100	0.88	67

retained prominence as fruit producers, they were beginning to lose ground to developing fruit districts in Sonoma, Solano, Santa Clara, and Alameda counties, which did not suffer from mining debris. The destruction caused by debris, together with population shifts away from the foothills to the valleys and the San Francisco Bay Area, enabled the fruit districts in the latter four counties to outproduce those in the foothills and along the Sacramento River by the 1870s. Sonoma County had two fruit-growing areas: the Sonoma Valley, which had become one of the major producers of grapes in the state, second only to Los Angeles County, and the Sebastopol area, a prime apple-growing area. The Vaca Valley in Solano County produced peaches and other stone-pit fruits in abundance; the Santa Clara Valley emerged as the premier fruit-growing area in California by 1880 and retained its preeminence until recent times; and a wide variety of tree and small fruits thrived in the plains of Alameda County.

Fruit districts that developed in the late 1880s in the Newcastle-Penryn area of Placer County and the Winters area in Yolo County were also free from the scourge of mining debris. Toward the end of the nineteenth century, areas in the San Joaquin Valley, especially the viticultural districts of Fresno, Tulare, and Kings counties, had also become so productive that they surpassed the northern districts in the value of orchard crops produced.

A small number of Chinese had begun to cultivate fruit in the late 1870s, but it was not until the 1880s that their fruit acreage expanded. In those years, 40 percent or more of the farms leased by Chinese in the Sacramento Valley and 35 percent or more of those they leased in the San Joaquin Valley grew fruit. Though the farms leased by Chinese tenants were quite large, fruit trees were grown only on a few acres of each of the ones with orchards. This pattern was similar to that exhibited by white farmers and fruit-growers. With the exception of those in the Vina district of Tehama County, most Chinese fruit-growers each cultivated only three or four acres of fruit—mostly apples, peaches, pears, plums, and grapes. As is shown in table 22, the most extensive acreages recorded by census takers in 1880 were operated by Ah Shew, with ten acres of apples and thirty acres of peaches and other fruit worth $3,400; Hop Sing and Company, with four acres of apples and thirty acres of peaches and other fruit worth $3,200; and Ah Chaw and Company, with two acres of apples and twenty acres of peaches and other fruit worth $3,000. Small as these acreages were, fruit brought in a disproportionate fraction of the Chinese tenants' total income. While most of them devoted no more than 5 to 20 percent of their land to fruit, fruit provided one-third or more of their total income from farming.

As is shown in maps 15 and 16, large concentrations of Chinese tenant farmers were found in the Sacramento Valley in four locations: the Vina district of Tehama County, the fruit belt along the Feather River, the foothills of Placer County, and land along the major bends of the Sacramento River in Colusa, Sutter, and Yolo counties. Fruit was grown in all these areas, although much of the land leased by Chinese along the upper Sacramento River was used for raising corn and other field crops.

Chinese tenant farmers were also numerous in the coastal valleys of central California: the Sonoma Valley, where they planted hops and tended vineyards and orchards; the Napa Valley, where they tended vineyards and grew vegetables; the Vaca and Suisun valleys of Solano County, where they cultivated deciduous fruit; the Santa Clara Valley, where they grew strawberries, fruit, and garden seeds; the Pajaro Valley, where they harvested and dried apples; and the Salinas Valley, where they practiced diversified farming, growing vegetables, fruit, potatoes, and sugar beets. (No attempt has been made to map the location of Chinese-leased tracts in the coastal valleys, because lease and chattel mortgage documents in these counties, where there were many Spanish and Mexican land grants, often designated tracts by lot numbers within subdivisions of such grants. Many of the old subdivision plats are no longer available, so it has not been possible to pinpoint the precise location of the lots.)

In the San Joaquin Valley during the nineteenth century, as is shown on map 17, Chinese tenant farmers were found in a ten-mile radius around the city of Fresno, along the Kings River, around the city of Visalia in Tulare County, and in the northeastern corner of Kings County. Map 18 indicates that Chinese tenants remained in the same locations—though in smaller numbers and scattered among Japanese tenants—during the

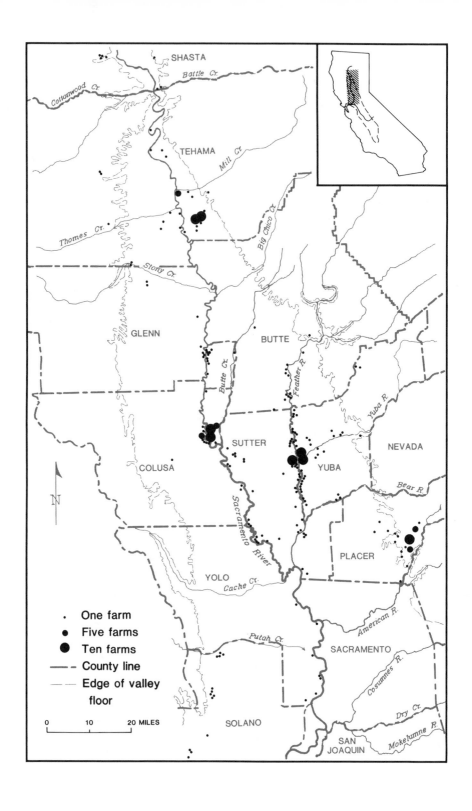

SHASTA

Battle Cr.

Cottonwood Cr.

TEHAMA

Mill Cr.

Thomes Cr.

Big Chico Cr.

Stony Cr.

GLENN

BUTTE

Butte Cr.

Feather R.

Yuba R.

SUTTER

NEVADA

COLUSA

YUBA

Sacramento River

Bear R.

PLACER

YOLO

Cache Cr.

American R.

• One farm
● Five farms
● Ten farms
– – County line
– – Edge of valley
    floor

Putah Cr.

SACRAMENTO

Cosumnes R.

0   10   20 MILES

SOLANO

Dry Cr.

SAN
JOAQUIN

Mokelumne R.

twentieth century. In these areas of the more arid San Joaquin Valley, Chinese were among the pioneer horticulturists and viticulturalists.

Chinese who cultivated fruit and other labor-intensive specialty crops followed different patterns of specialization. At least five such patterns can be discerned. First, in Sacramento, Sonoma, and Santa Clara counties, where a sizable number of orchards had already been planted by the time the Chinese began to lease land there, they worked in fruit orchards mainly as pickers and packers. When they did lease land in these areas to farm on their own account, they tended not to grow tree fruit as a general rule, specializing instead in crops such as hops, strawberries, or garden seeds which brought higher returns per acre than deciduous fruit. Second, in Tehama County, where they were the pioneer orchardists, few became harvest laborers. Here, the small number of Chinese farm laborers available worked for other Chinese who leased orchards or operated truck farms. Third, in Yuba and Sutter counties, Chinese did not enter the harvest labor force in large numbers either, because the orchards along the Feather River were already suffering from mining debris by the time large numbers of Chinese became available. Chinese took up fruit

Map 15   *The Sacramento Valley: Farms Leased by Chinese Tenants, 1870–1899*
SOURCES: *Tehama County, "Leases"; Butte County, "Leases"; Colusa County, "Leases" and "Personal Property Mortgages"; Sutter County, "Leases" and "Crop and Chattel Mortgages"; Yuba County, "Leases" and "Chattel Mortgages"; Placer County, "Leases" and "Personal Mortgages"; Yolo County, "Leases"; and Solano County, "Leases" and "Chattel Mortgages."*
NOTE: *Each dot shows the exact location of farms leased by Chinese tenants. Farms whose exact locations could not be determined have not been plotted on this map. The number of farms is underrepresented in those counties where the chattel mortgage record books have been disposed of. Glenn County was not established until 1891, so most of the farms in Glenn County shown above came from Colusa County records. Data for Sacramento County, shown in maps 9–14, are not duplicated here.*

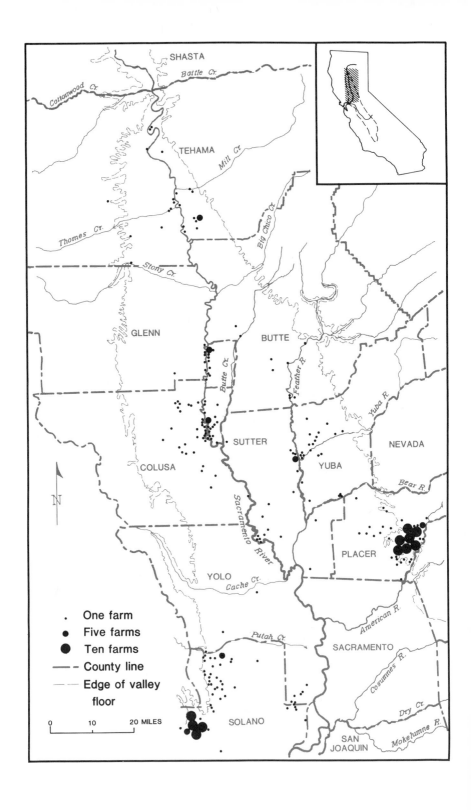

cultivation there by leasing old, dying orchards, which they dug up before planting new orchards in their place. By enabling the landowners to get some income from their holdings, the Chinese in effect played a scavenger role.

A fourth pattern was found in the fruit districts that developed relatively late: the Vaca Valley of Solano County, the Winters area of Yolo County, the Newcastle-Penryn district in Placer County, and the grape-growing areas of the San Joaquin Valley in Fresno, Tulare, and Kings counties. By the time these began to prosper in the late 1880s, the number of Chinese in California had begun to decline, but having established fine reputations for themselves as agriculturalists, they were much sought after as tenants and farm laborers—a fact that placed them in a more advantageous bargaining position when signing contracts. Furthermore, by the 1890s many shipping companies specializing in the export of fresh fruit to eastern markets had begun to dictate the pattern of fruit production in California. These companies preferred Chinese and Japanese tenants. In Placer County, in particular, several shipping companies leased land, which they sublet to Asians, stipulating in the leases that all the fruit harvested was to be consigned to the shippers in exchange

Map 16   The Sacramento Valley: Farms Leased by Chinese Tenants, 1900–1920
SOURCES: *Tehama County, "Leases;" Butte County, "Leases;" Glenn County, "Leases," "Crop Mortgages," and "Chattel Mortgages"; Colusa County, "Leases" and "Personal Property Mortgages"; Sutter County, "Leases" and "Crop and Chattel Mortgages"; Yuba County, "Leases" and "Chattel Mortgages"; Placer County, "Leases" and "Personal Mortgages"; Yolo County, "Leases"; and Solano County, "Leases" and "Chattel Mortgages."*
NOTE: *Each dot shows the exact location of farms leased by Chinese tenants. Farms whose exact locations could not be determined have not been plotted on this map. The number of farms is underrepresented in those counties where the chattel mortgage record books have been disposed of. Data for Sacramento County, shown in maps 9–14, are not duplicated here.*

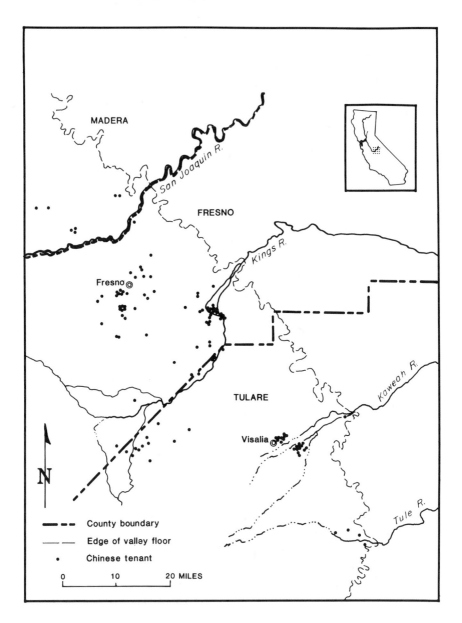

Map 17   *The San Joaquin Valley: Farms Leased by Chinese Tenants,*
1888–1899
SOURCES: *Madera County, "Leases"; Fresno County,*
"Leases"; Tulare County, "Leases"; and Kings County,
"Leases."
NOTE: *Kings County was not established until 1893, so some of*
the data for farms leased by Chinese tenants there came from
Tulare County records. The chattel mortgage record books for
all of these counties have been disposed of, so the number of
farms shown is underrepresented. Farms whose exact locations
could not be determined have not been plotted on this map.

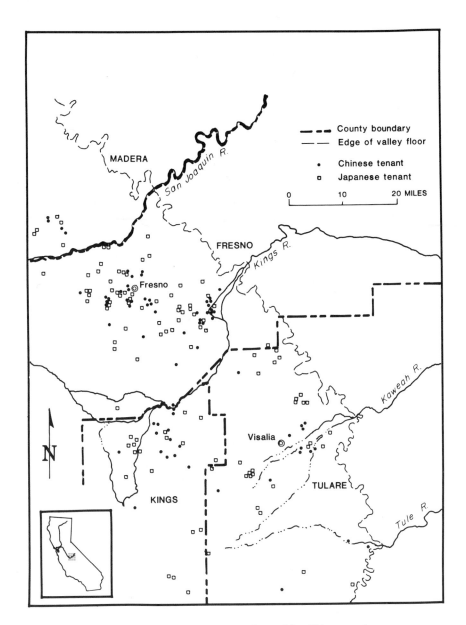

Map 18   The San Joaquin Valley: Farms Leased by Chinese and
Japanese Tenants, 1900–1920
SOURCES: Madera County, "Leases"; Fresno County,
"Leases"; Tulare County, "Leases"; and Kings County,
"Leases."
NOTE: The chattel mortgage record books for all of these coun-
ties have been disposed of, so the number of farms shown is
underrepresented. Farms whose exact locations could not be de-
termined have not been plotted on this map.

Chinese strawberry workers, the Pajaro Valley, California, 1896. (Photograph from the Florence Waugaman Collection, courtesy of Professor Sandy Lydon, Cabrillo College, Aptos, California)

for liberal cash advances at the modest interest rate of 7 percent a year.

Finally, a fifth pattern emerged when Chinese undertook cultivation of field crops on contract, such as sugar beets in the Salinas Valley and rice in the upper Sacramento Valley. In the former location, leasing fields of 60 to 200 acres from individual landowners or the Spreckels Sugar Company, they grew and harvested beets that were delivered to the sugar refinery in Watsonville; in the latter area they leased plots of 1,000 or more acres and used modern methods to grow a crop their people had known since ancient times.

Chinese Tenants in the Coastal Valleys

The Sacramento Delta, the coastal valleys, and the plains around San Francisco Bay were among California's most important early agricultural regions. In each locality, both extensive and intensive agriculture were practiced. Each area also had many Chinese agriculturalists. The role Chinese played in turning raw land into farms in the Sacramento Delta has already been described, so attention will be given here to how Chinese first came to work in the Sonoma Valley. A few words will also be said about them in the Santa Clara Valley.

Chinese first entered Sonoma County in the 1860s and 1870s when they were hired to plant new vineyards in the Sonoma Valley and to work as harvest hands in the apple orchards around Sebastopol.[3] When they began to lease land in the late 1870s, they cultivated hops, not grapes or deciduous fruit, on the leased acreages because the former brought a higher return per acre than the latter.[4] Chinese, however, were an integral part of the early grape-growing and wine-making industry that developed in the 1860s in the Sonoma Valley. The 1860 census of agriculture listed only 5 farms in the entire county that produced wine, but by 1870, 134 of the farms in Sonoma County did so. (The number growing grapes was doubtless larger.) Wine making was concentrated in the Sonoma Valley where, as was recorded in the manuscript schedules of the census of agriculture, ninety-one vineyardists cum wine-makers produced 271,875 gal-

The Chinese bottle washer, Buena Vista Vineyard, Sonoma County, California. (Photograph courtesy of the California Historical Society, San Francisco)

lons of wine in 1869–70. Colonel Agoston Haraszthy, a Hungarian immigrant often called the "father of California viticulture," was largely responsible for introducing Chinese workers into the valley's viniculture industry. He had purchased the Buena Vista Ranch in the Sonoma Valley in 1857 and employed Chinese laborers to clear land and to plant 70,000 vines in 1860 and 135,000 in 1861.[5] An account published in the *Daily Alta California* of July 23, 1863, stated: "Arriving at a place called Lovell Valley, we found the overseer, Attila Haraszthy, working thirty Chinamen, grubbing oak saplings on a pretty steep hill, facing south." Chinese also built the wine cellars and did "inside work" at Haraszthy's winery, as the same reporter observed:

> On the same floor we found four Chinamen filling, corking, wiring, etc. Champagne bottles, while the youngest son of the Colonel selected for them the corks and bottles.... There are now in progress three cellars, close to the press house. These are all being blasted and excavated by Chinese. They are to be twenty-six feet wide, thirteen feet in height, and three hundred feet long.[6]

Haraszthy expanded his operations by reorganizing the ranch into the Buena Vista Viniculture Society with additional capital from eight San Francisco financiers, and employed a hundred Chinese to help him make 100,000 gallons of wine and 5,000 gallons of brandy and to produce other agricultural products valued at $135,000 a year.[7] In the next two decades, other vineyardists and wine-makers in Sonoma County also began to use Chinese labor. Later, when vineyards were planted in the Napa and San Joaquin valleys in the 1880s, there were still enough Chinese around to be employed to clear land, plant vines, pick grapes, and build stone cellars and fences. With regard to picking, a reporter for the *San Francisco Chronicle* in 1883, writing about the Napa Valley, stated:

> It is tedious work, because the picker is compelled to squat on his haunches, the grapes seldom being more than a foot or a foot and a half from the ground.... The Chinese make good pickers on account of their stolid industry and genius for plodding, and they are largely employed in the valley.[8]

When Chinese labor became scarce, vineyardists had to prune their vines higher, leaving more room between the ground and the grapes so white pickers could work more easily.[9] In 1887 an article in a trade journal—which obviously was in favor of Chinese immigration—also commented on the value of Chinese grape-pickers:

> The prettiest and cleanest employment I ever saw is that of gathering grapes. But it is hard on the back, and in the end is not easy work.... The best hand in the grape field by all odds is the little Chinamen [*sic*]. He grows close to the ground, so does not have to bend his back like a large white man. Besides, he is very supple-fingered. And it does not take a John L. Sullivan to lift a bunch of grapes. And so when you have decided in your own mind that the grape is going to be a conspicuous figure in the political economy of this, the greatest State of the Union, and when you have further decided that the Chinamen [*sic*] is necessary to make it profitable, you can decide very certainly in your own mind as to whether the little brown men are to go or to stay.[10]

According to William F. Heintz, Chinese worked not only as laborers in the wineries and in the fields, but some were also engaged in the more technical aspects of wine making. Young Moon, a resident of Glen Ellen and later of Santa Rosa in Sonoma County, worked in the Chauvet winery and distillery in Glen Ellen and was recognized as "an experienced brandy distiller and blender" who learned his craft from the winery's founder, Joshua Chauvet, and his son.[11] Heintz claimed that Chinese cellar bosses were quite common in the Sonoma Valley in the 1880s because, according to one old resident Heintz interviewed, "You were able to trust them. They wouldn't bother your wine."[12] That is to say, Chinese were not known for drunkenness—a fact that many other employers of Chinese workers often cited in their favor.

Other accounts indicate that the Chinese were expert grafters. Carl Dresel of Napa recalled, "I remember the Chinese grafting vines, an entire field. These were field grafts, varietal grapes grafted to the resistant vines so they would not get phylloxera."[13] Grafting required great skill. Nursery grafts were easier to make,

Chinese grape pickers, probably Southern California. (Photograph courtesy of The Huntington Library, San Marino)

1372.

Chinese pruners in a vineyard, probably Southern California. (Photograph from the Pierce Collection, courtesy of The Huntington Library, San Marino)

but field grafts were preferable, since vines planted in the field itself grew better. When vines were improperly grafted, the losses could be considerable. That vineyard owners used Chinese grafters is an indication of the trust the former had in the latter.

Though Chinese were indispensable in tending the vineyards and in working in the wine cellars, the archival records of Sonoma County show that few of them leased vineyards there. Instead, Chinese tenant farmers in the Sonoma Valley specialized in hop cultivation: over 90 percent of the leases signed by Chinese in Sonoma County between 1880 and 1900 were for the cultivation and curing of hops. Often, hops were grown on a section of a tract that already contained an orchard or a vineyard, and the Chinese tenants were expected to take care of both kinds of crops. Typical of the leases signed in the 1880s for hop cultivation was one between James Stuart and Ah Foon and Sam Lum. The Chinese leased 212 acres containing an orchard, a vineyard, a hopyard, and a grain field for six years for $1,700 a year. Stuart provided four horses and various farm implements. The tenants' responsibilities included plowing and harrowing the vineyard twice a year, suckering the hopvines twice a year and resetting them whenever necessary, pruning both the grapevines and the hopvines as well as the fruit trees, tieing all the hops, and building a hop house. To insure the payment of rent, and in return for the advance of $700 of operational funds, the tenants mortgaged their crops to the landlord, with interest at 10 percent a year until paid.[14]

As elsewhere, landlords in the Sonoma Valley used the Chinese to convert unimproved or extensively used land to more intensive usage. Many leases specified the number of "hills" of hops that the Chinese tenants were required to plant each year. In some cases, leases stated that the Chinese also had to build the hop-drying houses. Chinese tenants helped to finance the commercialization of hop cultivation not only by paying rent and providing labor; frequently, they had to pay half of the insurance costs as well—fire being a constant danger in the drying houses.[15] During the nineteenth century, of the crops grown in Sonoma County, hops brought one of the highest returns per acre, even higher than grapes, so hops provided Chinese tenants with an acceptable income even after half of the proceeds had been drained off as rent. Hops were sold on the international market to beer brewers through commission merchants, among whom Philip Wolf and Company and Lilienthal and Company were most frequently used by Sonoma County hop growers.

In the Santa Clara Valley, Chinese also had a dual pattern of specialization. As was described in an earlier chapter, the first Chinese agriculturalists in Santa Clara County were strawberry growers. Like Delta potatoes and Sonoma Valley hops, Santa Clara County strawberries brought a higher dollar return per acre than deciduous fruit. Cultivation of this crop on a rental basis provided a viable living even after profits had been shared. The Chinese in Santa Clara County continued to grow strawberries well into the twentieth century. According to a Japanese immigrant who cultivated strawberries in the Florin area south of Sacramento city for sixty years, the Chinese had such a firm hold on the San Francisco Bay Area strawberry market that Japanese growers never succeeded in dislodging them there as they had managed to do elsewhere.[16]

Chinese began to work in the fruit orchards of the Santa Clara Valley as pickers and packers only some years after they became strawberry cultivators. A number of Chinese tenant farmers began to lease orchards in the valley later still, but the number of farms they leased in this area was never large, partly because Italian immigrants, who possessed some of the same advantages as the Chinese—such as having access to a large supply of laborers from among their fellow countrymen and, from the point of view of employers, the convenient habit of boarding themselves, in addition to being "industrious, attentive, and regular"—early entrenched themselves in this area, making it difficult for the dwindling number of Chinese to compete with them. A number of the farms that Chinese leased specialized in seed production.[17]

Pioneer Orchardists in Tehama County

A quite different pattern emerged in Tehama County, where the Chinese happened to be the first large orchardists—of any ethnicity—in the area. They were members of the same

Chinese on an onion seed farm, Santa Clara County, California. (Photograph courtesy of the California Historical Society, San Francisco)

companies that operated the first large Chinese truck farms in California. On these tracts, vegetables were intercropped with fruit trees.

There is no record of how the Chinese found out about this area. Two possible explanations may be offered. One possibility is that Leland Stanford, who owned a 3,000-acre vineyard in Vina, which at one time was probably the world's largest, may have introduced them to the region.[18] Another possible explanation is that several large landowners, including J. Granville Jones and Joseph Spencer Cone, made a conscious decision to recruit Chinese tenants. Jones signed the first recorded leases with Chinese tenants in Tehama County in 1882. Chinese involvement in fruit growing predated that first lease, however: the 1880 manuscript agriculture census already had listed eight large Chinese orchardists in the area.

Jones first leased two hundred acres for ten years to a group of Chinese partners operating as the Chung Quong Sang Company. No rent was charged the first five years; instead, the tenants had to cut down all trees with trunks measuring more than twenty inches in diameter and cut and grub all trees whose trunks were smaller than twenty inches. After clearing the land and grubbing the roots, the tenants were obliged to plant 5,000 fruit trees at their own expense. Beginning in the sixth year, the Chinese were to pay a rent of $3,500 a year. The landlord provided lumber for building fences, a house for the tenants to live in, and a building for drying fruit.[19]

Jones must have found the arrangement to his liking, for in 1883 he signed leases with two additional groups of Chinese tenants who promised to buy and plant at their own expense 10,000 fruit trees for him during the first two years of their tenancy.[20] The rent was calibrated to the life cycle of the trees. While the land was being cleared, no rent was charged. After saplings had been planted, when the land between the trees was used for growing vegetables—which, as the census indicates, the Chinese did on a large scale—the rent was moderate. As the saplings grew and less land could be plowed for fear of injuring their roots, the rent was reduced, only to be increased again when the trees began to bear fruit. In Tehama County many landlords allowed their tenants to market the early fruit on their own account. Only prime-quality fruit picked at the height of the harvest season was shipped out of state under the landlords' names. In 1884 Jones found two other sets of Chinese tenants, and in 1885 yet another group.[21] When he died at the end of the decade, his heirs continued their relationship with Chinese.[22]

Another landlord of the Chinese, Joseph Spencer Cone, came from Ohio and first mined for gold at Ophir. When he decided to go into farming, he tried his hand first at wheat cultivation and then at sheep raising. The 1870 manuscript agriculture census listed him as the resident owner of 15,000 acres in Antelope Township on the east bank of the Sacramento River in northern Tehama County. Cone's land was valued at $75,000, and he had 2,000 sheep, 500 pigs, and more than 300 other animals on it. The total value of his livestock was $21,000. He also grew 16,000 bushels of wheat, 300 tons of hay, $1,000 worth of fruit, and $1,000 worth of truck crops on his farm. Cone was one of the largest employers of farm labor in 1870, having paid $10,000 in wages in the preceding year.[23]

Cone increased the size of his landholding greatly when he purchased the Rio de los Molinos grant—a Spanish land grant that had been confirmed to A. G. Toomes for 22,172 acres and patented at the end of 1858[24]—on the east bank of the Sacramento River, with Deer Creek running through it. Cone's new property in Lassen Township was to the south of his holding in Antelope Township. Cone acquired the grant when the heirs of the grantee, who owed the government $50,000 in back taxes, had exclaimed in exasperation that their land was not even worth $50,000. With alacrity, Jones offered them $50,000 in cash for their land—an offer they accepted.[25]

In 1887 Cone began to rent out parcels to Chinese. He first leased 80 acres to Hop Lee and Company for vegetable cultivation.[26] Finding his relationship with them satisfactory, he asked these tenants to plant the tract in fruit. Then, in 1891 he leased 200 acres to ten partners in the Ung Wah Company on a graduated rent schedule. There were already apple, peach, and pear trees on the tract, but he asked this group to plant the uncultivated portion to plum trees whose fruit was to be dried into prunes. In addition to receiving a cash rent, Cone got 1,000 pounds each of peaches and pears and 4,000 pounds of apples per year. The Chinese could keep the rest of the fruit to be sold as they saw fit.[27] In the early 1890s Cone leased one tract after

another, ranging from 60 to 100 acres, to Chinese tenants, all of whom were obligated to plant the land in fruit.[28]

In the Chinese-operated orchards in Tehama County, a wide variety of deciduous fruit was grown, including apples, apricots, cherries, nectarines, peaches, pears, and plums. Since in these early years no one knew what fruit would grow best in which location, mixed orchards gave the orchardists some insurance in

Chinatown in Tehama, flooded by record high waters of the Sacramento River, 1909. (Photograph courtesy of the California State Library, Sacramento)

case of failure in any single species or variety. Also, having fruit that ripened in staggered periods brought a more even cash flow and utilized labor more efficiently. Most of Cone's leases were for ten years, which meant that he had fine bearing orchards by the time each company of tenants moved on.[29] In addition to getting his land planted in orchards, Cone seemed to have benefited in other ways from his dealings with the Chinese. For example, he required his tenants to haul supplies for him free of charge from the railway depot at Red Bluff to his ranch and asked them to perform miscellaneous tasks whenever he desired.[30]

Chinese were not just the pioneer orchardists in this section of Tehama County, but were virtually the *only* fruit-growers there. The main beneficiaries of their labor were men like Jones and Cone and some of their neighbors, including Joseph Leininger, L. O. and Minnie Carter, S. C. Dicus, and Loomis Ward, who, compared with Cone, owned only medium-sized farms, but who also found it profitable to lease a third to a half of their acreages to Chinese, who converted their unimproved land, pasture, or grain fields to orchards.[31]

A number of Chinese bought land in Tehama County, but the documentary records do not indicate whether they planted fruit trees on these purchased tracts. In 1876 a Chinese from San Francisco named Chong Chick Sing began buying land in Tehama County. He first acquired four tracts totaling 330 acres through the probate court by making the highest bid of $6,000 for these properties, which were part of the estate of Frank M. Peak.[32] The following year he purchased 67 acres for $750 from the Central Pacific Railroad Company.[33] In 1881 he made a small purchase of 12 acres for $230.[34] Then, in 1882 he sold everything he had bought to Lee Wan Gow, another Chinese resident from San Francisco, who paid $17,000 in cash for the six parcels.[35] During the next decade the same land passed through the hands of three different Chinese buyers,[36] until a company owned by Joseph Spencer Cone bought it all in 1892 from the last Chinese owner, Lee Ping, a prominent farmer and merchant who had several businesses elsewhere in California.[37]

Salvaging Old Orchards in Yuba and Sutter Counties

Being so far north in the Sacramento Valley, Tehama County did not suffer from mining debris. Farther south, however, many thousands of fertile acres were destroyed in the three decades when the hydraulic miners' powerful jets of water tore open hillsides and even mountains, washing down avalanches of "slickens," destroying good farmland downstream. Land along the Yuba, Bear, and Feather rivers, as well as smaller streams, was buried under the slimy waste. Even John Sutter's famous Hock Farm in Sutter County suffered the same fate as many other tracts in the Sutter Basin, when it was completely ruined by debris.

At the same time that landowners and farmers sought court injunctions to halt hydraulic mining, they looked for ways to recover their losses. In the late 1870s and early 1880s, landowners in Yuba and Sutter counties began to lease tracts—many with old orchards on them, some where the trees were still healthy, others where they were dying—to Chinese, who had first entered agriculture in the area as hop-pickers near the town of Wheatland in Yuba County.[38] Most of the tracts were within a ten-mile radius of the twin cities of Marysville and Yuba City. In many instances, the Chinese tenants had to dig out the stumps of the dead or dying trees before planting new ones in their place. The wood was cut and bound into cords to be sold as firewood. Some landlords demanded payment for the wood even though they paid no wages to the Chinese for the work of cutting and grubbing the trees and chopping the trunks and branches into firewood, while others allowed the Chinese to keep the proceeds from the sale. A few paid the Chinese wages for clearing the land, but most did not.

Some county documents provide evidence that "old" and "new" orchards existed on some tracts. Eli Teegarden of Yuba City, Sutter County, for example, leased both his "Old Orchard" and his "New Orchard" to Do Joe, who was told to cut down all nonbearing trees.[39] When Chinese had to plant new fruit trees, the landlords supplied them in some instances.[40] To earn some income for themselves, they planted vegetables between the saplings.[41] In this way, the Chinese who had not been involved in the first round of orchard planting in Yuba and Sutter counties were instrumental in resurrecting the bottomlands along the Feather and Yuba rivers as a productive fruit belt.

The clearest documentary evidence of the scavenger role that Chinese played in debris-ruined areas is found in the voluminous briefs of a case tried in the Circuit Court of the United States for the Ninth Circuit and the State of California. The owner of a tract in the Sutter Basin, Edward Woodruff, who lived in another state, filed suit on behalf of himself and many of his neighbors against the North Bloomfield Gravel Mining Company, which was engaged in hydraulic mining. An absentee landowner was chosen because this enabled the case to be tried in a federal court. The plaintiffs presented evidence that their land had been destroyed by debris, while the defendant called on witnesses to testify that the land allegedly destroyed was still arable because Chinese were renting strips outside of the levees to grow several varieties of corn, beans, barley, and alfalfa. Twenty-six references were made in the briefs to the presence of Chinese farmers.[42] However, the fact that these tenants paid rents of only six or seven dollars per acre was proof that the land was not fertile, since rents elsewhere were much higher during the same period. Besides, the kinds of crops the Chinese grew in these areas did not require very good soil. Low as the rents were, the owners of these tracts nonetheless managed to recover some value from their holdings by renting them to Chinese. Few other tenants were willing to cultivate such poor soil in locations subject to the constant risk of floods. But here, as elsewhere, some Chinese tried to earn a living in whatever way they could.

Preferred Tenants in the "Uplands" of Placer County

Several of the more prosperous agricultural areas in California developed in the 1880s and later decades—that is, in years after Chinese exclusion had reduced the number of Chinese in California. Moreover, according to conventional wisdom, that was a period when Chinese were being driven away from rural areas and herded into urban Chinatowns. Nonetheless, many Chinese participated in the development of the new fruit-growing areas, showing up in considerable numbers in such places as the Winters area of Yolo County, the "Uplands" of Placer County, the grape-growing regions of the San Joaquin Valley, the citrus- and celery-producing districts of southern California, and the rice lands of the upper Sacramento Valley to partake of the boom in the production and export of California fruit, nuts, and other specialty crops. Aside from the Sacramento–San Joaquin Delta, the San Francisco Bay Area, and many of the coastal valleys, where they continued to farm in this period of Chinese exclusion,

the Chinese presence was most notable in Placer County from
the 1890s through the 1910s; from the mid-1910s to the mid-
1920s, their resurgence in California agriculture was most apparent in the rice-growing areas of Colusa, Glenn, Sutter, Yuba,
and Butte counties in the upper Sacramento Valley.

During gold rush days, a number of orchards had been established in the foothills of Placer County, but the county did not
become prominent as a horticultural district until more than
three decades later. Chinese had begun to lease land for cultivation in Placer County from the early 1860s onward,[43] but they
did not became active as tenants there until the late 1880s. Their
entry into fruit cultivation in the county thus coincided with the
large-scale development of its "Uplands" into a prime fruit-
growing region. The main factor that led to the growth of this
area as a thriving horticultural district was its location along the
route of the transcontinental railroad. Although the trunk line of
the railroad had been completed in 1869, high freight rates pre-
vented shippers of California products from using it extensively
to export goods to other parts of the nation for almost two decades after its completion. Freight rates were finally reduced in
the late 1880s; at the same time, refrigerated railroad cars had
come into use. However, "icing" the cars was still cumbersome
and problematic. When it was discovered that the "Uplands"
of Placer County on the western slopes of the Sierra Nevada
that lay alongside the transcontinental railroad was suitable
for the cultivation of a wide variety of tree fruits, that district
immediately enjoyed a boom because it took three days less
time for fruit grown and loaded there to reach midwestern and
eastern destinations.

A lease signed in 1888 between Wong Suey, Wing Suey, and
Wong Sing with H. Reinecke, S. N. Blanchett, Will Reinecke,
and Bertha Reinecke gives an indication of the terms under which
Chinese took up fruit cultivation in this newly developing district. The fifty-two acres leased was part of the Reinecke Ranch,
located three-quarters of a mile below Newcastle. The rent was
$40 per acre. The landlords provided a dwelling house, two
horses valued at $80 each, a wagon, a set of double harnesses,
four plows, one harrow, one cultivator, two shovels, two hoes,
one ax, one sledge hammer, one cross-cut saw, and pipes for
water. The lease did not state clearly whether the waterpipes had

Chinese weeding irrigated orchard, California. (Photograph by Carleton E. Watkins, courtesy of the California Historical Society, San Francisco)

already been laid in place, but a clause stating that the tenants were responsible for keeping the pipes in good repair implied that they were. The tenants had to plant an unspecified number of fruit trees to be provided by the landlords. If any of the saplings should die, the landlords would supply new ones at no extra cost. The tenants were not held responsible for the loss of or damage to any of the tools so long as they kept up their rent payments. However, if the horses died, the landlords and tenants had to split the cost of buying new ones.[44]

Another lease signed in 1889 between Ah Lue and Joe Sing and C. and A. Walker provides insight into the terms of those leases which were for land that already had growing orchards. The rent for the forty-five acres leased was to be $36 per acre the first year, $40 the second year, and $44 the third year. In addition to providing tools similar to the ones described above, the landlords also supplied the equipment and materials for spraying the existing fruit trees. The bills for the water, packing boxes, and other materials used would be paid out of the profits from the first sale of the fruit, which was to be sold only through a shipping house or a commission agent chosen by the landlords.[45]

Shipping companies played a key role in the horticultural development of the Placer County "Uplands." More than a dozen companies were active in the area. Among them, The Producers' Fruit Company (the article "The" being part of its full name) recorded the largest number of chattel mortgages with Chinese tenants. It was first incorporated in Illinois in 1896, with Chicago as its principal place of business, to buy, sell, and deal in all kinds of fruit, fruit products, nuts, and vegetables. In 1908 it reincorporated in South Dakota, with Sioux Falls as its principal place of business, changing its name to Producers' Fruit Company (without the "The.") The company first appointed H. A. Fairbank of Sacramento city, one of its vice-presidents, as its California agent—an appointment that was revoked in 1921, Fairbank being replaced by L. H. Godt. In the first decades of the twentieth century, the company had connections with other fruit-shipping firms, of which the Baltimore Fruit Exchange, whose president at that time was Joseph DiGiorgio, was the most notable. The Producers' Fruit Company extended its first chattel mortgage to a Chinese fruit-grower in 1898, two years after its establishment. Ah Dick, who had leased ten acres of the Perry Fruit Ranch two miles east of Loomis, received $27 from the company—with the possibility of an additional advance of $1,000 should he so desire—using all the deciduous fruit he grew as collateral. The advance was repayable in one month at 8 percent interest. From this small beginning, the company became one of the major creditors of Chinese tenant horticulturalists in Placer County. More than fifty chattel mortgages—about half of the total number recorded in the Placer County Recorder's office from the late 1890s through the mid-1910s—were signed between Chinese and the company.[46] These probably represented only a fraction of the mortgages in effect. Other fruit-shipping firms that dealt with Chinese included local companies, such as the Schnabel Brothers and Co. and George D. Kellogg, both based in Newcastle, the Penryn Fruit Company based in Penryn, and the Loomis Fruit Growers Association in Loomis, as well as out-of-state companies, such as the Porter Brothers' Company with headquarters in Chicago.[47]

Investigators for a congressional committee that studied immigrants in industries between 1907 and 1909 found that landowners and shipping firms in Placer County greatly desired Chinese tenants, whose numbers were dwindling by this time. The investigators commented that "the fruit houses will take a Chinese tenant in preference to a Japanese every time," because "the Chinese are entirely honest in all contractual relations. The confidence in them is so great that they usually pay no rent until the crops are harvested. The fruit-shipping houses frequently make loans to them on their personal unsecured notes. They do not abandon their leases."[48] The agents estimated that the Chinese were leasing about 500 acres at that time, some 15 percent of the total acreage in fruit. This figure is corroborated by data computed from information in county lease documents that indicate that in 1909, sixteen Chinese tenants leased a total of 589 acres in Placer County. The congressional investigators estimated that the more numerous but less liked Japanese operated about 60 percent of the fruit acreage, which would amount to some 2,000 acres. Computations from information in county documents show that in fact fifty Japanese were renting 3,325 acres in Placer County in 1909, but not all of these were used for growing deciduous fruit, since the Japanese also grew a lot of strawberries in the area. Italian and Portuguese tenants and a few resident owner-operators cultivated the rest of the fruit acreage.[49] No doubt, had exclusion not kept the Chinese population from replenishing itself, the Placer County "Uplands" might have become a center of Chinese settlement.

Chinese in Extensive Agriculture

Lest the discussion of Chinese cultivation of specialty crops gives the impression that Chinese grew only labor-intensive crops, a brief examination of Chinese involvement in extensive forms of agriculture is in order. Though Chinese tenant farmers who cultivated wheat, barley, corn, buckwheat, and hay were relatively few in number, they did exist. As table 23 shows, the 1880 manuscript agriculture census listed over twenty Chinese grain farmers, four of whom produced wheat, eight of whom grew barley, and fourteen of whom grew different varieties of corn, including Indian, Egyptian, and broom corn. These Chinese tenants were scattered throughout the Sacramento, the San Joaquin, and the coastal valleys. The acreages farmed by these tenants were not large, but their very existence proves that Chinese did grow cereals, the most important of which eventually turned out to be rice.

Some of the Chinese farmers who first engaged in extensive agriculture had purchased the land they farmed. Two of the largest farms ever owned by Chinese in California belonged to Chin Shin and Ah Wing. In 1893 Chin Shin made a surprisingly large purchase of 663 acres in El Dorado County from Philip Monroe, an absentee landowner who resided in Berkeley, Alameda County.[50] Three years later, Chin Shin made two other purchases to add to his holdings—one for 760 acres from John Phillips,[51] the other for 74 acres from Woo Hong Seck.[52] The year after that, he bought 560 acres from a widow named Annie Simas for $3,200.[53] He lost part of his holdings in 1896 because he owed the state delinquent taxes, but he managed to redeem his property two years later.[54] As though the amount of land he owned were not enough, Chin Shin bought a final 1,120 acres from the Farmers and Mechanics Savings Bank in 1899 with a

down payment of $4,000. The archival records do not show if
he took out a mortgage on this last purchase.[55] For unknown
reasons, he sold all 3,177 acres in 1903 to three prominent landowners from San Joaquin County: Cyrus Moreing, W. C. White,
and R. B. Oullahan, men who had many previous dealings with
Asian immigrants around Stockton, to whom they had leased
many tracts of land.

Chin Shin used these thousands of acres to raise cattle and to
carry out mining. Raising cattle was hardly a traditional Chinese
agricultural enterprise, so undoubtedly Chin Shin had learned
the business only after he came to California. When he sold his
properties, he sold along with the land 191 cows and heifers over
two years old, 5 Durham stud bulls, 54 two-year-old steers, 106
yearling calves, 14 horses, 2 farm wagons, 2 spring wagons, 1
buggy, 1 cart, many sets of saddles and harnesses, 100 cowbells,
various blacksmithing tools, many plows and hayforks, 1 cream
separator, and 25 tons of hay. On the different tracts was also
some leftover mining equipment.[56]

Perhaps the most astonishing farm ever owned by a Chinese in
California was one of 978 acres—of which 580 acres were under
cultivation—belonging to Ah Wing and Hong Tong Vey in
Suisun Township, Solano County. The 1880 manuscript agriculture census showed that Ah Wing cultivated 80 acres of barley
producing a crop of 2,200 bushels, 500 acres of wheat producing
4,000 bushels, with 40 acres of grasslands producing 50 tons of
hay in 1879–80. The total value of his products was $5,000, so
the yield per acre of his crops was not high. He had $800 worth
of farming implements and $600 worth of livestock, consisting of
16 horses, 30 pigs, 3 milk cows, 15 cattle, and 3 calves. He
employed 72 man-weeks of labor and paid $600 of wages to farm
laborers.[57] Certainly, nothing that has been written in the entire
corpus of Chinese American history would lead one to expect
that any Chinese might have owned and operated a wheat and
barley farm of such size anywhere in America.

As though with poetic justice, Chinese tenants finally got a
chance in the late 1910s and early 1920s to grow a field crop they
were familiar with: rice. However, the conditions under which
they grew this crop differed as greatly from those they had known
in China as night differs from day. In California, during its heyday, rice was cultivated on such a grand scale that airplanes

Table 23 Chinese Grain Farmers Listed in the 1880 Manuscript Census of Agriculture by County

County and Township	Name of Person or Company	WHEAT		
		Acres	Bushels	Yield/Acre
SACRAMENTO VALLEY				
Colusa County				
Grand Island Twp.	Ah Toy	133	2,000	15 b/A
Sutter County				
Sutter Twp.	Ah Fiesan	—	—	—
	Tung Ling	—	—	—
Nicolaus Twp.	Ah Sing	—	—	—
Yuba Twp.	Joe Do	—	—	—
Solano County				
Suisun Twp.	Ah Wing and Hong Tong Vey	500	4,000	8 b/A
Vacaville Twp.	Ah Hing	—	—	—
Sacramento County				
Dry Creek Twp.	Ah Ing	40	8,000	200 b/A[c]
COASTAL VALLEYS				
Napa County				
Napa City	Ah Jim	—	—	—
Santa Clara County				
Gilroy Twp.	Moon Fung & Co.	—	—	—
San Mateo County				
5th Twp.	Ah Pan (Oats)	10	100	10 b/A[d]
	Sun Lee	—	—	—
SAN JOAQUIN VALLEY				
San Joaquin County				
Douglas Twp.	Ah Gee	—	—	—
Merced County				
Twp. # 14	Ah Wong & Co.	—	—	—
Kern County				
5th District	Ah Jo	—	—	—
	Ah Got	—	—	—
	Sing Lee	—	—	—
	Ah Ki	—	—	—
	Ah How & Ah Sow	—	—	—
	Ah Jim	—	—	—

SOURCE: U.S. Bureau of the Census, "Census of Productions of Agriculture" (manuscript), 1880.
NOTE: Other products grown on the above farms have not been included in the table. Most farms had at least one other crop. Owing to the difficulty of allocating relative value to each crop, no attempt has been made to compute the average value per bushel or per acre of grain grown.
[a] Indian corn. [b] Broom corn. [c] Buckwheat. [d] Oats (placed under Wheat column for lack of space).

BARLEY			CORN		
Acres	Bushels	Yield/Acre	Acres	Bushels	Yield/Acre
92	3,732	41 b/A	—	—	—
—	—	—	5	200	40 b/A[a]
—	—	—	15	200	13 b/A[a]
—	—	—	70	2,400	34 b/A[a]
			20	20,000	1,000 b/A[b]
—	—	—	1	30	30 b/A[a]
80	2,200	28 b/A	—	—	—
20	400	20 b/A	—	—	—
—	—	—	—	—	—
—	—	—	2	20	10 b/A[a]
—	—	—	12	480	40 b/A[a]
25	700	28 b/A	—	—	—
25	300	12 b/A	—	—	—
—	—	—	100	2,000	20 b/A[a]
—	—	—	40	1,000	25 b/A[a]
6	120	20 b/A	15	690	46 b/A[a]
12	240	20 b/A	12	500	42 b/A[a]
—	—	—	2	80	40 b/A[a]
—	—	—	20	1,500	75 b/A[a]
—	—	—	10	1,000	100 b/A[a]
—	360	—	—	—	—

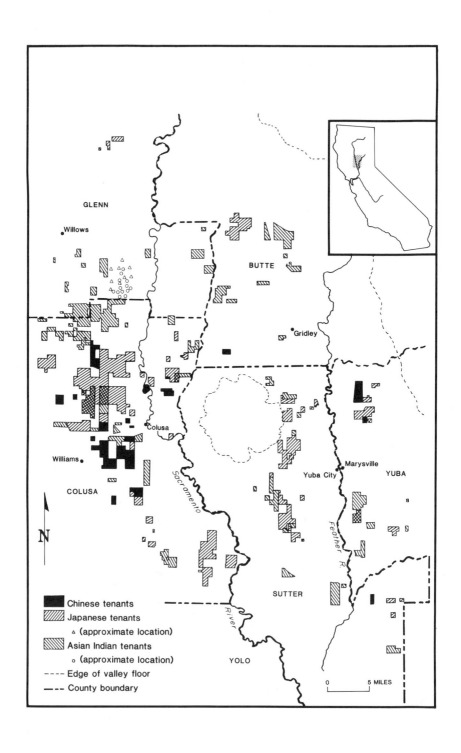

GLENN

Willows

BUTTE

Gridley

Colusa

Williams

Marysville

Yuba City   YUBA

COLUSA

N

Sacramento

Feather R.

Chinese tenants
Japanese tenants
△ (approximate location)
Asian Indian tenants
○ (approximate location)
Edge of valley floor
County boundary

SUTTER

River

YOLO

0   5 MILES

were used to sow the seeds, although it is not known whether Chinese used airplanes for this purpose.

Rice was cultivated successfully in California only after sixty years of experimentation and failure. As early as 1856, it was reported that a Chinese had observed that it must be possible to grow rice in the swamp lands of the Sacramento–San Joaquin Delta.[58] American entrepreneurs also wished to grow rice because they saw a large market for it, given the increasing number of Chinese in the state. Supposedly, over a million pounds of rice were consumed each month by the Chinese resident population. The California State Legislature also tried to encourage rice cultivation by offering premiums for its production. (Offering premiums for various crops was a common practice that state agencies in those days followed to encourage agricultural development in the state.)[59] However, these aspiring rice-growers also realized that Chinese preferred to eat only long-grain rice, so the early experimenters used exclusively the long-grain varieties as seed. Therein lay the cause of their failure.

In experiments carried out in the 1880s by private parties in Los Angeles and Sonoma counties, the heads of the rice plants failed to fill out. Similarly, in experiments carried out by the United States Department of Agriculture on Union Island in the San Joaquin Delta in the early 1890s, again heads did not form. Eventually, it was discovered that coastal and Delta fog and the cool night temperatures in these localities made it impossible for the rice plants to develop properly before autumn frosts arrived. Long-grain varieties of rice simply were not hardy enough to grow properly in the cool nights and short growing season of the Delta and along the coast. Moreover, these early rice-growers did not seem to know how to irrigate the rice plants properly.[60]

It was not until the first decade of the twentieth century that

Map 19   *Farms Leased by Chinese, Japanese, and Asian Indian Tenants for Rice Cultivation, 1915–1924*
SOURCES: *Butte County, "Leases"; Colusa County, "Leases" and "Personal Property Mortgages"; Glenn County, "Leases," "Crop Mortgages," and "Chattel Mortgages"; Sutter County, "Leases" and "Crop and Chattel Mortgages"; and Yuba County, "Leases" and "Chattel Mortgages."*

attempts were made to use Japanese short-grain varieties. W. W.
Mackie of the United States Bureau of Soils, Charles E. Chamb-
liss of the Office of Cereal Investigations of the Bureau of Plant
Industry, I. Ikuta, a Japanese immigrant, and various landowners in Butte County all contributed efforts to find a way to plant
rice successfully in the Sacramento Valley. Experimentation
reached a high point in 1909 when some 300 varieties were
tested. Success was achieved when 180,000 pounds of rice were
grown on 55 acres of adobe soil in Butte County in 1911. Chamb-
liss persuaded some local farmers to organize the Sacramento
Valley Grain Association, which then donated 56 acres near
Biggs in Butte County, buildings, and tools to establish a formal
experimental farm under the supervision of Ernest L. Adams, a
trained agronomist. From 1913 on, rice acreage doubled almost
every year, increasing from 6,000 acres in 1913 to 16,000 in 1914
to 32,000 in 1915, 65,000 in 1916, 79,000 in 1917, 112,000 in
1918, and 142,000 in 1919.[61] Bankers, land companies, and professional men all rushed into the rice-growing business, though
few undertook this task themselves. Almost all the rice grown
during the boom years was produced by Chinese, Japanese,
Korean, and Asian Indian tenants. The acreage distribution
of each is shown in map 19. It can be seen that even at this late
date, there were enough Chinese around to lease and farm a
considerable proportion of the rice land in this region.

The tracts leased by Chinese tenants, among whom one
Jew Foo was most prominent, were large—most about 1,000
acres, rented at an average of $10 per acre in 1919–20.[62]
Landowners—whether individuals or corporations or irrigation
districts—subdivided their holdings into tracts and built the
main ditches of the irrigation works to bring to the fields the
large amounts of water needed. The tenants, however, had to
pay for the cost of hiring engineers to mark the contours of the
land, build the dikes and feeder ditches, the headgates and
sluices, and prepare the soil for cultivation. Large sums of operating capital were required; most Asian tenants obtained these
from the landlords or third parties by mortgaging their crops at
an interest rate of 7 or 8 percent a year. Two major problems
were weeds and the inability to get sufficient water to submerge
the fields for the period when such flooding was required.
Tenants were responsible for keeping the ground free of weeds,

while landlords agreed to refund all rents paid should insufficient water be available when it was needed.[63] There were many success stories of various Asian "rice kings," as well as of failures. In retrospect, Chinese participation in rice growing in the upper Sacramento Valley represented—like asparagus cultivation in the Delta in the same period—a triumphant last effort to reap some returns from the land in exchange for years of toil.

EIGHT

"Cheap Labor"

CHINESE TENANTS managed to accomplish what they did only because they had plenty of help from fellow Chinese who worked for them and for white farmers as farm laborers. Before discussing the contributions of Chinese farm laborers to the growth of California agriculture, it is necessary first to evaluate how scholars have depicted them, since they have been the only group of Chinese agriculturalists who have received any scholarly attention. Interestingly, it is students of California's farm labor problems and, with the exception of Ping Chiu, not historians of the Chinese in America, who have written about Chinese farmworkers.

In the literature on farm labor problems in California, the works of Varden Fuller, Carey McWilliams, Paul Taylor, Lloyd Fisher, Cletus Daniel, and Linda and Theo Majka are the most important because each offered a different theory on how the Chinese presence has affected the historical evolution of California agriculture.[1] The use of a seasonal, migratory labor force has been one of the salient characteristics of agricultural production in the golden state. California Indians, Chinese, Japanese, Asian Indians, Filipinos, and Chicanos in turn have served as the backbone of California's migratory labor force. Only briefly during the 1930s did large numbers of white people enter California's migratory stream, when refugees from the Dust Bowl—the Okies and the Arkies—came to find work in the state's fields and orchards and thereby engendered great public concern.[2] But after many of them were absorbed into the industrial labor force during World War II, public consciousness of migrant farm

Chinese field hands, California. (Photograph from the Pierce Collection, courtesy of The Huntington Library, San Marino)

laborers waned once again, not to be awakened until more than two decades later when César Chavez made their plight into a cause to shame the nation.

Varden Fuller's Study

While almost all studies of California farm laborers make an obligatory mention of the Chinese, most of them contain little original research and rely mainly on the exposition contained in Varden Fuller's Ph.D. dissertation, "The Supply of Agricultural

Labor as a Factor in the Evolution of Farm Organization in California," and on Carey McWilliams's book *Factories in the Field*. These two pieces of writing have dominated the historiography of the field for more than forty years. Fuller's work is authoritative because it made extensive use of primary sources and expounded a carefully researched and reasoned thesis which has been extensively cited by all those who have written on the topic after him. McWilliams's book, which reads very well and lends the subject a tone bordering on the epic, is compelling and has been especially popular with radical critics of California's agribusiness and race relations.

Varden Fuller, an agricultural economist, whose Ph.D. dissertation was entered as evidence in the hearing held in 1939 by the Subcommittee on Violations of Free Speech and Rights of Labor of the Senate Committee on Education and Labor chaired by Senator Robert La Follette, has been considered the leading authority on the subject of California's farm labor supply. His work has been influential in studies not only of the Chinese in California agriculture but of all other minorities in the farm labor supply as well. In examining the natural and so-social factors that have influenced the evolution of California's farm structure, Fuller considered the size of farms and the degree of crop specialization to be most important. In his view, four factors—two economic, one political, one social—have influenced the size of landholdings and farms. The two economic factors have been the technique of production and the external economies of scale inherent in large-scale production and distribution. Policy on public land disposal at the time agriculture began to develop in a particular area has been the political factor impinging on size, since such policy most likely set arbitrary limits on the size of tracts available for sale. The degree of homogeneity in the agricultural population has been the social factor affecting the size of landholdings. The first three factors are self-explanatory, but the fourth—a key to understanding Fuller's thesis—requires comment.

Fuller measured the degree of homogeneity in the farming population by the differential entrepreneurial ability that different members of the farming population possess and their relative access to capital resources. He argued that the greater the degree of homogeneity, the more likely are farms to be similar in size, especially if one disregards internal economies of scale. Fuller believed that "equally endowed" persons should be equally able to operate farms requiring the same amount of labor, capital, and entrepreneurship. On the one hand, the less homogeneous the population, the more likely certain individuals or groups are to organize large farms and to obtain financing while others remain wage laborers. Some of the latter may even have to accept seasonal work because numerous obstacles bar them from alternative employment elsewhere. When the farming population is not homogeneous, Fuller argued, a greater proportion of large farms is likely to exist.

As other agricultural economists have noted, Fuller also believed that soil and climate are the most important influences on the degree of crop specialization. Even if farmers may want to grow certain crops, they cannot do so if the natural conditions are not right. In addition, the accessibility of markets and the ability of the transportation network to distribute highly perishable products efficiently determine whether it is feasible to grow certain specialty crops profitably.[3]

Fuller then traced how the farm population in California became nonhomogeneous. He pointed out that in the pre-American period, California Indians were turned into peons to work in the fields and gardens of the Spanish missions. Between 1833 and 1846, when the missions were secularized, rancheros found various means—such as arresting Indians for vandalism or reducing them to debt peonage—to keep their now legally emancipated Indian labor force intact.[4] According to Fuller, the idea that being field hands was the natural lot of nonwhite persons had already become so ingrained in the minds of Californians by the time large numbers of Americans came that the latter simply adopted the practice of employing nonwhite persons for field work. But, Fuller was careful to insist, although Americans adopted the practice, they did not initiate it.

Fuller further believed that the mentality of California's early immigrants was also crucial in perpetuating the racial division of labor: the white population that came during the gold rush was so intent on making quick fortunes that when gold ran out, the ex-miners attempted to become independent entrepreneurs and had no predilection whatsoever for work as wage laborers, especially not in agriculture. During the first two decades of statehood, therefore, California had a population made up of active, avaricious, ambitious persons, with a preponderance of males in their prime, and was "devoid of a laboring class available for wage work."[5]

Although Fuller recognized that the scarcity of labor was definitely a limiting factor in the development of large-scale, specialized, and diversified agriculture, he carefully divided his analysis into two time periods: before 1870, and after it. He said that during the 1850–70 period, the slow growth of large-scale, labor-intensive, commercial agriculture was due to many reasons unrelated to the labor supply. These included (1) a lack of knowledge on the part of eastern, midwestern, and foreign farmers on how to farm successfully under California's peculiar climatic conditions; (2) the need to experiment to find crop varieties that were suitable for different regions of the state; (3) no access to a national market, owing to the lack of transportation linking California to consumer markets elsewhere; (4) the primitive technology available for preserving highly perishable fresh fruit and vegetables; and (5) the slow development of irrigation, which was required for the successful cultivation of specialty crops in most areas of the state. These, rather than the scarcity of labor or the nature of the labor supply, were the primary influences on the pattern of farm organization which emerged in the early years.[6]

After 1870, however, the labor supply became the most critical factor in transforming California agriculture from extensive cereal cultivation to intensive fruit growing, because some of the earlier obstacles were slowly being overcome: growers had accumulated knowledge about how to farm successfully in California; the transcontinental railroad had been completed; fruit-canning and -drying technology had been developed; and more and more irrigation works were being built. With regard to a labor supply, Fuller argued that growers did not purposely go out of their way to look for or to create a cheap one. Rather, with the large number of discharged Chinese railroad-workers milling around the cities and countryside looking for work after the completion of the transcontinental railroad, a ready-made labor supply was simply seized upon by those who stood to benefit from one.

Chinese farm labor was ideal from the growers' point of view because it was cheap, reliable, and convenient to engage. Fuller's definition of "cheap labor" was far more sophisticated than that used by other writers. He argued that Chinese labor was cheap, not only because it was available at low wages, but because the workers could be hired only when their labor was needed. Since they were not kept on the farms during the slack seasons, their wages did not become part of the farmers' year-long overhead costs.[7]

The availability of a cheap labor supply had an important effect on land use patterns and consequently on the price of land. At the very time that a large number of Chinese laborers became available, Fuller believed that three alternatives were open to large landholders: they could have continued to use their land for grain or livestock; they could have held it idle, subdividing it bit by bit as settlers became more numerous; or they could have converted such land into more intensively used orchards. The last alternative was only possible with an adequate, affordable labor supply. Fuller was convinced that had Chinese labor not been available, large landowners would have been compelled by market forces to subdivide and sell their land sooner or later. But since the needed labor could be found, and since labor-intensive crops brought much higher returns per acre, the change to intensive agriculture increased the returns to land investment. Land prices in California became so high that the average aspiring farmer could not afford to buy large enough plots to grow crops that produced a sufficient income for a viable living. Moreover, small farmers also found it difficult to compete against the large landowners and growers who, given their profitable returns, had no incentive to subdivide or sell their land. As Fuller put it:

> Wherever intensive cultivation had already begun or was in prospect, land values were capitalized on the basis of actual or anticipated returns from the employment of the cheap and convenient Chinese labor supply. To the prospective small operator, this meant paying so high a price for land as to permit him a labor return approximately equal to the wages of Chinese. Since wages of Chinese were approximately equal to those prevailing in Europe and below the general level in the United States, such a prospect did not encourage either European immigrants or people from [the] East to come to California. Established farm operators in

the eastern states would have had to sell their less highly capitalized land in order to buy the more highly capitalized land of California, and pay the high costs of migration as well. And there were other factors discouraging to settlement of small farm operators. One of these was the relative heterogeneity of the soils and climate and the differences in the systems of cultivation ... such variability in conditions contributed to many failures.... The already high price of land was frequently augmented by the speculative possibilities of minerals, of oil, and of exotic forms of agriculture.[8]

Fuller attributed such importance to the Chinese presence— even though he pointed out that estimates that Chinese constituted three-quarters to seven-eighths of the farm labor force were too high—because they made up much of the seasonal labor force. In his view,

probably the maximum proportion [of Chinese to white farm laborers] between 1882 and 1884 did not exceed one half. Before this time the absolute number and the proportion of Chinese had been increasing; subsequently they both declined. By 1910, the Chinese had declined to a position of relative insignificance in the aggregate farm labor supply.... Even though the Chinese probably never constituted a large majority, and were in fact throughout the period 1870 to 1900 much of the time a small proportion of the total agricultural labor supply, it is significant that they did supply a large proportion of the demand for casual and seasonal labor. There are no data to permit a quantitative measurement of this; it is, however, the consensus of innumerable comments by employers, the agricultural press, and contemporary observers. Moreover, Chinese employment was restricted principally to the northern part of the state and to intensive crops—fruits, truck crops, hops, and later, sugar beets.[9]

Unlike other sections of his study, which were carefully documented, Fuller unfortunately gave no examples of the comments of contemporary observers upon which he based his thesis.

After the Chinese Exclusion Law was passed, according to

Fuller, Californians had to think about choosing between depreciating their land values sufficiently so that small operators could afford to buy farms, or finding a new source of labor to enable the large units to continue in operation. But the effects of Chinese exclusion were not immediately felt—first, because many Chinese still remained in California, and second, because during much of the 1880s and 1890s, economic depression caused white workers to lose their jobs in the cities and to flock to the countryside in search of work. By the early years of the twentieth century, when the number of Chinese had greatly dwindled, Japanese immigrants had begun to arrive in large numbers to work in California's fields and orchards. Thus growers never really were forced to make a choice. Fuller claimed that the groups who came after the Chinese were available through no special action taken by California's growers: a series of fortuitous historical accidents made them available. Only in the 1920s—when there were few Japanese farm laborers left in general circulation because of curbs on Japanese immigration, and those still around chose to work mainly for fellow Japanese—did growers finally take concerted action to find a new tractable labor supply.[10]

In depicting nonwhite groups as cheap labor, Fuller was careful to point out that he was not simplistically arguing that they were willing to be a source of cheap labor; motivation did not enter into his analysis. He recognized that there were many instances of upwardly mobile Asian immigrants.[11] However, he said that, on the whole, because the employment opportunities open to nonwhite groups were severely limited by forces external to the agricultural sector, they became "available" to agricultural employers, who conveniently used them to perpetuate an objectionable pattern of social, economic, and political inequality. Despite such qualified statements, Fuller created the impression that if Asian and Chicano farmworkers in California have been oppressed, they themselves—by their very presence—were responsible for creating some of the conditions that have oppressed them. Though he never explicitly said so, his line of argument implied that the presence of cheap labor led to the perpetuation of land monopoly.

In contrast, he was quite firm in exculpating large growers from any responsibility for bringing such inequality about, even

though he noted that "many employers declared their willingness to incur the losses which would be involved in making such
an adjustment [to land subdivision]. But other elements in the
employer group were determined to protect their immediate
interests, by importation of labor, if necessary."[12] He never indicated who these "other elements" were. In the end, though
Fuller did not approve of the undemocratic conditions facing
California's farm laborers, he felt that no one could be held
responsible for their evolution.

Carey McWilliams's Study

Carey McWilliams, crusading journalist, author of more than
half a dozen books about California, and for many years editor
of *The Nation*, has argued in *Factories in the Field* that the development of an oppressive farm labor system was the consequence of
land monopoly, and land monopoly, in turn, was the result of
deliberate, conscious action on the part of avaricious individuals,
whom he called "land barons."[13] Though he recognized the importance of Spanish-Mexican land grants in enabling a small
number of owners to hold tens of thousands of acres, McWilliams apparently thought the activities of American "land grabbers," such as Henry Miller, Charles Lux, William S. Chapman,
Ben Ali Haggin, Isaac Friedlander, Thomas Fowler, Abel Stearns,
John Foster, the Murphy family, and the Bixby family, and public
officials such as Surveyor General B. F. Houghton, who dishon-
estly snatched up large tracts of prime California land, were
more important in creating land monopoly.[14]

McWilliams agreed with Fuller that California agriculture
long ago took on an industrial character—hence the apt title of
his book, *Factories in the Field*—but he disagreed with Fuller on
where the responsibility for initiating such a pattern lay. In
McWilliams's view, it was conscious human action, not fortuitous historical accidents, that led to the use of large armies of
migrant farmworkers in California agribusiness. Because land
monopoly existed, landowners and growers did whatever they
could to obtain cheap labor. Industrial agriculture in California
has followed a typically capitalist pattern of operation, using
migrant farm laborers as an "agricultural proletariat."[15]

According to McWilliams, besides the adoption of oppressive labor practices, the monopolization of land in California had other far-reaching consequences, including attempts by speculators to expropriate land that had originally been claimed by bona fide settlers, some of whom had put in much energy and expense to improve it; a retardation of the social, political, and economic growth of the state; and the perpetuation of a wasteful pattern of farming.[16] Each undesirable characteristic became more firmly entrenched as California agriculture changed rapidly from one type of farming to another. The changes, first from bonanza wheat farms to fruit orchards and then to sugar-beet fields, each not only brought about changes in production methods but led to many unwelcome social consequences as well.

In McWilliams's view, the first transition from wheat to fruit had several important social ramifications: growers established much closer relations with merchants in the cities; heavy capitalization was necessary with the development of irrigation; crop diversification increased rapidly; the average size of farms decreased; and labor requirements increased sharply. Because so much more hand labor was needed, McWilliams said that "only cheapening the labor" made intensive cultivation possible.[17] Moreover, the ratio of temporary, seasonal workers to permanent employees increased greatly, since fruit rots on trees if not picked quickly during the harvest. Finally, though the labor had to be cheap, it also had to be skilled: if the fruit were not picked and packed properly, its salability would be diminished.[18]

Chinese farm laborers became important at this juncture because they were both cheap and skillful. In emphasizing the numerical importance of the Chinese in the farm labor supply, McWilliams used a figure that had appeared in the 1886 report of the Bureau of Labor Statistics of the state of California. In April 1886, John S. Enos, the Commissioner of Labor Statistics, had called in a number of owners of employment agencies to testify on "the supply of white labor which will be available for the fruit harvest of the present year, 1886, and from whence a supply of such labor can be had for such purposes."[19] J. P. Johnson, owner of an employment agency in San Francisco which specialized in the supply of Chinese labor, had indicated that he had "never estimated how many Chinese are employed on farms, but about seven eighths of the help on farms are

Chinese." Later, Johnson reiterated that "at least seven eighths
of the labor in the vineyards and orchards is done by Chinese."[20]
W. D. Ewer, another witness, who kept a "white employment
office" in San Francisco, testified that "the farm work heretofore,
and now, has principally been done by the Chinese; at least seven
eighths of the men on the farms are Chinese."[21] McWilliams
simply quoted the figure "seven eighths" without assessing its
reliability, and other writers who have cited McWilliams have
never asked where he got this number. Those who knew that the
estimate came from the Bureau of Labor Statistics assumed it
must have been accurate, since it was published in an official
source. No one has investigated who J. P. Johnson or W. D.
Ewer were, how many years they had been in business, how
familiar they might have been with contracting Chinese workers,
how many Chinese they had found employment for, what employers they supplied, what the basis of their impressionistic
estimate was, and consequently, how reliable the figure was.

McWilliams also cited an estimate by Bertram Schrieke, who
had stated that in 1870, 90 percent of the agricultural labor in
California was performed by Chinese, and that a decade later,
the figure had decreased to 75 percent. Schrieke was a Dutch
sociologist who had been invited by the Board of Trustees of the
Julius Rosenwald Fund to visit the United States in 1934 to conduct a study of race relations. He traveled through the eastern
and southern states and California for seven months to compile
data. His conclusions were based on "personal observation,
upon conversations . . . and wide reading of authoritative literature." However, his estimate of the number of Chinese farm
laborers, along with another statement that half a million acres
of farmland had been put out of cultivation as a result of the
expulsion of the Chinese from rural California, were undocu-
mented.[22] McWilliams himself admitted that since the available
estimates varied widely, it was impossible to say accurately how
extensively Chinese had been employed in California agriculture,
although he was "inclined to believe" that they were more widely used than "some historians" have realized. Unfortunately,
later writers who have cited him have ignored his qualifying
statements and have relied on his book, which had no footnotes,
as an authority on the subject.

According to McWilliams, growers considered the Chinese to

be "well-nigh perfect": they could be hired at "sub-subsistence wages ... had no families ... when the season was over, they vanished into San Francisco and obligingly re-appeared when required ... [and] were extremely efficient workers."[23] Not only were they expert fruit-pickers but they were the "only consider-able body of people who [understood] how to pack fruit for east-ern shipment."[24] Despite the growers' fondness for the Chinese, they were eventually "expelled" from rural California. Accord-ing to McWilliams, an "amalgamation of social forces—or-ganized labor, small manufacturers, small farmers"—and the arrival of many immigrants as the result of lowered railroad fares in the mid-1880s, enabled rural dwellers to "take matters into their own hands; they expelled the Chinese and gave them rough notice not to return."[25] A series of anti-Chinese riots that broke out in rural California in the early 1890s allegedly removed the Chinese so completely from California agriculture that they were not involved in the second transition from fruit to sugar-beet cultivation during the last years of the nineteenth century. This assertion of McWilliams will be evaluated in a later chapter.

The second transition—from fruit to sugar-beet cultivation—also had a number of social ripple effects. McWilliams probably has been the only observer who has attached revolutionary social implications to the introduction of sugar-beet cultivation. The organization of the sugar-beet industry, in his view, is based on a true application of the industrial pattern of factory production to California agriculture. The industrial character of sugar-beet cultivation and sugar refining is visually apparent: the smoke-stacks of the beet-sugar refineries stand out tall and stark against the sky, their dense, black smoke marring the rustic landscape. Radiating out from this symbol of industrial power are fields of beets cultivated by farmers on contract. The power of the factory is such that small independent farmers in the vicinity go out of business and are forced to become the company's tenants. Initially, these contract farmers were white Americans; by the turn of the century, the Japanese had taken over.[26]

The contract beet-growers and tenant farmers hired migratory labor during the weeding, thinning, topping, and harvest sea-sons. Since beets will not keep more than a week after they are taken from the ground, the demand for seasonal labor is even greater than in fruit orchards. In the opening years of the twentieth century, almost all the beet-field laborers in northern California were Japanese; in southern California, one-fifth were Japanese and four-fifths Mexicans. Inside the refinery itself, however, the labor was white. The beet-sugar industry in California, like its cane-sugar counterpart in Hawaii, was and still is both organized and run like a plantation. As a matter of fact, McWilliams pointed out, members of the Spreckels family helped to found both.[27]

Writing in the muckraking tradition in the years following violent outbreaks in the Imperial Valley and elsewhere, McWilliams was trying to warn the reading public about the emergence of an incipient form of fascism in rural California. Tracing the state's vigilante tradition and race riots to its pattern of land-holding, McWilliams was trying to make a case for agrarian socialism under which land monopoly would be ended, production cooperatives would flourish, and farmworkers' welfare would be promoted by unionization.

Paul Taylor's Analysis

Fuller's study and McWilliams's book, as well as John Steinbeck's novel *The Grapes of Wrath*, were published in 1939. The simultaneous appearance of these books, in conjunction with the hearings of the La Follette Committee, following almost a decade of intermittent violence in California's agricultural valleys, focused much public and governmental attention on the nature of California agriculture. But it was left to Paul Taylor, an economist who had written extensively about Mexican labor in the United States, to analyze the impact of agribusiness on California rural society and how it departed from the American agrarian ideal. This he did in a long article published in 1945, entitled "Foundations of California Rural Society."[28]

According to Taylor, on at least five occasions during the nineteenth century, great public concern over the foundations of California rural society had been expressed. During the 1849 constitutional convention, delegates debated whether California should enter the union as a free or a slave state. Although the discussion was over the nature of a society founded on mining rather than on agriculture, everyone understood that in the Southern states, free and slave labor were primarily alternative forms of agricultural labor. What the miners feared, according to Taylor, was that "manual work such as they performed would become dishonorable ... also that economic competition from unremunerated slave labor would drive out free working miners like themselves." Antislavery spokesmen wanted no free blacks to come to California, either, because for them the question was one not simply of legal status but of the existence of a "caste system of labor whether the subordinate group was enslaved by law or not." They did not wish to have a "permanent labor class."[29] It seemed to matter little that it was white slaveholders and capitalists who brought such conditions of enslavement or peonage about; in the minds of the miners and white workers, the black victims themselves were somehow responsible for allowing themselves to be made permanently subservient. The conclusion was that "if we wish to avoid placing them in a position of servitude, we must exclude them."[30]

Such a line of argument was reiterated in the second great debate over the nature of California rural society when, from the 1850s through the 1880s, Californians contended over the issue of Chinese immigration. This question was related to the state's agricultural development because it was alleged that the availability of cheap Chinese labor enabled landowners to hold on to large tracts of land without subdividing them. Such a pattern of landownership was said to impede the immigration into California of farm families from the East and the Midwest and to retard the state's overall social development.

The third debate took place during the second constitutional convention in 1879 when delegates argued about whether measures should be written into the new constitution to curb land monopoly—the outcry against land monopoly being part of the general struggle against trusts and monopolies, which were considered to be antithetical to a genuine free enterprise system. A fourth occasion for discussion arose during the famous *Lux* v. *Haggin* case in 1886 when the issue of water monopoly was joined to the question of land monopoly, for it was recognized that in California agriculture, control over water was more crucial than control over land per se. Finally, during the passage of the 1902 federal reclamation law, debate took place over whether land irrigated by water from projects built with federal funds should have acreage limitations placed on them. The desire to impose an acreage limitation represented yet another attempt to counter land and water monopoly in the state.[31]

Paul Taylor has been an indefatigable modern proponent of the agrarian ideal. He has been concerned with the Chinese only because their presence seemed germane to an understanding of why "California has not placed at the head of her agricultural goals, the achievement of a satisfactory relation between land and the families who labor upon it." He believed that "this divergence from national policy has not survived because Californians generally willed their agriculture to be different. On the contrary, they have condemned it repeatedly, and struggled against it over and over again ... but they have never achieved more than limited success."[32] In his article Taylor never satisfactorily identified all the factors that undermined the agrarian ideal in California; he only concluded that Chinese exclusion represented a partial victory of the "popular will" in its attempt to establish agrarian democracy in California. This is an ironic statement from a scholar who has been a great champion of the workingman and the small farmer.

In Taylor's opinion, dichotomous visions of California rural society were never more clearly expressed than at the 1877 congressional hearings on Chinese immigration when "wealth" and "welfare" were counterposed. The question of which value should be more important was discussed in terms of "cheap versus dear labor" and "large estates versus working farmers." Colonel W. W. Hollister, a large landowner and employer of Chinese, not only did not care what color or race his laborers were, but wanted as many of them as possible so that they would "create competition among themselves for positions," thus reducing "the price of labor to such a point so that its cheapness will stop their coming."[33] If California were to prosper, "work must be obtained, and it must be obtained at prices that will leave a fair margin for brains and capital." He found Chinese workers to be honest, very intelligent, never drunk, kindly, and submissive—in great contrast to white men, whom he called "bummers," who got drunk and desired to "live off [their] wits and not work."[34]

Other witnesses felt that it was better to attain material progress more slowly in order to ensure the creation of a "casteless society," which was best represented by farming communities with schools, churches, pleasant houses, fine yards, and roads. A desirable rural society was one without the evils of "absentee ownership, tenancy, and a large laboring population, the land-lords pressing upon the renters and the renters pressing upon the laborers and upon the land."[35]

The Chinese were objectionable because they allegedly created a labor caste, contributed to the exhaustion of the soil, deterred the entry of "more highly paid white laborers and even of boys and girls into agricultural labor," and "divided rural California sharply into classes with an inferior caste at the bottom."[36] In short, the Chinese presence bolstered all the characteristics of an undesirable rural society. According to Taylor:

> The alignment of California opinion on the issue of the desirability of more Chinese in order to provide cheap labor for agriculture showed a decided class character. Large landholders and operators of large tracts were conspicuous among those who favored ample supplies of Chinese laborers. The Grangers, and small farmers generally, opposed cheap Chinese labor strongly. The Workingmen's Party, led by Dennis [sic] Kearney, was the most vehement organization in opposition, as well as representative of the most numerous interest.

In comparison, "middle class opinion was not unified. Ministers of the gospel were divided.... Merchants were divided."[37]

Regardless of whether the Chinese in fact served as a "caste at the bottom," it never occurred to anyone—including Paul Taylor himself—that neither the Chinese in California nor blacks in the plantation South had the power to determine the structure of the society in which they found themselves. Opinions such as those that were aired during the various public debates on the so-called Chinese question were classic examples of blaming the victims for the oppression they suffered. Be that as it may, in Taylor's view the anti-Chinese forces ultimately succeeded in pressuring Congress to pass the Chinese Exclusion Law because the economic depression and social instability of the late 1870s moved "middle class opinion to the side of American laborers and small farmers."[38] After this period, though the issue of monopolies was to crop up several more times in the following decades, at least the Chinese were no longer considered to be the major cause of an undemocratic rural society in California.

Lloyd Fisher's Study

The next important work on the subject of Chinese farm laborers to appear was *The Harvest Labor Market in California* by agricultural economist Lloyd Fisher, published in 1953.[39] A liberal "deeply concerned that no man or group of men should enjoy unopposed power, that men should be free to clash their wills and minds against each other on terms approaching equality," Fisher "regarded the harvest labor market as one skirmish— albeit a revealing one—in this perpetual struggle."[40] Fisher did not believe, as Fuller, McWilliams, and Taylor did, that the large size of farms was the main reason that a seasonal, migratory labor force has been used in California agriculture. He stated:

> There are crops which cannot be handled by family labor, either with or without a hired man. Even on the smallest farms the labor requirements of the harvest far exceed the family supply available. Ten acres of peaches, tomatoes, or apricots cannot be harvested without a force of seasonal farm laborers. The characteristics of the agricultural labor market in California are created only in part by the prevalence of large-scale farming enterprises or by the corporate structure of some farm businesses. Small farms are as dependent as large farms upon seasonal hired labor, resident owners as dependent as absentee corporate owners. The harvest labor market is primarily a function of specialization in labor-intensive crops, and the intense seasonal demand for labor that results.[41]

While accepting Fuller's thesis that "the present character of California agriculture is a direct response to the volume and character of the supply of agricultural labor," and agreeing that "the coincidence of continuing immigration of the Chinese, their violent expulsion from the mines ... and the completion of the transcontinental railroad combined to provide the first substantial group of casual laborers for California agriculture," Fisher pointed out that the harvest labor market had been a "structureless market," one that was—at least at the time he wrote—"without any structure of job rights or preferences."[42] Growers, who derive advantages from such "disorganization," have used various mechanisms to maintain the status quo, including preventing the formation of unions; maintaining impersonal relationships between employers and employees; differentiating skilled from unskilled tasks; using piece rates for the unskilled workers so that "differences in skill, age, and sex become matters of relative indifference"; and avoiding—at least until recent decades—mechanization.[43] In short, since there was "literally no relationship between employer and employee upon which any claims to recurrent employment might be built,"[44] almost anybody at all could be hired. The laborers, who could fully substitute for each other, were easily replaceable, making it virtually impossible for them to find any basis on which to organize to demand any rights.

One way growers have established a division of labor has been to treat the harvest operation itself as a separate economic enterprise, so that decisions regarding it are made independently of other decisions in the agricultural production process.[45] Since the major cost of the harvest is the cost of labor, growers purposely have tried to insure that more laborers are available than they can actually use, because an oversupply makes it possible to keep wages as low as possible, discourages strikes, and makes it possible for the harvest to be completed as quickly as possible in order to minimize the risks of inclement weather. That is why throughout much of California's agricultural history, growers have persistently reported "labor shortages" when there were in fact none, because a "chronic 'flooding'" of the labor market was advantageous to them. Since "the cost of recruiting additional labor [has been] negligible, the cost of unemployment [has been] borne by the community," and the cost of transportation was the responsibility of the farmworkers themselves, farmers have had little incentive to change their wasteful pattern of labor usage.[46]

It has been possible for growers to perpetuate such labor practices because they have primarily used workers who had little

"Chinese camp" on A. A. Lohens's Ranch, showing the kind of dwellings that Chinese farm laborers lived in, Alameda County, California, 1897. (Photograph courtesy of the California State Library, Sacramento)

opportunity to move from low-paid, seasonal, agricultural work to better-paid, more stable, industrial employment. According to Fisher, most of the migrant farm laborers, who have been disenfranchised minorities or immigrants, have faced two major obstacles in occupational mobility: general prejudice based on racist stereotypes about how various racial groups are most suitable for particular kinds of work, and barriers erected against their entry into the industrial labor market by labor unions. However, though nonwhite farm laborers have not been able to move from agriculture to industry, whenever there has been an economic depression, unemployed white workers have been able to seek jobs in agriculture, thus reducing the number of jobs open to the subordinate groups when they have needed work most badly.[47]

In Fisher's view, the Chinese have been significant in the harvest labor market, not so much because they were cheap, but because they initiated the system of labor contracting that has persisted ever since. "The farm labor contractor, Chinese, Japanese, Filipino, and Mexican, performed a function which has been found consistently to be necessary" because this practice has been "one of the few organizing influences in a disorganized market. It continues to bring workers and jobs together, to provide an element of stability and regularity in a chaotic market."[48] Thus, the manner in which Chinese laborers had organized themselves enabled agricultural employers to enjoy the advantages of a structureless market while simultaneouely being assured of a labor force when one was needed.

In the last twenty years, with increasing mechanization, the introduction of hourly wages in place of piece rates, and farm labor unionization, some of Fisher's analysis no longer applies. But the conditions he described held true for almost a century.

Ping Chiu's Study

Ping Chiu's 1963 study, *Chinese Labor in California, 1850–1880*, provides the only available exposition on Chinese farm laborers set in the context of nineteenth-century Chinese American economic history. Like Fisher, Ping Chiu observed that the economics of grain farming that most Americans were familiar with simply were not applicable to fruit growing. The labor requirement of even small orchards is far greater than the public assumes it to be. The labor difficulties faced by aspiring fruit-growers in nineteenth-century California were exacerbated by the fact that wages in general were abnormally high in the state, based originally as they were on the average income of miners. So, "unable to compete with high-priced labor, the agricultural-ists [initially] tried to regulate their production according to the size of their labor supply."[49]

Ping Chiu made two points that differed from those made by other writers on the topic. Whereas others have thought that their discharge by the Central Pacific Railroad Company in 1869 forced the Chinese to become farm laborers, Ping Chiu believed that since "the initial attraction for the Chinese was mining, [it was when] the placer mines gave out in the mid-1860s [that] Chinese laborers [became] available in any quantity."[50] This observation is consistent with the thesis presented in chapter 2 that mining—its availability or lack thereof—influenced the rest of the occupational distribution of the Chinese. In addition, Ping Chiu argued that fruit growing was not all that lucrative in the early years. Given the high cost of labor—which often consti-tuted one-quarter to one-third of the total estimated income—as well as high interest and freight rates, Ping Chiu cited data from the manuscript agriculture census for Los Angeles County to show that only about half of the vineyards and orchards had a "surplus" (revenue minus labor cost, taxes, interest payments, marketing costs, and the growers' own living expenses) of over $1,000 in 1880. In Ping Chiu's view, "because of a low profit margin and high borrowing charges, wages remained the only cost that could be kept at a minimum level.... Direct competi-tion for labor with industry was out of the question [so] cheap Chinese labor was indispensable."[51] This argument echoed faithfully the statements of many contemporary growers them-selves. Moreover, Chinese labor was desirable because it "repre-sented a constant unit labor cost; that is to say, an increase in demand for labor had little or no effect on wage rates."[52]

Ping Chiu also tried to evaluate where big and small farmers stood on the Chinese question. Relying on episodic evidence, he concluded that farmers "as a class were not committed to the expulsion of the Chinese." However, because of the myth of the family farm and the "divergence of interest between large and small land owners," some small farmers "began to join the workers in the anti-Chinese movements." He conjectured that some of the

> more calculating farmers might have realized that by supporting the urban anti-Chinese movement, and by limiting the job opportunities of the Chinese in the cities, more of them would have been made available on the farm. That, in turn, would have achieved a reduction of labor cost by lowering alternative cost. Others might have conceived the support of the anti-Chinese movement to be a small price for a farmer-labor alliance against the "monopoly," and the unity of the farm movement.
>
> Consciously trying to maintain a common front against the railroad, or intimidated by the threat of violence, farm producers began to replace Chinese with white laborers in the mid-1870s.[53]

Ping Chiu's speculation about what motivated some farmers to become anti-Chinese was not substantiated by any sort of historical evidence; moreover, his assertion that whites began to replace Chinese in the farm labor force in the mid-1870s was simply erroneous. For these reasons, this section of Ping Chiu's study, one of the weakest parts of his book, though focused on the Chinese as protagonists, failed to illuminate their situation any more clearly than other studies had done.

Cletus Daniel's Study

Until the publication of Cletus Daniel's finely crafted book *Bitter Harvest* in 1981, most writers had treated farm laborers only as pawns in the interplay of larger forces.[54] Using many hitherto untapped sources, Daniel has resurrected the history of farmworkers' struggles to better their own lot against overwhelming odds. As a labor historian who relied primarily on written sources, perhaps it was inevitable that his tale focused primarily on white farmworkers, who left more written records of their activities, than on nonwhite farm laborers, who left few documents. In particular, Daniel's discussion of the Chinese was brief and offered little fresh analysis. But the book is very important because the author presented a new theory on why agribusiness developed in California,[55] highlighting the role that local, state, and federal governments played in adjudicating and controlling labor unrest. Furthermore, he explained why a liberal, reformist approach to improving the lives of an oppressed group can, at best, bring only marginal results and why radical approaches based on an inaccurate understanding of concrete conditions were also bound to fail.

Daniel's most important insight was his assertion that California agribusiness was not a departure from the American agrarian ideal of small family farms, because "California was never part of that tradition."[56] Though many factors led to the creation and perpetuation of large landholdings, Daniel favored the explanation that such holdings resulted from the existence of Spanish-Mexican land grants. Furthermore, he pointed out that many of the early American landowners and farmers in California differed from most agrarian idealists because the "gold mania" that brought them to the Pacific Coast imbued them with a speculative tendency, and led them to view farming less as a way of life than as a business. Moreover, since agricultural development in California occurred during an era of rapid industrial expansion in the nation as a whole, the temper of the times influenced farmers, who were enjoying the benefits of concrete economic gains, to ignore an increasingly anachronistic agrarian ideal. As the businessman emerged as the new role model of the successful American with an attendant change in social values, California's farmers were more than happy to become part of the new age.[57] Unlike Taylor, Daniel did not consider California agriculture to be a stage upon which a great conflict in values was fought, though considerable lip service was paid to agrarian idealism as it slowly and imperceptibly eroded.

Daniel argued that "the highly commercialized, even industrialized farms or ranches that emerged by the end of the nineteenth century ... had begun as a counter to bonanza farming."[58] Reiterating the by now familiar litany that owners and operators of huge bonanza wheat farms were criticized by their contemporaries for "giving no thought to the long-run developmental needs of the state," for perpetuating "the tradition of monopolistic landholding," for causing a "relatively slow growth of the state's population," for making California too dependent on a single crop "as to make the entire farm economy ... vulnerable to the vagaries of a notoriously unstable international grain market," and for using "cheap, seasonal wage labor [that] introduced class divisions,"[59] Daniel claimed that, ironically, advocates of agricultural diversification who led the move toward intensive agriculture had acted out of agrarian motivations. But, benefiting themselves from large-scale operation and market control, they soon became leading apologists for practices that have come to characterize agribusiness. Though there was indeed a considerable reduction in the size of farms in California during the 1880s, the labor practices that had been initiated by large growers were retained. Even owners of small family farms began to argue that given their lack of control over commodity prices and their inability to "successfully compete for labor with their counterparts in the cities," they had no recourse except to use low-paid, seasonal labor—labor "regarded as not suitable for whites" because the work was "onerous" and the attendant economic and social status was "degraded."[60]

Farm owners had no qualms about employing Chinese, however, because it was said that they "did not, indeed could not, have aspirations or expectations comparable to those that white workers might reasonably harbor or consider a matter of right on the basis of their whiteness alone.... [Chinese] were immune to the democratizing forces of tradition, circumstance, and social contract that afforded the lowly just enough opportunity for advancement to keep the popular expectation of upward movement alive." As a "work force whose estrangement from the social and cultural mainstream was so profound and unalterable as to render it captive economically," the Chinese were just the kind of laborers farm employers sought.[61] Although he recognized that "the vocal assaults launched by agrarians [against the Chinese] were blatantly racist and chauvinistic," Daniel's analysis of the Chinese indicates that he basically agreed with the contemporary assessment of them.

Daniel pointed out that "if urban agitators tended to regard Chinese exclusion as an end in itself, agrarians saw it only as an important first step toward the eventual dismantling of California's large-scale agricultural industry."[62] But such dismantling never occurred, because "newly evolved personal and social values," "a very rapid market expansion brought about by new and improved methods of transportation," and the "nearly simultaneous development of new and improved canning and drying techniques permitted California farmers not only to sell all of the fresh produce they could grow, but to profit as well from an equally strong and enduring demand for processed fruits and vegetables."[63] Thus, intensive agriculture, instead of countering the undesirable features of bonanza farming, in fact "reinforced and perpetuated" the "large-scale commercial agriculture established by wheat farmers in the 1850s."[64] Daniel answered the question that Paul Taylor raised but never answered: why did agrarianism fail to triumph in California? According to Daniel, it failed because, being nothing more than a nostalgically recalled abstraction, it lacked "the force and durability" that had buttressed it in older parts of the country— it was "simply not strong enough to retain the allegiance of a farming population to whom the imposing benefits of horticulture were being demonstrated in real terms on every side."[65]

But the existence of demonstrable benefits was not enough. Agribusiness could not have evolved so rapidly had various government agencies not contributed to its growth. More than any other author writing on California agriculture, Daniel has revealed the crucial role that the state has played. He not only chronicled the violent means that local law enforcement agents used to control farm labor unrest, but showed that even well-meaning reformist agencies of the state government—such as the Commission of Immigration and Housing, and the Commission on Land Colonization and Rural Credits, created in 1913 and 1915, respectively—as well as New Deal federal agencies, served to dampen farmworker militance because these agencies were prolabor but antiunion. Since the major thrust of the National Industrial Recovery Act of 1933 was to rehabilitate capitalism, national planning and paternalistic management rather than worker militancy were considered the proper means to harmonious labor relations. The National Recovery Administration worked to divest unions of independent economic power, while the National Labor Relations Act—though recognizing that relations between workers and capitalists were inherently in conflict and that the interests of labor had to be protected by guaranteeing its right to collective bargaining—did not benefit farmworkers at all, because they were left out of the Act's provisions.

A tragic tone was inevitable in Daniel's exposition, since in his view neither individual reformist efforts by men like Simon Lubin, Carleton Parker, and lawyers from the American Civil Liberties Union, nor radical action by organizers from the Industrial Workers of the World (the Wobblies), the Agricultural Workers Industrial League/Union, and the Cannery and Agricultural Workers Industrial Union could do much to improve farmworkers' lives in the long run. And the dozens of strikes— some spontaneous, some organized—that farmworkers themselves engaged in from the late 1920s through the 1930s brought few gains because the farmworkers' "powerlessness was so profound and engulfing as to have become at once the cause and effect of [their] wretchedness."[66] To this day, farmworkers remain among the most oppressed workers in the United States.

Despite the book's many insights based on meticulous research, it failed to broaden or deepen our understanding of nonwhite farmworkers, even though these have provided the bulk of the farm labor force in California. Daniel devoted many pages to Mexican, Chinese, Japanese, and Filipino farmworkers, but his account essentially concentrated on the attempts of leftist, white unions to organize workers in the fields and canneries. Though Daniel was unquestionably sympathetic to the plight of California's farmworkers, like some other liberal white scholars, his study has unwittingly perpetuated the stereotype that nonwhite workers are powerless and virtually unorganizable. Thus, even as he explained "the reluctance of Chinese farmworkers to challenge prevailing conditions" because of the limits imposed by the labor contracting system and the risks facing an alien, disenfranchised group, he inadvertently confirmed the impression that the Chinese were indeed cheap, docile laborers, and little more.[67]

Linda and Theo Majka's Study

Daniel's detailed analysis of how various levels of government have shored up agribusiness in California has been taken one step further by Linda C. and Theo J. Majka, two sociologists who had worked as organizers for the United Farm Workers and who placed the history of California farmworkers in an explicit framework based on a "structuralist" Marxist theory of the state.[68] In *Farm Workers, Agribusiness, and the State,* published in 1982, the Majkas argued that the conditions faced by California farmworkers can best be understood as a series of struggles between two groups with grossly unequal power, and that the outcome of each incident was heavily influenced by actions taken by the state.

The "structuralist" Marxist theory of the state—based primarily on the exposition of Nicos Poulantzas—views the state differently from both the pluralist conception of government and the Leninist theory of the state. The former views governmental units as neutral mediators in the democratic competition among various interest groups for power and resources, whereas the latter considers the state to be a tool of the dominant capitalist class. In "structuralist" Marxist theory, a more complex role is attributed to the state, which must enable the dominant classes to accumulate capital while acting to ensure its own legitimacy by appearing to serve society as a whole. Since the two goals, accumulation and legitimacy, are at times contradictory, sometimes the state functions in a way that is detrimental to the short-term interests of capitalists, but it does so to insure the ultimate reproduction of capitalist class relations. In structuralist theory, class conflicts, which can be regulated by bureaucratic procedures, are not seen to lead inevitably to revolutionary upheavals. By channeling militancy, the state can guarantee reforms and insure the survival of "insurgent" subordinate groups while simultaneously undermining their political strength and ability to win further concessions.[69] Depending on the politicians and bureaucrats in office, the relative strength of agribusiness vis-à-vis that of the farmworkers, and the general social and political climate of a period, governmental agencies can vary, and have varied, the role they played in particular episodes of the struggle between agribusiness and farm laborers, but in the long run they have acted mainly to insure growers a dependable, cheap, and malleable labor supply.

The second through eighth chapters of the book, which traced the history of California farmworkers before the United Farm

Workers entered the scene, were culled almost entirely and quite
uncritically from secondary works. Consequently, the Majkas incorporated into their own study many of the historiographical
problems inherent in the nature of the subject matter: the lack of
documentary sources, the multiplicity of ethnic groups that have
worked in California agriculture, and the difficulty any single
author faces in attempting to master the subject matter sufficiently to write a good general history. The chapter on the Chinese,
only three pages of which dealt with Chinese farm laborers,
contained many half-truths, only the most important of which
will be discussed.

The linchpin in the Majkas' theory was that throughout most
of its history the California farmworkers' movement had not succeeded in making permanent gains, because it lacked "external
resources," owing to racism against nonwhite workers and the
existence of a dual labor market. Unlike some dual labor market
theorists, who simply distinguish between a primary market,
with permanent, well-paid, unionized jobs, and a secondary market, with seasonal, poorly paid, nonunionized jobs, the Majkas
claim that in California the first can be found in urban centers
and the second in rural areas. They believed that the presence of
a "passive and tractable," and thus exploitable, Chinese labor
force made the emergence of the dual labor market possible.[70]
Basing their thesis on only eight secondary sources,[71] the Majkas
recounted how, first, the Chinese were driven out of the mining
regions and entered urban trades, but later, as Chinese competition threatened white workers as well as manufacturers, they
were forced to seek agricultural work in rural areas.[72] Unfortunately, such a view is too simplistic: the Chinese were not driven
entirely away from the mining regions in the 1850s, as the Majkas (and Fisher before them) claimed, nor did they take up agricultural work solely because they were driven out of the urban
areas. As I discussed in chapter 2, throughout the nineteenth
century Chinese engaged in both urban and rural occupations.
Moreover, while anti-Chinese legislation and activities restricted
the kind of work Chinese could find, their own perception of
where their comparative advantage lay also determined what
kind of work they sought.

The main contribution the Majkas made to the history of
farmworkers in California lay in their recognition that relative-

ly powerless as they were, farmworkers—like their industrial
brethren—have struggled to gain some control over their working conditions, to have a larger role in making decisions about
the production process, and to change the balance of power between employers and workers. It was possible for them to succeed from time to time because their labor has been, and continues to be, crucial to the prosperity of California agriculture,
given the central place that agriculture occupies in the state's
overall economy. Since disruptions could not, and cannot, be
tolerated for long periods, the state has had to intervene to enable production to continue and to prevent the farmworkers'
movement from turning into a wider, antiestablishment movement. Those who are sympathetic to the farmworkers' struggles
see hope in the fact that the farmworkers themselves have come
to understand more clearly their own power. But like everyone
else who has written on the subject, the Majkas did not believe
that the Chinese ever had such understanding.

It is obvious that all the authors whose works have been reviewed here considered the Chinese presence central to the development of intensive agriculture in California, though their
analyses differed in subtle ways. Fuller, Taylor, Daniel, and the
Majkas laid some responsibility at the feet of the Chinese for the
perpetuation of land monopoly and the emergence of large-scale
agriculture in California; McWilliams saw them as pawns and
victims of "land grabbers"; Fisher and Ping Chiu thought that
impersonal forces, such as the labor requirements of intensive
crops and high interest and freight rates, were more important
than the "willingness" of Chinese to work at low wages, in determining the pattern of land and labor usage. This study has
shown that the Chinese were indeed crucial to the development
of intensive agriculture in California, but it should also be clear
by now that they played multiple roles in this endeavor and were
by no means only providers of "cheap labor." The following
chapter will consider the extent to which the Chinese can be
"blamed" for the existence of land monopoly in California, and
the nature of their cheapness and docility.

NINE

Working Hands

THE QUESTION of whether the Chinese presence in nineteenth-century California was responsible for the perpetuation of land monopoly and the growth of a large-scale, industrialized form of agriculture cannot be answered directly, simply because there is insufficient reliable information to answer such a question. The issue can only be examined indirectly by asking what proportion of the farm labor force was composed of Chinese, and whether this proportion was large enough to justify the claim that the Chinese were mainly responsible for the undemocratic nature of California agriculture. If the answer is no, then what other factors were at work? Moreover, even if the Chinese were sufficiently numerous in the farm labor force to have affected the character of California agriculture, were they as "willingly" cheap, docile, and servile as contemporary observers charged them with being?

The Number of Chinese Farm Laborers

The authors whose works were reviewed in chapter 8 accorded the Chinese a central role in the evolution of large-scale specialty agriculture in California, but none attempted to obtain an accurate count of the actual number of Chinese in the farm labor force. To be sure, given the large number of migrant seasonal workers in California's farm labor supply, such a task is well-nigh impossible because individuals enter and leave the farm labor force from week to week, or even from day to day. For this reason, Fuller believed that the sources he relied on, such as the *Pacific Rural Press*, the *California Farmer*, and the *California Fruit Grower*, contained statements as accurate as could be had on the relative importance of the Chinese; at least, such statements showed how farmers perceived the situation.

It is important to get a better estimate of the number of Chinese farmworkers than those figures given in the existing literature, because unless it is known how numerous the Chinese were relative to other groups, it cannot be claimed that their presence as "cheap labor," more so than any other factor, caused "land monopoly" to be perpetuated in California. Unfortunately, even the census, the most systematic source available, is not fully reliable, because although it counted most of the permanent farm hands, it included only a fraction of the seasonal workers. Too, statistics culled from the census schedules for the different years must be interpreted differently. Chinese farm laborers first became noticeable in the 1870 census; the figures for that year accurately reflected the total number of year-round and seasonal farm laborers in California because the count was taken between June and August. Since in most counties the peak months for hiring seasonal farmworkers were August and September, census takers who worked through August "caught" not only the permanent workers but also a large proportion of the seasonal ones in their count. The 1880 and 1900 censuses, on the other hand, did not reflect the total number of farmworkers accurately, because the enumeration was made in the single month of June prior to the peak harvest season. Thus many of the seasonal workers were not included. To get more accurate estimates, statistical adjustments must be made for these years.

Table 24, showing the number of farm laborers who had been born in different places, indicates that in 1870 Chinese ranked first in Alameda, Sacramento, and San Mateo counties. These locations coincided partially but not entirely with those where Chinese had first leased land to farm as tenants. Unlike the latter, who often entered areas where they were among the pioneer farmers, Chinese farm laborers first appeared in areas which white farmers had already developed. Grain, deciduous and small fruit, and vegetables had been grown since the Spanish period in the western part of Alameda County, a fertile plain

Table 24   *Farm Laborers and Laborers in Farmers' Households by*
Nativity and County, 1870

Place of Birth	Alameda	Sacramento	San Joaquin	San Mateo	Santa Clara	Solano	Sonoma
CHINA	350	622	50	180	145	163	42
U.S.A.	274	398	625	132	582	1,084	588
California	(32)	(23)	(26)	(11)	(91)	(34)	(49)
Other Western States	(5)	(12)	(2)	(0)	(5)	(15)	(8)
New England/ Mid-Atlantic	(114)	(144)	(183)	(81)	(171)	(285)	(137)
North Central States	(93)	(159)	(320)	(23)	(227)	(635)	(276)
South Atlantic States	(13)	(28)	(62)	(3)	(43)	(43)	(28)
South Central States	(17)	(32)	(32)	(14)	(45)	(72)	(90)
ENGLAND/ SCOTLAND/ WALES	27	44	36	29	52	71	43
IRELAND	159	79	77	146	182	417	142
GERMANY	101	75	48	71	70	128	72
PORTUGAL	314	36	1	47	31	41	7
ITALY	6	28	16	6	14	0	14
CANADA	56	40	51	33	83	99	75
MEXICO	10	0	0	7	0	0	0
OTHER PLACES	130	47	26	62	96	42	41
TOTAL	1,427	1,369	930	713	1,255	2,045	1,024
% Chinese	24.5	45.4	5.4	25.2	11.6	7.9	4.1
% American	19.2	29.1	67.2	18.5	46.4	53.0	57.4
% European & Other	55.6	25.5	27.4	55.2	42.1	39.0	38.5
% Mexican	0.7	0	0	1.0	0	0	0
Rank of Chinese	1st	1st	4th	1st	3rd	3rd	6th

SOURCE: Tallied and computed from U.S. National Archives, Record Group 29, "Census of the U.S. Population" (manuscript), Alameda, Sacramento, San Joaquin, San Mateo, Santa Clara, Solano, and Sonoma counties, 1870.

several miles wide between the salt marshes of San Francisco Bay and the foothills of the Coast Range. Chinese worked in these fields and orchards of the coastal plains from the 1860s on. In Sacramento County, Chinese farm laborers were concentrated along the Sacramento River in the northern Delta, where a thriving fruit belt had developed from the beginning of the gold rush, and along the American River in Center Township, which white farmers had discovered was suitable for the cultivation of hops. In San Mateo County, they worked on farms in the lowlands just south of San Francisco which had also been developed by white farmers. Chinese were also relatively important in the Santa Clara Valley and the Vaca Valley of Solano County, nascent fruit-growing areas, where they ranked third in number of farm laborers.

Another way of looking at the relative importance of the Chinese as farm laborers is to examine what percentage of the farm labor force they constituted. Chinese were 45 percent of the farm labor force in Sacramento County, approximately 25 percent in Alameda and San Mateo counties, 12 percent in Santa Clara County, and 8 percent in Solano County. However, if only the townships in which they predominated are taken into account, they made up far larger percentages: 67 percent in Washington Township in Alameda County; 39 and 72 percent, respectively, in Franklin and Georgiana townships in Sacramento County; 48 percent in First Township in San Mateo County; 22 and 13 percent, respectively, in Santa Clara and San Jose townships in Santa Clara County; and 26 percent in Vacaville Township in Solano County. Overall, in the counties included in table 24, there were some 1,600 Chinese farm laborers out of approximately 9,000 counted, or about 18 percent of the total. Since the published census gave a figure of only 1,637 Chinese out of a total of 16,231 farm laborers in California in 1870,[1] there were apparently only a handful of Chinese farmworkers in those counties not included in table 24.

By 1880, as is shown in table 25, the Chinese had dispersed geographically and increased both in absolute numbers and in relative percentage. There were now many of them throughout the Sacramento Valley and around the San Francisco Bay Area. They ranked first in Alameda County and second in Contra Costa, Placer, Sacramento, San Joaquin, Santa Clara, Solano, Sonoma, Tehama, and Yuba counties. The largest clusters were in Alameda, Sacramento, San Joaquin, and Santa Clara counties, with 455, 537, 471, and 763, respectively. In three of these counties, the Chinese were approximately one-third of all farm laborers, but considering only those townships where they were concentrated, they made up far larger proportions: 68

Table 25   Farm Laborers and Laborers in Farmers' Households by Nativity and County, 1880

Place of Birth	Alameda	Contra Costa	Colusa	Placer	Sacramento
CHINA	455	168	169	150	537
U.S.A.	323	345	1,416	367	630
California	(147)	(176)	(407)	(111)	(212)
Other Western States	(9)	(5)	(59)	(8)	(21)
New England/Mid-Atlantic	(111)	(64)	(227)	(80)	(133)
North Central States	(45)	(86)	(590)	(134)	(211)
South Atlantic States	(5)	(6)	(59)	(14)	(21)
South Central States	(6)	(8)	(74)	(20)	(32)
ENGLAND/SCOTLAND/ WALES	27	29	53	22	38
IRELAND	102	81	192	26	55
GERMANY	103	45	125	34	72
PORTUGAL	362	135	24	2	21
ITALY	13	14	7	8	6
CANADA	25	23	108	18	37
MEXICO	11	8	2	0	0
OTHER PLACES	107	41	39	42	23
TOTAL	1,528	889	2,135	669	1,419
% Chinese	29.8	18.9	7.9	22.4	37.8
% American	21.1	38.8	66.3	54.9	44.4
% European & Other	48.4	41.4	25.7	22.7	17.8
% Mexican	0.7	0.9	0.1	0	0
Rank of Chinese	1st	2nd	3rd	2nd	2nd

SOURCE: Tallied and computed from U.S. National Archives, Record Group 29, "Census of the U.S. Population" (manuscript), Alameda, Colusa, Contra Costa, Placer, Sacramento, San Joaquin, San Mateo, Santa Clara, Solano, Sonoma, Tehama, Yolo, and Yuba counties, 1880.
NOTE: The figures for Sacramento County are an undercount because in Brighton township no farm laborers were enumerated, and it was also not possible to count the laborers living in farmers' households, because the enumerator failed to separate the individuals by households.

and 30 percent, respectively, in Brooklyn and Eden townships in Alameda County; 52 and 92 percent, respectively, in Franklin and Georgiana townships in Sacramento County; 83 percent in Union Township in San Joaquin County; 68 and 22 percent, respectively, in Santa Clara and San Jose townships in Santa Clara County; and 54 percent in Vacaville Township in Solano County.

In addition, several new localities now also had significant numbers of Chinese farm laborers: along the Feather River north and south of Marysville, around the hopyards of Wheatland in southern Yuba County, and in the newly planted fruit belt be-

San Joaquin	San Mateo	Santa Clara	Solano	Sonoma	Tehama	Yolo	Yuba
471	89	763	240	202	53	30	124
1,373	161	904	525	821	451	684	290
(419)	(70)	(467)	(158)	(463)	(130)	(199)	(116)
(22)	(1)	(10)	(36)	(23)	(18)	(5)	(1)
(302)	(53)	(181)	(135)	(127)	(61)	(141)	(58)
(528)	(32)	(196)	(151)	(153)	(193)	(271)	(93)
(49)	(1)	(24)	(27)	(21)	(16)	(31)	(7)
(53)	(4)	(26)	(18)	(34)	(33)	(37)	(15)
80	17	61	22	32	13	37	10
139	73	149	131	99	29	103	50
135	57	100	57	135	18	83	17
21	38	70	33	19	9	26	12
69	243	50	14	48	0	28	7
132	6	80	53	66	14	38	16
0	1	0	0	0	3	1	0
91	47	150	70	51	20	47	20
2,511	732	2,327	1,145	1,473	610	1,077	546
18.8	12.2	32.8	21.0	13.7	8.7	2.8	22.7
54.7	22.0	38.8	45.9	55.7	73.9	63.7	53.1
26.6	65.7	28.4	33.2	30.5	16.9	33.4	24.2
0	0.1	0	0	0	0.5	0.1	0
2nd	3rd	2nd	2nd	2nd	2nd	7th	2nd

tween Rocklin and Auburn in the foothills of Placer County. Chinese farm laborers were present not only in areas with labor-intensive crops but in grain-growing areas as well: about 8 percent of the farm laborers in Colusa County, which ranked as California's foremost wheat-producing county in 1880, were Chinese.[2]

Chinese farm laborers never became as numerous in the drier San Joaquin Valley as they were in northern California, because irrigation works had to be built before intensive agriculture could be undertaken. By the time that irrigated agriculture became possible, the number of Chinese had declined because of

Chinese orange pickers, Santa Ana, Orange County, California, ca. 1895. (Photograph by Blanchard, courtesy of the California Historical Society, San Francisco)

Chinese orange pickers and packers, Riverside, California. (Photograph by Jarvis, courtesy of the California Historical Society, San Francisco)

their exclusion. Where Chinese farm laborers did show up in the San Joaquin Valley, they worked mainly as grape pickers in the Fresno and Visalia areas, which were planted with grapevines only in the 1880s.

In the counties shown in the tables, in both 1870 and 1880, next to the Chinese, native-born white Americans from the northeastern seaboard and the north central states were the second most numerous. Among European immigrants, Irish, Germans, and Portuguese predominated. There were as yet very few California-born farm laborers in 1880, but by 1900, Californians, many of whom were the children or younger relatives of farmers, had become the largest group of farm laborers in the state.

Since the 1890 manuscript schedules were lost in a fire, there is no information for that year. However, there are some figures gathered by the California Bureau of Labor Statistics which give an idea of how numerous Chinese farm laborers were in the mid 1880s. In 1886 the California Bureau of Labor Statistics received responses from twenty-four counties with information on Chinese farm laborers and they are shown in table 26. Excluding Mendocino and Yolo counties, where the 4,000 and 2,000 Chinese workers in the orchards, vineyards, and hopyards greatly exceeded the number of Chinese normally in those counties, the largest number of Chinese farm laborers were found in Los Angeles County, with 2,000; Sacramento County, with 1,700; Santa Clara County, with 1,600; San Joaquin County, with 900; and Colusa County, with 800. The Commissioner of Labor Statistics had sent out his letter requesting information from the county assessors in April 1886, but it is not known whether the numbers given by the assessors were for the peak harvest season or not. The figures seem reasonably accurate, however, because for the most part they are in line with those tallied from the 1880 and 1900 manuscript census schedules.

Table 27, with figures for 1900, shows that the absolute number of Chinese remained largest in those counties where they had been traditionally important—Sacramento, San Joaquin, and Santa Clara counties, with 1,165, 917, and 670. In addition, Chinese had entered the farm labor force in the San Joaquin Valley, particularly in Fresno and Kern counties, which had 403 and 473. Some Chinese also showed up along the central Califor-

*Table 26   Estimated Number of Chinese and Chinese Farm Laborers
in Selected Counties, 1886*

County	Estimated Total Number of Chinese	Estimated No. of Chinese in Orchards, Vineyards & Hopyards	Estimated No. of Chinese in Other Agricultural Work	Percent in Farm Work
Butte	3,000	100	150	8.3
Colusa	1,500	0	800	53.3
Contra Costa	500	75	125	40.0
El Dorado	400	200	15	53.8
Los Angeles	6,000	2,000	—	33.3
Marin	325	25	50	23.1
Mendocino[a]	1,000	4,000 [sic]	100	—
Monterey	500	0	50	10.0
Napa	650	200	—	30.8
Plumas	500	0	50	10.0
Sacramento	3,000	1,500	200	56.7
San Bernardino	500	100	—	20.0
San Joaquin	2,500	300	600	36.0
San Luis Obispo	150	0	20	13.3
Santa Barbara	500	0	100	20.0
Santa Clara	2,500	1,000	600	64.0
Siskiyou	1,800	0	50	2.8
Sonoma	1,500	200	0	13.3
Sutter	550	100	—	18.2
Tulare	1,000	125	125	25.0
Tuolumne	500	0	10	2.0
Ventura	300	50	150	66.7
Yolo[a]	400	2,000 [sic]	0	—
Yuba	2,000	500	0	25.0

SOURCE: *Second Biennial Report of the Bureau of Labor Statistics of California for the Years 1885–1886*, Sacramento: State Printing Office, 1886, pp. 53–54.
[a] Mendocino County produced hops, so presumably the 4,000 orchard, vineyard, and hopyard workers were seasonal laborers who came just during picking season. Yolo County produced fruit, so the 2,000 workers were probably seasonal laborers who came only during the harvest. A dash means no estimate was given, though there were certainly Chinese doing other agricultural work in these counties.

nia coast, mainly in the Salinas Valley in Monterey County, which had 411. Since the total number of farmworkers had increased dramatically, however, Chinese now made up smaller percentages of the total farm labor population than twenty years earlier. Only in Sacramento County did their numbers reach about one-quarter of the total. Elsewhere, they ranged from under 10 percent to about 20 percent. However, they still

Table 27   *Farm Laborers and Laborers in Farmers' Households by Nativity and County, 1900*

Place of Birth	Alameda	Butte	Contra Costa	Fresno	Kern	Monterey	Placer
CHINA	360	223	180	403	473	411	334
JAPAN	398	145	163	262	1	610	104
U.S.A.	1,069	887	786	3,848	841	974	693
California	(897)	(475)	(592)	(1,015)	(429)	(754)	(312)
Other Western States	(9)	(27)	(12)	(83)	(20)	(19)	(24)
New England Mid-Atlantic	(76)	(84)	(62)	(434)	(80)	(46)	(121)
North Central States	(78)	(238)	(105)	(1,755)	(228)	(115)	(174)
South Atlantic States	(3)	(15)	(5)	(160)	(24)	(15)	(22)
South Central States	(6)	(48)	(10)	(401)	(60)	(25)	(40)
ENGLAND/ SCOTLAND/ WALES	17	10	33	218	31	22	29
IRELAND	54	28	42	124	36	23	11
GERMANY	93	41	53	315	31	40	36
PORTUGAL	643	11	175	107	31	71	9
ITALY	62	9	39	232	47	11	12
CANADA	20	23	17	143	42	27	12
MEXICO	3	1	7	28	63	9	1
OTHER PLACES	124	27	93	740	57	133	33
TOTAL	2,843	1,405	1,588	6,420	1,653	2,331	1,274
% Chinese	12.7	15.9	11.3	6.3	28.6	17.6	26.2
% Japanese	14.0	10.3	10.3	4.1	0.1	26.2	8.2
% American	37.6	63.1	49.5	59.9	50.9	41.8	54.4
% European & other	35.6	10.6	28.5	29.3	16.6	14.0	11.1
% Mexican	0.1	0.1	0.4	0.4	3.8	0.4	0.1
Rank of Chinese	4th	2nd	2nd	2nd	2nd	3rd	2nd

SOURCE: Tallied and computed from U.S. National Archives, Record Group 29,"Census of U.S. Population" (manuscript), Alameda, Butte, Contra Costa, Fresno, Kern, Monterey, Placer, Sacramento, San Joaquin, Santa Clara, Solano, Sonoma, Sutter, Tehama, Yolo, and Yuba counties, 1900.

ranked second in Butte, Contra Costa, Fresno, Kern, Placer, Sacramento, San Joaquin, Santa Clara, Sutter, Tehama, and Yuba counties.

To compensate for the failure of the 1880 and 1900 censuses to include a sufficient number of seasonal laborers because they were taken in June alone, adjustment ratios were used to compute a set of hypothetical figures. In his study of the labor requirement of various crops, R. L. Adams, an agricultural economist working for the University of California's College of Agriculture, collected detailed statistics in the late 1930s on fluctuations

Sacramento	San Joaquin	Santa Clara	Solano	Sonoma	Sutter	Tehama	Yolo	Yuba
1,165	917	670	417	166	52	302	93	25
635	226	117	700	48	21	50	188	0
1,777	2,707	3,226	1,870	3,175	615	520	1,077	374
(924)	(1,282)	(1,385)	(968)	(1,504)	(348)	(235)	(593)	(246)
(25)	(42)	(59)	(43)	(91)	(10)	(34)	(29)	(7)
(245)	(422)	(671)	(275)	(454)	(44)	(61)	(102)	(34)
(455)	(730)	(940)	(456)	(885)	(166)	(160)	(273)	(72)
(47)	(105)	(79)	(51)	(80)	(27)	(5)	(24)	(1)
(81)	(126)	(92)	(77)	(161)	(20)	(25)	(56)	(14)
105	120	209	108	195	10	10	32	11
99	165	228	142	172	16	19	48	14
273	461	394	256	550	46	14	86	17
100	20	300	149	181	1	3	28	6
331	399	256	37	272	1	1	41	2
66	111	199	103	103	12	9	16	5
0	0	0	0	0	0	2	1	0
127	163	368	179	181	33	23	68	15
4,678	5,289	5,967	3,961	5,043	807	953	1,678	469
24.9	17.3	11.2	10.5	3.3	6.4	31.7	5.5	5.3
13.6	4.3	2.0	17.7	1.0	2.6	5.2	11.2	0
38.0	51.2	54.1	47.2	63.0	76.2	54.6	64.2	79.7
23.5	27.2	32.7	24.6	32.8	14.8	8.3	19.0	14.9
0	0	0	0	0	0	0.2	0.1	0
2nd	2nd	2nd	3rd	7th	2nd	2nd	3rd	2nd

in the number of seasonal agricultural workers on a month-by-month and county-by-county basis.[3] In table 28, Adams's figures have been used to compute an adjustment ratio for each county by taking the number of seasonal workers required in the peak harvest month over the number of seasonal workers required in June (column 3). This ratio was then used to compute the adjusted total number of farm laborers for the peak month for each county (column 4). The difference between the peak month and June is then calculated (column 5). Since contemporary observers and most scholars have claimed that Chinese made

Table 28   Maximum Possible Number and Maximum Possible Percentage of Chinese in the Total Farm Labor Supply, 1880

County	(1) Total No. in June	(2) Peak Month	(3) Adjustment Ratio[a]	(4) Adjusted Total	(5) Difference June/Peak	(6) No. Chinese in June	(7) Max. Possible No. of Chinese	(8) Max. Possible % of Chinese
Alameda	1,529	April	2.0	3,058	1,529	350	1,879	61.4
Colusa	2,135	Sept.	1.4	2,989	854	169	1,023	34.2
Contra Costa[b]	889	July	1.1	978	89	168	257	26.3
Los Angeles	1,156	June	1.0	1,156	0	24	24	2.1
Placer	669	Aug.	1.4	937	268	150	418	44.6
Sacramento[b]	1,419	Sept.	4.4	6,244	4,825	537	5,362	85.9
San Joaquin[b]	2,511	Oct.	1.3	3,264	753	471	1,224	37.5
San Mateo	732	June	1.0	732	0	89	89	12.2
Santa Clara	2,327	Sept.	1.5	3,491	1,164	763	1,927	55.2
Solano[b]	1,145	Aug.	1.4	1,603	458	240	698	43.5
Sonoma	1,473	Sept.	2.6	3,830	2,357	202	2,559	66.8

Tehama	610	July	1.6	976	366	53	419	42.9
Yolo[b]	1,083	Aug.	1.8	1,949	866	30	896	46.0
Yuba	546	Aug.	5.2	2,839	2,293	124	2,417	85.1
(Sacramento–San Joaquin Delta)	(1,978)	April	(1.1)	(2,176)	(198)	(981)	(1,179)	(54.2)

SOURCE: Computed from U.S. National Archives, Record Group 29, "Census of U.S. Population" (manuscript), 1880; and R. L. Adams, *Seasonal Labor Requirement for California Crops* (Berkeley: University of California, College of Agriculture, Agriculture Experiment Station, 1938), Bulletin 623, pp. 22–23.

a Ratio of number of seasonal workers needed in peak harvest month over number needed in June.

b Includes both Delta and non-Delta lands. For Delta lands, see the last row. (Figures for the Delta are the sums for Merritt Township, Yolo County; Rio Vista Township, Solano County; Township No. 5, Contra Costa County; Sutter, Franklin, and Georgiana Townships, Sacramento County; and Union Township and half of O'Neal Township, San Joaquin County.)

Col. 4 = Col. 1 × Col. 3
Col. 5 = Col. 4 – Col. 1
Col. 7 = Col. 5 + Col. 6
Col. 8 = Col. 7 ÷ Col. 4

up "most" of the seasonal work force, the maximum number of
Chinese was calculated by assuming that all the seasonal workers added to the labor force between June and the peak month
were Chinese. Column 7 is the sum of the number of Chinese
counted in June plus the differential between June and the peak
month. Column 8 then shows the maximum percentage that
Chinese might have been in the overall farm labor force in each
county. It appears that in 1880 in Sacramento and Yuba
counties, Chinese might indeed have constituted three-quarters
to seven-eighths of the farm labor force, as is claimed in some of
the writings that have been reviewed, but nowhere else could this
have been true. In most counties, the maximum possible percentage of Chinese was about 45 percent. There is some corrobora-
tive evidence that the hypothetical figures shown in column 8 are
accurate. In response to an inquiry by the Commissioner of
Labor Statistics, seven farmers said that between 1,500 and
3,000 Chinese were expected to work in the fruit harvest in
Alameda County in 1886, with 2,000 the most commonly cited
figure.[4] The figure 1,879 for Alameda County in column 8 is
therefore compatible.

Were there enough Chinese in these counties or the adjacent
areas for the hypothetical additional number of workers to be
recruited? Table 29, which shows farm laborers as a percentage
of the total Chinese population, indicates that in 1870, with a
single exception, farm laborers ranged from 3 percent to less
than 20 percent of the total Chinese population; in 1880 they
ranged from 5 to 28 percent; in 1900 they ranged from 4 to 52
percent in the major agricultural counties. In other words, the
majority of the Chinese in these counties were not farmworkers,
so it was entirely possible to recruit many Chinese who normally
had other occupations to work for a few weeks during the harvest
to supply the additional workers needed.

Was Varden Fuller justified, then, in arguing that the presence
of abundant and cheap Chinese farm labor helped to perpetuate large landholdings in California? At least one nineteenth-
century employer of Chinese farmworkers would not have
thought so. William W. Hollister, a large landowner, testified
before the 1877 congressional committee that "as a rule the
large farmers of California employ the least number of Chinamen. They are generally employed upon the smaller farms. You

Table 29   Farm Laborers as a Percentage of the Total Chinese
Population, 1870, 1880, and 1900

| | 1870 | | 1880 | | 1900 | |
County	Chinese Population	% Farm Laborers	Chinese Population	% Farm Laborers	Chinese Population	% Farm Laborers
Alameda	1,939	18.1	4,386	10.4	2,211	16.3
Butte					712	31.3
Colusa			970	17.4		
Contra Costa			732	23.0	627	28.7
Fresno					1,775	22.7
Kern					906	52.2
Monterey					857	48.0
Placer			2,190	6.8	1,050	31.8
Sacramento	3,195	19.5	4,892	11.0	3,254	35.8
San Joaquin	1,626	3.1	1,997	23.6	1,875	48.9
San Mateo	519	34.7	596	14.9		
Santa Clara	1,525	9.5	2,695	28.3	1,738	38.6
Solano	920	17.7	993	24.2	903	46.2
Sonoma	473	8.9	904	22.3	599	27.7
Sutter					274	19.0
Tehama			774	6.8	729	41.4
Yolo			608	4.9	346	26.9
Yuba			2,146	5.8	719	3.5

SOURCE: Computed from U.S. Bureau of the Census, *Census of U.S. Population*, 1870, 1880, and 1900; and U.S. National Archives, Record Group 29, "Census of U.S. Population" (manuscript), Alameda, Butte, Colusa, Contra Costa, Fresno, Kern, Monterey, Placer, Sacramento, San Joaquin, San Mateo, Santa Clara, Solano, Sonoma, Sutter, Tehama, Yolo, and Yuba counties, 1870, 1880, and 1900.

can understand that very readily, because the large farms require teamsters, men who can drive teams, plow, and harrow, and all that kind of work, which the Chinamen cannot do."[5] To explore Fuller's thesis further, many factors other than the Chinese presence must be taken into consideration.

Chinese Farm Laborers and Land Monopoly

That the Chinese were significant in the farm labor supply is obvious, but it is less certain that their presence was the key or the sole factor that enabled landowners to change from extensive grain cultivation to labor-intensive agriculture and to do so without subdividing their land. As this study has repeatedly shown, there is no question that the Chinese played a crucial role in converting land to more intensive usage, but they did so not simply as harvest laborers but as tenants. The point to understand is that when landowners leased out only part of their farms to Chinese tenants, they were in fact subdividing their land into lots of manageable size. So it is not true that subdivision did not take place, as Fuller argued; it happened that in California subdividing the land did not necessarily lead to its sale. There is little evidence to support the argument that the landowners would have sold such subdivided plots to their tenants had these been white and not Chinese. Thus tenancy, more than the availability of Chinese labor, enabled the large landholders to keep what they owned. And this situation was not peculiar only to California.[6]

Furthermore, the effect that the transition to fruit culture had on land values was less definitive than Fuller had believed. Fuller had correctly pointed out that land values were (and are) dependent on many factors, some of which are unrelated to agriculture per se, and because of this, he thought it pointless to attempt to measure how different crops had affected the value of the land on which they were planted.[7] Yet he did not hesitate to assert that the use of Chinese labor had "capitalized" land values to such an extent that easterners and immigrants could not afford to buy land in California. He believed this to be true even though he could not prove empirically that those landowners whose farms had increased most in value were indeed employers of Chinese labor.

Data on the relationship between crop pattern and farm value, as well as the relationship between crop pattern and product value, can be used to evaluate Fuller's thesis further. The manuscript schedules of the 1860, 1870, and 1880 censuses of agriculture contained detailed information about individual farms which can be used for this purpose. Since few farms grew fruit in 1860, and few Chinese worked as farm laborers then, the 1860 census will not be considered. The relationships between land use (or crop pattern) and farm value, and between crop pattern and total product value, for 1870 and 1880 have already been examined for Sutter, Franklin, Georgiana, and Union townships in the Sacramento–San Joaquin Delta. It was shown that there was

little significant difference between diversified and all-grain
farms in terms of both sets of relationships, and that potatoes
and beans, rather than fruit and vegetables, brought higher returns in the 1870s and 1880s in the Delta. To investigate this
question further, Sonoma and Analy townships in Sonoma
County, Santa Clara and San Jose townships in Santa Clara
County, and Vacaville Township in Solano County will also be
analyzed because these were the most rapidly developing fruit
districts in California in those decades.[8]

In 1870 in Sonoma Township, 113 farms were already growing fruit along with cereals, and only 37 remained purely grain
farms. Yet the distribution curves for the farm value of the two
kinds of farms did not differ significantly. That is to say, there
was not a larger percentage of farms containing orchards that
were valued more than grain farms. Similarly, in the other
townships for both 1870 and 1880, where fruit and vegetable
cultivation did push farm values somewhat toward the higher
end of the scale, differences in the pattern of farm value distribution between the farms with orchards and those without were
only moderate. Only in San Jose Township was there a noticeable divergence in farm values between the two kinds of farms.
This divergence, however, might have been due as much to the
township's proximity to the growing city of San Jose as to the
intensification of land use—a situation similar to the pattern
found in Sutter Township just south of Sacramento City.

Though fruit obviously was worth more than grain, most of
the early orchards did not seem to be great economic successes.
Correlating crop pattern with the total value of products showed
that those farms producing fruit indeed received slightly larger
revenues, but once again, differences were not very significant. It
cannot be claimed, as Fuller did, that fruit cultivation increased
the total value of a farm's products so dramatically that land
became too expensive for prospective immigrants to buy, thereby
impeding their settlement in California. However, in fairness to
Fuller, these findings pertain only to the early years of the transition from extensive to intensive agriculture. Since no schedules of
the agricultural censuses taken after 1880 have been released by
the U.S. Bureau of the Census, it is not possible to make similar
computations for later years when fruit cultivation became truly
important in California. In any event, in the later period, when

the number of Chinese in the farm labor force was declining, even if land values had shot up dramatically, it is difficult to attribute such changes to their presence.

There are some accounts that did tout the fabulous sums to be earned from fruit, but they did not represent average conditions, since they had been cited as success stories. In 1883, for example, the *Pacific Rural Press* reported that Judge W. C. Blackwell of Hayward had earned $1,000 from each acre of apricots in his orchard in Alameda County, A. T. Hatch had got $800 from each acre of currants from his vineyard in Solano County, one "O. G." had received $1,500 per acre from Bartlett pears in Solano County, Henderson had made $400 per acre from raisins in Riverside County, Wolfskill hauled in $2,000 per acre from oranges in Los Angeles County, Thomas had earned $1,280 from each acre of peaches in Tulare County, and Barter had received $1,440 from each acre of peaches in Placer County.[9] In contrast, good wheatland was producing only about 20 bushels per acre at the time, and wheat was selling for $1.75 per hundred pounds, bringing the wheat farmer an income of only $31.50 per acre (20 bushels = 1,800 pounds).[10] However, the examples cited were the exceptions rather than the rule: the manuscript schedules of the 1880 census of agriculture showed that such large earnings were very rare.

To get a satisfactory answer to the question of how the presence of Chinese farmworkers might have affected farm size in California, it would be necessary to compare those farms employing Chinese with those farms that did not. Unfortunately, such a comparison cannot be made, because in every county investigated, few Chinese farm laborers lived in the households of white farmers, so there is no sure way to tell which farmers used Chinese labor and which did not. The best that can be done is to see which size farms used the most labor, and whether farms growing labor-intensive crops used more labor than those growing grain and hay.

In each of the five townships examined, large farms did use more labor, as measured by the total amount of wages paid. However, it was not the smallest farms (those under 40 acres) but medium-sized ones (those with 40 to 160 acres) that used the least labor. Size seemed to have little relationship to the amount of wages paid for hired help. Moreover, a significant proportion of farms in all size categories paid no wages for hired help. Of those farmers paying wages, the majority of the ones owning less than 320 acres spent under $500 a year. In most instances, farm operators—whether owner-operators or tenants—paid only $100 to $200 in wages. The 1870 manuscript agriculture census did not contain information on the number of man-weeks of labor hired, but the 1880 census did, and it was found that on the average, hired hands received $6 to $7 a week, so $200 would have bought approximately thirty man-weeks of labor. If it is assumed that the harvest period was one month, the average farmer hired perhaps seven persons to help him. Or if he kept only one hired man, that person could have been employed for thirty weeks of the year. For those who spent more, $500 would have bought between seventy and eighty man-weeks of labor. In those instances, some twenty persons might have been employed for a month, or one person for the entire year plus another for half a year.

The data are inconclusive with regard to what kinds of farmers employed more labor. Surprisingly, in all five townships, farmers producing fruit spent about the same amount in wages as grain farmers. Regardless of what crops were grown, a certain number of individuals owning farms in each size category used hired help. What the analysis revealed was that in California, regardless of the crops grown, 30 to 50 percent of the farmers employed hired hands for part of each year. Thus, the available evidence indicates that the causal relationship among Chinese "cheap labor," labor-intensive crops, and "land monopoly" that Fuller and others have posited simply was not as clear-cut as they would have their readers believe.

Other Factors Leading to Large Farms

Though Chinese tenants and farm laborers did enable farm owners to convert their land to intensive use, which in turn raised farm and product values to some degree, the high cost of land is not the only reason that farms have been big in California. There were other important forces that led to bigness. The cost of land reclamation and irrigation, the need to overcome the problems posed by distance from markets, changing consumer preferences, and the general temper of the times that promoted the values of a capitalist society, all had an influence on the size of California farms.

With the exception of certain regions along the coast and in the northern part of the Sacramento Valley, intensive agriculture in California has not been possible without irrigation. Like reclamation projects, irrigation works were expensive, requiring large capital investments, so farmers who desired to engage in irrigated intensive agriculture had to find financial capital. Farmers encountered great difficulty in this regard because in the first three decades of statehood, interest rates in California, particularly for nonmanufacturing ventures, were very high. Irrigation projects could be built only by big capitalists, or on a cooperative plan, or with government aid. Up to 1887 all irrigation developments in California were privately financed. The passage of the Wright Act in 1887 marked an important milestone in the development of irrigated agriculture.[11] Under the terms of the act, landowners could join together to form irrigation districts and to pool their resources for the construction of dams, canals, ditches, head gates, and so forth. After 1877 irrigated acreage in California increased at an extraordinary rate. In 1880 it was estimated that there were some 350,000 acres of irrigated land in the state, of which 292,885 acres were in Los Angeles and San Bernardino counties and in the Central Valley. In 1890 there were 1,004,233 irrigated acres; in 1900 there were 1,446,114; by 1910 there were 2,664,104.[12]

The movement to irrigate had all the fervor of a moral crusade. For example, William E. Smythe, author of *The Conquest of Arid America*, argued that irrigation had social and moral consequences. He believed that irrigation would make the establishment of small farms possible:

> The essence of the industrial life which springs from irrigation is its democracy ... irrigation entails expense, either of money or of labor ... where water is the foundation of prosperity it becomes a precious thing ... men cannot acquire as much irrigated land, even from the public domain, as they could acquire where irrigation was unnecessary. It is not only more difficult to acquire in large bodies, but yet more difficult to retain ... Only the small farm pays.[13]

But Smythe's prophetic vision did not materialize. Irrigation did little to break up large landholdings because once water was made available, it became more, rather than less, possible to cultivate large tracts. Before irrigation, the lack of water frequently had constrained the size of farms; after irrigation was introduced, tracts of land may indeed have increased greatly in value, but this did not mean that owners were eager to sell, because now they could use their irrigated fields for profitable intensive agriculture. Since irrigation and reclamation projects both cost so much, small farmers who could not afford to build their own irrigation works were at a disadvantage. Even the pooling of resources sometimes did not provide sufficient funds. Only when government funds became available for irrigation projects were small farmers theoretically placed on the same footing as large ones, but by then large landowners in California had acquired so much political power that they could ignore or circumvent acreage limitation laws without fear of reprisal.[14] Irrigated agriculture, therefore, actually helped to perpetuate rather than break up the holding of large tracts even though its early promoters had hoped it would make family farming possible.

California's geographic isolation has also affected its landholding pattern. From the 1870s onward, California has specialized in growing and exporting highly perishable fruit and vegetables, only a small part of which is consumed locally. To sell such products in a distant national, indeed, world, market, improved methods of preservation, transportation, and market control were needed. Improvements in canning technology came at about the same period when the transcontinental railroad finally became available to farmers as a feasible means of transportation. (Though the railroad was completed in 1869, it was another decade before bulk export of fresh produce was possible, when refrigerated railroad cars were introduced and freight rates were reduced.) Large growers were more able than small farmers to take advantage of these changes. At the same time, the need to find economies of scale in marketing also propelled California agriculture toward bigness. California has been a pioneer in organizing marketing cooperatives that control national and worldwide markets through the use of sophisticated ways to gather and relay market information and to adjust supplies instantaneously in response to fluctuations in demand in various

places as the fresh produce is being moved across the country.
For example, several carloads of lettuce originally destined for
Chicago may be rerouted to St. Louis if it is discovered that the
Chicago market will be glutted on the day the lettuce is scheduled
to arrive there. The first truly successful fruit exchange was
developed by citrus growers. Subsequently, similar organizations have been established for a wide variety of other crops,
each set up to cater to special marketing needs.[15]

Aside from breakthroughs in food processing, transportation
technology, and marketing systems, urbanization and the emergence of a wage-earning middle class have also helped to expand
the market for California's specialty crops. Until rather recently,
fresh fruit and vegetables (unless they were grown on the family
farm) were considered to be luxury items and were not part of
the common people's everyday diet. But with rapid urbanization
and the rise of a middle class of consumers who possessed purchasing power, fresh and preserved fruit and vegetables began to
be eaten more regularly. In the last quarter of the nineteenth
century, the urban population of the United States increased
twice as rapidly as the rural one, with immigrants accounting for
about 40 percent of the overall increase. Eastern and midwestern
urban wage-earners who did not grow their own food became
consumers of California's agricultural products. Consumers
around the world also soon learned to appreciate the same superior quality, so that today in some remote areas of Asia, Africa,
or Latin America, California brand names such as "S and W,"
"Del Monte," and "Hunts"—besides Coca Cola—may be
the only English words that non-English-speaking villagers
know.

Finally, as Cletus Daniel has argued so persuasively, the
rise of the capitalist ethic helped California growers to justify
the pattern of their operations. No longer constrained by old-
fashioned agrarian idealism, they experienced few qualms in becoming successful businessmen, and California never became a
land of family farms. But their critics did not approve. Although
reformers such as Henry George have condemned monopolies
for inhibiting democracy, novelists such as Frank Norris have
dramatized the rapacity of railroads, and Progressive politicians
such as Hiram Johnson have won office by promising to combat
the Southern Pacific Railroad's stranglehold on California,

monopolies have persisted in California. The tension inherent in a society that attempts to be democratic, and yet protects the right of individuals to make as much profit as possible within a capitalist economic system, is by no means a uniquely California phenomenon, but the fact that monopolies existed in its rural sector made them more difficult to accept, given the hold that agrarian ideals still retained. In the late nineteenth century, mergers, trusts, and monopolies were evils that were associated with the nation's burgeoning urban industries, so that amid the widening social divisions created by industrial growth, it was still possible for beleaguered city dwellers to dream of a simpler, cleaner life in a bucolic countryside. But in California where factories sprang up in the fields, no such idyllic vision was possible. In defense of themselves, large landowners and growers have argued that large landholdings were part of the natural—or at least the inherited—order of things. By invoking the sanction of natural law, they have tried to excuse themselves from blame and have tried to demonstrate the need to continue using migrant farm laborers under oppressive conditions because they claim that the cost of labor is the only one they can control. Besides, they feel that by organizing the state's agriculture for maximum profit, they have only been good American entrepreneurs, and entrepreneurship, it is said, is surely as important as democracy and equality. The familiar cliché that farming is a business and not a way of life in California misses the point: agribusiness *is* a way of life. In the eyes of the large growers, it is and should be the California way. Ironically, it is not growers but liberal scholars who have held the Chinese responsible for the emergence of "undemocratic" social relations in California agriculture.

Cheapness and Docility

All the available scattered evidence showed that the Chinese did indeed receive lower wages than others doing comparable work. But whether such discrimination against them was the result of their "willingness to accept low wages," and whether such "willingness" was the culmination of "four thousand years of enslavement"—as many contemporary observers alleged—are different questions altogether.

During the last three decades of the nineteenth century, employers seemed to believe that approximately $1 a day (without board) was the appropriate wage for Chinese workers regardless of what they did. As a rule, Chinese laborers worked six days a week or twenty-six days a month. The 1872 *Transactions of the California State Agricultural Society* stated that Chinese were receiving $22 a month during the winter season and $30 a month during the summer, while white workers were being paid $30 a month during the winter and between $40 and $50 a month in the summer. In addition, white laborers generally received board.[16] White workers, therefore, were paid at least one and a half times as much as their Chinese peers. Moreover, Chinese farm wages remained static. The 1884 report of the California Bureau of Labor Statistics stated that Chinese farm laborers were paid $20 to $25 a month with board, and calculated that Chinese board was usually worth $5 a month.[17] So those without board persumably received about $25 to $30 a month. Only after the turn of the century did Chinese wages seem to exceed $1 a day.

No doubt the Chinese knew they were being paid less than white workers, but it is difficult to tell in retrospect whether they considered themselves "cheap labor." Those who lived frugally and did not gamble could save a third to half of their wages. The Bureau of Labor Statistics estimated that Chinese paid a monthly rent of $2 to $4 per person and spent $10 to $12 a year (or an average of $1 a month) for clothing.[18] The Bureau did not state how much the Chinese farm laborers might have spent on opium, prostitutes, and gambling. If it is assumed that the average man spent perhaps $2 a month for such recreation and that he had no other expenses, then a Chinese farm laborer could save from $8 to $15 a month in the 1870s and 1880s. Such amounts, or even a fraction of them, when sent to China, represented considerable sums. Those who charged that the Chinese were able to accept low wages because they had no families in the United States to support had a point: low wages earned in California translated into good pay in Kwangtung, where the cost of living was much lower. By maintaining trans-Pacific families, with the wage earners on one side of the Pacific and

Chinese gathering mulberry leaves, Curtis Silk Farms, Los Angeles, California, 1907. (Photograph courtesy of the California Historical Society, San Francisco)

their dependents on another, the cost of what Marxists call the reproduction of labor was greatly reduced. Thus California's employers reaped the benefits of having laborers who came full grown, the cost of whose nurturance had been borne by families and villagers in China and not by communities in America.

California farmers had pointed out to their critics that although the Chinese were paid less, they were still making more than what white farm laborers in other parts of the country earned. According to information collected by the Division of Statistics of the United States Department of Agriculture, the wages of farm laborers fell continuously from the peak reached in the late 1860s until the end of the century, but throughout this period those in California earned higher wages than their counterparts elsewhere. In 1869 white farm laborers earned $46.38 a month in California; they earned $32.08 in the eastern states, $28.02 in the midwestern states, $17.21 in the southern states, and $27.01 in the western states (excepting California). In 1875 California farmworkers earned $44.50 per month, while those in the eastern, midwestern, southern, and western states earned $28.96, $26.02, $16.22, and $23.30, respectively. In 1879 Californians earned $41.00 a month, whereas those in the same four regions earned $20.21, $19.69, $13.31, and $20.38, respectively. Though California farm laborers also got paid less, the decline in their wages was far smaller than declines in other regions. The figures for 1882 were $38.25 for California, $26.61 for the East, $22.24 for the Midwest, $15.30 for the South, and $23.63 for the West.[19] The numbers most likely represented the average wages of farmhands who were employed year round. It is obvious that Chinese farmworkers in California who earned about $26.00 a month were doing as well as white farm laborers in the eastern states, and better than people in the other three regions, though in making these comparisons, it must be kept in mind that the cost of living was higher in California.

Though their wages were unquestionably lower than those of whites, there was a "floor" below which Chinese farm laborers would not go. The *San Francisco Post* of October 12, 1877, published an article that reported the views of a farmer's wife whose husband preferred not to employ Chinese laborers except when he had to during the height of the fruit season. She explained how the Chinese had come to dominate farm work in the Sacramento Delta:

One Chinaman rents a place; he hires two or three to help him. Then a lot more hang on to them, and the Chinamen that came there first won't let the new ones work at less than a dollar a day; they would rather board them until they get work at that price. And that's the way a whole colony of Chinese is brought on a place by one man leasing to them.... White men are getting $1.50 a day. There are lots of white men in the state, but they are driven away from here by the Chinese. If the Chinese were not here, white men would be; but the Chinese are here, so the white men can't be. Our fruit will rot; we must have help to pick it at once, and so we are compelled to hire Chinese at that time, against our will, because our neighbors lease to them when they have no need.[20]

Thus, it seems that the Chinese had a firm idea of what was an acceptable wage and saw to it that their countrymen did not violate it. All things considered, the Chinese were "cheap" only relatively speaking.

Neither were Chinese farm laborers as docile as they were said to be. Stuart Jamieson, in his study of labor unionism in American agriculture, claimed:

Labor unionism and organized conflict in a primitive form first appeared on a significant scale in rural California when the Chinese became an important part of the agricultural labor supply.... As a racial minority excluded from other industries and subject to considerable intimidation from the white community, the Chinese in agriculture were not in a position to organize unions for collective bargaining.... However, the Chinese developed an indigenous form of labor organization which they transplanted to rural areas. Quite early, in San Francisco and other urban centers, they had formed native "brotherhoods" or "protective associations" known as tongs. The California Bureau of Labor Statistics ... described these as a type of "trade-unions" [*sic*] which are "rarely heard of, but nevertheless exist and are very powerful. In case of a strike or boycott they are fierce and determined in their action, making a bitter and prolonged fight." ... The tong became, instead of a labor union, a type of employment agency which facilitated the recruiting and hiring of Chinese for seasonal jobs requiring considerable mobility. It was a forerunner of the labor contractor system.... The system constituted a type of collective bargaining in a semi-union form of organization.[21]

Jamieson did not document his assertion properly, but it was highly plausible that the "system" he described existed, given the fact that Chinese guilds (in America as well as China) effectively set standards of acceptable workmanship, wages, and prices, and carved out territory for each enterprise in a particular line of business.[22]

Though the existing literature is replete with descriptions of Chinese laborers being "reliable," "industrious," "sober," "docile," "servile," and "plodding," there *are* records of Chinese farmworkers going on strike and engaging in other acts of resistance. The *San Francisco Bulletin* of August 18, 1880, reported one such event. Apparently, Chinese pickers in Santa Clara County had been demanding

> increased compensation from the value of one-half of the fruit to two-thirds of it. When the trouble first became manifest those not affected thought it was some misunderstanding between one or two growers and their men, but the affection spread until nearly all were involved, and gangs of Chinamen were striking everywhere. Where any gang had left it was found impossible to secure other Mongolians to replace them, and the growers were compelled to treat with their old hands before they could get any at all.[23]

Another instance of Chinese going on strike occurred in 1884, when a "large force" of Chinese hop-pickers reportedly struck on the Haggin Grant in Kern County. The farmers tried to replace them with blacks, but found the latter inexperienced.[24]

After Chinese exclusion went into effect in the latter part of 1882, the Chinese quickly recognized their improved bargaining position. As early as 1883, a writer in the *Pacific Rural Press* commented on the changed behavior of the Chinese who were

> demanding higher pay and doing less work. The railroads and factories have taken the great surplus, leaving but a small minority for house servants and field hands, and those seem to be well posted on the short supply of white labor as

well as their own; and as house servants they have become
absolutely unbearable, asking all the impertinent questions
usually asked by an astute servant girl. They fully realize
the advantages they hold under the operation of the exclusion act in checking any further supply.[25]

In 1884 it was said that Chinese farm laborers, who realized they
were in demand, were holding themselves "at high rates."[26] By
1890 a Chinese "labor union" was reportedly making a "$1.50
per day demand for work in the orchards and vineyards."[27] The
1900 report of the Bureau of Labor Statistics observed:

> Relieved, by the operation of the Exclusion Acts, in great
> measure from the pressing competition of his fellow-
> countrymen, the Chinese worker was not slow to take
> advantage of circumstances and demand in exchange for his
> labor a higher price, and, as time went on, even becoming
> Americanized to the extent of enforcing such demands in
> some cases through the medium of labor organization.[28]

A few other accounts revealed that Chinese farm laborers
were concerned not just with raising their wages but with their
rights as workers. For example, the *Pacific Rural Press* of August
28, 1884, reported that some Chinese hop-pickers had begun to
demand an advance for their labor.[29] Had this not been a
departure from expected behavior, no comment would have
been made. An item in the *Wheatland Graphic* of March 27,
1886, claimed that the Chinese had actually introduced the
boycott into California. One incident recounted was of

> a French hop raiser of Sacramento [who had] hired his pick-
> ers through a Chinese labor contractor, but being in a hurry
> to get done took a few men from the outside. A neighboring
> hop raiser who had got done before Mouton was going to
> help his friend dry, but the Chinese refused to work so long
> as he dried any hops picked by the outsiders.[30]

Such behavior would seem to indicate that if the Chinese had
been "docile," it was only because they realized they possessed
little power, and their docility was but a well-considered response to the situation they found themselves in rather than
an inbred quality they could not transcend.

Other Agricultural Work

Picking fruit and hops was by no means the only farm labor Chinese engaged in. They were also in demand as tree pruners, fruit packers, and cannery workers. All of these jobs require experience and skill. While many opponents and some supporters of Chinese immigration emphasized that they were cheap, experienced fruit growers stressed that the Chinese were invaluable because they were skilled. Only one year after Chinese exclusion went into effect, growers were already nostalgic for the days when Chinese could be "ordered" in gangs and "require[d] no further coaching."[31] Earl Ramey, a Yuba County historian, recalled that orchardists in the area all longed for the days when Chinese pruners were available because "they were expert at it.... They were so intelligent and smart and could catch a trade within a few weeks."[32] Even today, a few aged Chinese living in the Sacramento Delta, who are for all intents and purposes retired, are beseeched by orchard owners to work in the winter as pruners.[33] How well fruit trees are pruned greatly affects the quality of fruit they produce, and it is understandable that growers want to employ pruners who know how different varieties of trees at particular stages of their life cycles should be pruned.

Chinese had also made a reputation for themselves as fruit packers. It was said that

> in the matter of picking and packing fruit it seems to be with difficulty that we can find any desirable white help who will do this as satisfactorily to the consumer as the trained Chinamen, who, by tact peculiar to themselves mainly, seem to have reduced it to a science, and those versed in the marketing of fruits know how useless it is to send the same to market unless the greatest care has been exercised in the picking and packing of the same.[34]

The Chinese perfected a special way of packing known as the "China pack." One of the first deciduous fruits to be grown successfully on a commercial scale was the Bartlett pear. Being softer than many other fruits, Bartlett pears have to be packed with great care to avoid being bruised. One orchard owner in the Newcastle-Penryn area reminisced that the "China pack" was a work of art. "Every wrapper was smooth; not a wrinkle, and the tissue triangled to a point on top so that when the box was opened it was something to display in a grocer's window."[35] The Chinese fruit packers also made their own boxes: "from among these workers there always developed a lightning fast fruitbox maker."[36] The pears so wrapped and boxed reached their destinations uninjured and were attractive to behold, besides. The skill of Chinese fruit-packers proved once again that work performed by Chinese immigrants was reliable work. Regardless of the actual wages paid, then, Chinese labor was cheaper because losses due to careless work were minimized.

Only a part of California's fruit crop was packed, shipped, and sold fresh. Much of the rest was either dried or canned, which greatly increased the market for California fruit and vegetables because the fresh produce that could not be consumed or sold during the season was thus preserved. Waste was drastically reduced, and the canned products could be transported easily. The Cutting Fruit Packing Company established the first cannery in San Francisco in 1857.[37] In 1876 the pack of canned goods in California was 270,833 cases; by 1888 it had increased to 1,223,000 cases. (There are 24 cans to the case.) The largest volume of canned goods was shipped out of San Francisco, followed by San Jose, Sacramento, and Marysville. Canned goods were shipped by sea and by rail to every part of the United States, Europe, and Australia.[38]

The work crews in canneries were cosmopolitan, but there was generally a division of labor, with white men dominating the "floor work," which involved cooking the fruit or vegetables in huge vats and carting the cans to the warehouses, white women doing the pitting, cutting, and paring, and Chinese soldering the cans. Until the 1920s Chinese who worked in canneries were men. The 1880 manuscript census counted 150 Chinese workers in the Oakland factory of A. Lusk and Company, which, together with its San Francisco operation, was the largest cannery in the state at that time. Unfortunately, little is known about these workers other than what the census provided: 22 of them were under twenty years of age, 75 were in their twenties, 34 were in their thirties, 13 were between forty and sixty, and 2 were over sixty. Sixty-nine of the workers were married, though none had his wife living with him. These 150 cannery workers lived in several households, and their food was prepared by nine

Chinese cooks.[39] In 1884 Chinese were reported making cans for
the Sutter Canning Company and the Marysville Cannery.[40]

The 1890 report of the Bureau of Labor Statistics provided
information on twenty canneries in different parts of the state.
Chinese were found to be working in fourteen of them. Canneries
employing the most Chinese were the A. Lusk and Company's
plant in San Francisco, which had fifty-three; and the Martinez
Fruit and Canning Company, with sixty, the Carquinez Packing
Company in Benicia, with sixty, and the Black Diamond Canning Company with fifty, all in Contra Costa county. In all but
one place, the Chinese received the second highest pay among
the workers. White men received between $1.75 and $2.75 a day,
Chinese received between $1.00 and $1.67 a day, while white
women and girls received between $0.90 and $1.33 a day. Some
canneries paid by the day, some by the week, and others by the
month. A few used piece rates.[41]

By 1909 there were few Chinese workers left in canneries in
the San Francisco Bay Area, but many were found in canneries
located in the Sacramento–San Joaquin Delta, where they
earned between $1.50 and $4.00 a day, with the majority receiving
about $2.50.[42] Canneries in the Delta did not practice the same
ethnic and sexual division of labor that canneries elsewhere did.
Here, Chinese workers were employed in many job categories,
including handling heavy equipment—work that many employers normally reserved for white men. The 1910 manuscript
census listed three groups of Chinese cannery workers in the
Delta, one large contingent in Georgiana Township, and two
smaller ones on Roberts Island in O'Neal Township. Since the
census in this district was taken in the month of May, there might
have been other canneries employing Chinese, which had not yet
begun operation. Though the information from the manuscript
schedules of the census of population is skimpy, at least it
provides a glimpse into the composition of the households of
cannery workers.

In Georgiana Township, in a single boardinghouse run by
a fifty-five-year old immigrant, Leong Sam, there lived 140
asparagus-cannery workers served by three cooks, three waiters,
one dishwasher, and a laborer doing odd jobs. The group was
composed of 42 sorters, 67 canners, 7 traymen, 4 truckers, 4
solderers, 3 can inspectors, 4 steamers, 2 oven men, and 4 dis-

tributors. They were supervised by two young foremen, Leong
Sing, a twenty-four-year-old immigrant, and Chang Tom, a
nineteen-year-old Chinese born in California. A bookkeeper,
Leong Yung, kept accounts for the group. In this establishment,
the workers were relatively young recent immigrants, 28 percent
of whom were men in their teens and twenties. The majority (54
percent) were between the ages of thirty and fifty. Of the 126
who had been born in China, 75 had arrived in the United States
after 1900, but it is not known under what auspices they came.
Only 9 had come before the 1882 exclusion law went into effect,
with 13 arriving during the remainder of the 1880s decade, and
29 coming in the 1890s. Forty of the workers (29 percent) could
not speak English; among these, 34 were post-1900 immigrants.
With only one exception, the 14 workers born in California were
young; 10 were still in their teens, with fourteen-year-old Leong
Chip being the youngest in the group.

Age and the ability to speak English seemed to have little relationship to the kind of job assigned, although most of the native
teenagers and over 30 percent of the non-English speakers
worked as sorters. The literacy rate was high among the group,
only 10 of whom could not read and write. Most of them were
employed throughout the year, with 107 reporting no period of
unemployment, 16 being out of work for fewer than twenty weeks
during the preceding year, 9 unemployed for twenty to twenty-
nine weeks, and 6 for thirty or more weeks. (Two provided no
information to the census taker.) It was perhaps fortunate that
most of the men were steadily employed, for 86 of them were
married, 67 for more than ten years. Unlike Chinese listed in
other areas in other years, those in the Delta in 1910 were, for the
most part, shown with both their family and their given names.
There were seventeen Leongs, thirteen Wongs, and thirteen Lees
among the men, which suggests that many of them might have
been relatives or at least village mates.

The workers in the two canneries on Roberts Island were
much older immigrants who had been in the country for decades.
In one group of 76 men who boarded with Ah Ding, a sixty-two-
year-old farm laborer, there were 57 men (75 percent) who were
over fifty, 23 of whom were in their fifties, 31 in their sixties, and
3 in their seventies. Only one member of this group was a California native. Of the 75 immigrants, 1, Ho Tom, aged seventy-

seven, had arrived in 1859, 8 had come in the 1860s, 37 in 1870s, 17 in the 1880s, 8 in the 1890s, and only 4 after 1900. In all, 79 percent of the men had immigrated before 1882. Perhaps as a reflection of their long residence in the United States, only 12 percent could not speak English, although 29 percent were illiterate. A single cook, twenty-eight-year-old Yum Lee, born in California, served the entire group, all of whom were listed as single. (This was probably an error made by the census taker, who must have assumed that since they lived with no wives, they were single.) No information was provided on the number of months each individual in this household was unemployed. The job categories in which they were employed were less varied: there were 14 truckers, 36 canners, and 26 sorters.

The third cannery with Chinese workers, which also processed asparagus, had 47 workers who were also very old, with 15 in their fifties, 15 in their sixties, and 7 in their seventies. There were only 4 native Californians among them, all quite young. Except for 7 men, the other immigrants had all come before 1882, which may account for the fact that only 4 of them could not speak English, while the illiteracy rate among them was 23 percent. This group had 10 laborers, 22 canners, 8 sorters, 1 peeler (a fourteen-year-old boy born in California), 3 truckers, and 3 "general helpers." Like the second group, all of them were listed as single, though again this was probably an erroneous assumption on the part of the census taker. Likewise, no information was given on the period of unemployment that members of this group experienced.

That there were so many older workers suggests that while cannery work was by no means pleasant, it was nevertheless preferable to migratory farm labor. Working at a rapid pace for as many as eighteen to twenty hours a day, standing in damp surroundings, breathing air made foul by the odor of rotting fruit and vegetable peelings, cannery workers had to possess real stamina. Men in their seventies able to withstand such strain must have been individuals toughened by a lifetime of hard labor. Paying relatively better wages than farm work, canneries were able to attract Chinese workers even in a period when relatively few remained available.

Chinese agricultural laborers not only were skilled and enterprising but were surprisingly adaptable. In the semiarid San

Joaquin Valley, a small number became shepherds, cowboys, and irrigators. Intensive irrigated agriculture did not really develop there until the late 1880s and 1890s. Till then, most agricultural land was used for ranching, and a few Chinese found employment in the cattle industry. The 1880 manuscript census listed 7 Chinese sheepherders and 27 cattle herders in Merced County, 9 shepherds in Tulare County, and 6 shepherds in Kern County. Chinese cowboys? Yes, indeed! Their ages ranged from twenty to almost sixty, with the majority in their forties and fifties. If only more could be known about them!

When irrigation works were built, landowners needed workers who could man them. Not only did the correct amount of water have to be turned on or off into particular fields at the right time, but the irrigators had to keep an eye out for gophers and other rodents that bore holes in the levees, causing breaks. There is an account of at least one Chinese who was employed for many years as an irrigator in Kern County. Hugh S. Jewett, father-in-law of J. S. DiGiorgio, one of the most prominent fruit growers and wine producers in California today, himself was a prominent landowner and banker in the Bakersfield area, and had a Chinese male nurse named Hop when he was a baby. Jewett's father and uncle had come to Bakersfield in 1865 to settle, and in 1874 they had organized the Kern Valley Bank. When Hugh was about two years old, the family decided to move to San Francisco, but kept much of their land and business enterprises in Kern County. Hop begged to be taken along to the city because he had "evidently developed a great affection and devotion" to the child, but Hugh's father refused. In his reminiscences, Hugh Jewett recalled "the tall Cantonese" nurse breaking down in tears at the railway station as the child was removed from him.

During Hugh's boyhood, he heard that his father had hired a Chinese ranch-laborer named Ah Sam, alias "Yellow Sam." After he finished school, Hugh Jewett returned to take charge of his family's landholdings near Bakersfield in Kern County. One day, when he was inspecting the fields in one of his family's properties named Emory Ranch, he noticed gopher holes in the side of the ditch. Just then, a Chinese who turned out to be Ah Sam (whom he apparently did not recognize) came by, and Jewett, on discovering that the latter was the irrigator, pointed out the danger of letting the holes remain. Ah Sam listened quietly and then burst out vehemently, "What for you talk this way to me? I hold you in my arms"—stretching his arms out—"when you were so big. I work for your fadder [*sic*] long, long time. No water going to damage crops."[43] Apparently, Hop and Ah Sam were the same person. As an irrigator, Ah Sam knew what he was doing and did not appreciate his former charge, a city-bred greenhorn, telling him what to do.

Some years later, the Jewett family decided to sell the Emory Ranch, and Ah Sam was told he had to leave. Ah Sam responded, "All light [*sic*], you buy another ranch and I work just the same." Hugh Jewett said he did not intend to buy another one, but he would arrange for Ah Sam to work for his cousin. Ah Sam refused, saying, "I go to Chinatown and stay, you buy new ranch." Indeed, he went off to live in Chinatown for several months. Then Hugh Jewett's cousin, Philo Jewett, Jr., became ill and asked Hugh to take care of his property for him. An irrigator was badly needed for the property, so Hugh Jewett went to Chinatown to seek out Ah Sam and to offer him a job. Ah Sam told him, "No, I only work for your father and his family." Stymied by Ah Sam's utter devotion to his own family, Hugh Jewett went to visit him again in a few days and used a different approach. He told Ah Sam, "You are a member of the Jewett family. When anyone in the Jewett family gets sick some other member helps him. So you being of the Jewett family, you must go and help Philo Jewett." Thus persuaded, Ah Sam went to work for Philo Jewett, where he stayed until his retirement.[44]

Ah Sam's sense of vocation as an irrigator is a far cry from the image of "cheap labor" that many of the writings on Chinese Americans have projected. Not only were the Chinese not as cheap and docile as has been suggested, the skills they possessed were far more varied than what might be expected of a hired hand. Like Ah Sam, many probably took great pride in their work. And for those few who served as middlemen between Chinese and whites, their work even brought a little wealth and power.

TEN

Managers and Entrepreneurs

CHINESE farm laborers were an efficient migratory work force because their movement from one locality to another was orchestrated by Chinese tenants, labor contractors, and merchants, all of whom played managerial roles. These groups, together with the farmworkers, were integral parts of the Chinese production unit in California agriculture. The important work performed by tenants and farm laborers has already been examined. The relationship between them, as well as the functions performed by two other groups of Chinese, who did not engage directly in agricultural production but who nevertheless facilitated its smooth operation, will now be analyzed.

Chinese Social Organization in Agricultural California

According to Pardow Hooper, a white farm manager and adviser who began working in the Sacramento Delta in the early 1920s when Chinese tenant farmers and farm laborers were still in that area, white landowners never had to seek Chinese tenants. Rather, Chinese who acted as scouts traveled from farm to farm, inquiring about the possibility of leasing all or part of each, assessing the fertility of the soil, and offering rents they felt the parcels of land were worth. Hooper said that such individuals could speak some English and seemed keenly knowledgeable about farming conditions in many different areas. Those Chinese who ended up leasing land sometimes brought families

Chinese grape pickers, J. de Barth Shorb Vineyards, San Marino, Los Angeles County, California. (Photograph by A. D. Marchard, courtesy of the California Historical Society, San Francisco)

with them and lived the year around on the farm in houses that were provided them or that they built during their tenure.[1] When help was needed, the tenants recruited fellow Chinese from the rural Chinatowns, or if they needed a larger number than was locally available, they sent requests for workers to labor contractors, who often were local merchants who probably had contacts with merchants in the larger cities, such as San Francisco, Sacramento, Marysville, Stockton, or Fresno. Thus, the Chinese had their own built-in grapevine for relaying news of where work was available and how many workers were needed and when. This set-up explains how Chinese labor contractors could bring large gangs of men to a locality with a few hours' notice—a fact that many white employers had marveled at with obvious relief. The merchant-contractors charged the laborers a small commission for finding them employment, but made most of their money by provisioning them.[2]

In this well-organized division of labor, tenants established fixed bases of operation and acquired social standing and economic importance in the community through their managerial ability. Having learned how best to grow and harvest various crops, the tenants, some of whom might have worked for years as farm laborers themselves, showed the new laborers what to do and often worked alongside them. In the eyes of white landowners, one of the major advantages of leasing to Chinese was that it insured them a labor supply during the harvest. Chinese farmworkers did most of the seasonal work, not only on farms leased by Chinese but on neighboring farms operated by white owners and tenants as well.[3] While working in a particular district, the laborers usually lived in bunkhouses or in tents that were supplied (probably for a fee) by either the resident Chinese tenants or the peripatetic labor contractors. Food was prepared by hired Chinese cooks who quite likely traveled with the gangs, although there is no definitive information on this. One reason that white farmers prefered Chinese laborers was that unlike white workers who had to be fed, the Chinese boarded themselves, thereby saving the farmers' wives the bother of cooking.[4] Merchants and labor contractors maintained the social and economic network that made it possible to feed and house dozens, and perhaps even hundreds, of migrant farm laborers. In the larger towns, these merchants made additional profit by supplying the laborers with opium and women (some of whom, after Chinese exclusion went into effect, were white),[5] and by running gambling dens. They also sent and received letters, forwarded remittances, and served as "banks" for the migrant laborers, providing a familiar point of contact for men far away from home.

Such organization gave the workers a measure of protection, and in this sense the Chinese migrant farm laborers were better off than the "white tramps" who competed with them for work and who, though white and therefore relatively more "privileged," did not help each other. Several employers had noted that the Chinese laborers all seemed to "share in the contract," whether they were doing reclamation work or picking fruit. This sharing in the contract was a way to impose collective responsibility so that all did their share and none sloughed off. In densely populated China, where the margin of subsistence was very thin, cooperative effort and mechanisms for sustaining collective responsibility (such as the *pao-chia* system) enabled people to survive.[6] Chinese immigrants in California apparently used similar methods of social organization in many of the enterprises they undertook.

But the system, while providing protection, also engendered abuses. In many instances, laborers doubtless were overcharged for their food and entertainment, but they probably accepted this practice, knowing full well that without the tenants, labor contractors, and merchants, they would not be able to find work so effortlessly. Perhaps the most reprehensible activity the merchants engaged in was operating gambling dens, because losses in gambling, rather than low wages or exorbitantly priced food, probably kept more laborers in debt bondage for far longer periods than they had envisioned. Such practices led contemporary observers and a number of scholars to the conclusion that all Chinese were inalienably bound to their "bosses," who kept them under perpetual bondage.

Whatever resentment Chinese laborers might have felt toward their superiors, they recognized that organization was a necessity in an unfamiliar and hostile environment. Their membership in the community enabled them to cope with their social and physical environment and to deal with strangers whose language they did not speak. One advantage, for example, of allowing labor contractors to channel them from one job to another was

that it became more possible to be employed for a greater part of
each year. It was the labor contractors who found them work
wherever it might have been available and arranged for them
to be taken there. In a study of Chinese workers in the salmon-
packing industry of Alaska, Robert A. Nash listed one Chin
Lung as a labor contractor for the Alaska Packers Association in
1894. Sing Kee and Company was shown to be a contractor in
1900, Chin Wing was a contractor from 1910 to 1913, and Chew
Mock was involved from 1895 to 1922.[7] These and other names
given by Nash were identical to the names of prominent Chinese
tenants in the Sacramento–San Joaquin Delta. If the men and
the companies were indeed the same, then it was highly probable
that these salmon-cannery labor contractors cum Delta tenants
employed the same men in both enterprises. Since many of the
Delta's major field crops were harvested by May, it was entirely
possible that as soon as the harvest was completed, the men
headed north to join the summer crews in the salmon canneries.
Having more than one job increased the total annual income of
many a Chinese worker.

Agricultural Compradors

Chinese tenants, merchants, and labor contractors in agricultural California played a role similar to that played by
Chinese compradors in China. When Europeans attempted to
set up factories or conduct businesses there, they were hampered
by their inability to communicate in Chinese. Unable to speak
the language and not understanding Chinese culture or society,
they found it difficult to secure clerks and laborers on their own.
Accordingly, they relied on Chinese merchants to serve as middlemen or go-betweens on their behalf. Such middlemen, called
compradors—a term borrowed by English-speaking traders
from the Portuguese—not only helped the Europeans to recruit
the needed workers, but handled many aspects of their businesses as well. Known by their fellow Chinese as *mai-pan*, these
men "recruited and supervised the Chinese staff, served as trea-
surer, supplied market intelligence, assumed responsibility for
native bank orders, and generally assisted the foreign manager

in transactions with the Chinese."[8] According to Yen-p'ing Hao, the comprador had a precursor in the traditional Chinese broker, but whereas the latter was an independent commission merchant, the former was "in the main contractually employed by the foreign merchant."[9] Many compradors simultaneously ran their own businesses and served Western firms.

Before the 1880s the majority of China's compradors came from Canton and its environs. Three of Kwangtung's major emigrant districts—Heungshan, Namhoi, and Panyu—provided most of the Cantonese compradors. Many Cantonese emigrants to the United States, therefore, even if not compradors themselves, must have been familiar with the role they played. The Chinese compradors in the United States differed in one major respect from their counterparts in China: whereas many of the latter frequently owned capital and had independent businesses, most of those in California did not. To be sure, merchants owned or rented premises for conducting their business and for stocking merchandise, but tenants and labor contractors needed very little capital to get established. Their main assets were a working knowledge of English and ready access to laborers. Without the ability to communicate, they would find it impossible to transact business in America, and without access to Chinese laborers, they were not useful to white employers.

The Chinese who performed comprador functions in agricultural California made things convenient for the white landowners. Instead of keeping a separate account with each worker, the employers had only to pay the contractors one lump sum and had neither to board the men nor worry about providing them with acceptable housing. Almost like magic, once a contract had been made (usually orally) with the Chinese "boss," everything got done in time and with proper care. The observation by the Dillingham Commission that Chinese tenants were no longer sought after by landowners in the Vaca Valley in the early 1900s because "they could no longer obtain the necessary number of men to work for them during the harvest season"[10] underscored the fact that tenants and laborers were mutually supportive: each needed the other for survival, and landowners needed them both. If Chinese farm laborers were considered to be reliable, their dependability was in large measure guaranteed by their Chinese "bosses," whether these were resident tenants and mer-

chants, foremen who traveled with their gangs, or labor contractors who went from place to place scouting for work.

The Chinese tenants needed little capital of their own because
from an early date on, creditors were willing to grant them advances at the going interest rates. Table 30 shows that almost all
the creditors that Chinese tenants borrowed from during the
nineteenth century were white individuals. Later, commission
merchants, American banks, and a small number of Asians also
entered the picture. Ivan Light has argued, with little substan-
tiation, that Chinese and Japanese immigrants have been successful in business because they had access to capital through
traditional rotating-credit associations.[11] While such institutions

Chinese picking and hauling grapes, Fair Oakes Rancho, San Gabriel Valley, Los Angeles County, California. (Photographs courtesy of The Huntington Library, San Marino)

no doubt existed and were used by Chinese agriculturalists in California, the more important point is that there were ample other sources of credit available. One Japanese tenant farmer stressed that he never participated in a *tanomoshi* because he never needed to: commission merchants came to him, offering credit for crops he would grow. He said he never had to look for people willing to lend him money.[12] Labor contractors also did not need much initial capital, because white employers often advanced them the fare to bring the workers to a locality,[13] and Chinese merchants gave them provisions on credit. They did not need money to pay the men's salaries either, because the laborers got paid only after the employers had paid the contractors.

White farm employers and creditors who interacted only with these Chinese "compradors" seldom had any direct financial dealings with the farmworkers. Thus Chinese farm laborers, though a critical element in agricultural production, rarely became a part of the social world of rural California. Gangs of Chinese laborers appeared to perform certain tasks when needed and conveniently departed when the work was done, their presence orchestrated by Chinese tenants, labor contractors, and merchants who, unlike the migrant workers, were a more permanent part of the larger rural California society.

The life of Lee Bing, one of the most important Chinese merchants in rural California, illustrates the multifaceted talents that members of the merchant elite possessed.[14] Born in 1873 in the village of Namtong in Heungshan district, Lee Bing was the

Chinese weighing and crushing grapes, Fair Oakes Rancho, San Gabriel Valley, Los Angeles County, California. (Photographs courtesy of The Huntington Library, San Marino)

sixth of eight children. His father had gone to Australia to mine for gold in 1855 but returned penniless. His mother died when he was seven years old. His eldest brother emigrated to the United States in the 1880s, and Lee Bing himself sailed there when he was twenty-one years old as one of fifty Chinese workers recruited to build the Chinese pavilion at the Columbia Exposition in Chicago, paying his own way with borrowed money. For unknown reasons, after his arrival in Chicago, he was kept locked in a hotel for four days and then put on a train for San Francisco, where fortunately, upon disembarking, he was befriended by a Chinese merchant, one Mr. Leong. Lee Bing worked for this merchant for awhile, and eventually the latter took him to Walnut Grove, a town in the Sacramento Delta with a flourishing Chinatown that served the large Chinese population in the area.

Table 30 Creditors of Chinese Tenants by Amount Borrowed and Kind of Creditor, 1860–1920

	Total Number	White Individuals	Asian Individuals	Commission Merchants	White Banks	Asian Banks
1870–1879						
up to $200	2	2	0	0	0	0
$201–$500	1	1	0	0	0	0
$501–$1,000	1	1	0	0	0	0
$1,001+	3	3	0	0	0	0
no $ given	0	0	0	0	0	0
1880–1889						
up to $200	4	2	0	2	0	0
$201–$500	12	7	3	2	0	0
$501–$1,000	9	7	1	0	1	0
$1,001+	14	11	1	2	0	0
no $ given	0	0	0	0	0	0
1890–1899						
up to $200	40	32	2	5	1	0
$201–$500	76	59	5	10	2	0
$501–$1,000	56	45	4	7	0	0
$1,001+	44	38	2	4	0	0
no $ given	4	4	0	0	0	0

1900–1909

up to $200	100	42	0	58	0	0
$201–$500	86	61	2	21	2	0
$501–$1,000	120	94	6	16	4	0
$1,001–$5,000	144	94	6	37	7	0
$5,001–$10,000	19	8	2	9	0	0
$10,001+	4	4	0	0	0	0
no $ given	1	0	0	1	0	0

1910–1920

up to $200	45	13	1	31	0	0
$201–$500	74	32	5	35	2	0
$501–$1,000	91	56	7	24	4	0
$1,001–$5,000	210	92	18	83	14	3
$5,001–$10,000	91	49	0	39	1	2
$10,001+	71	26	3	31	3	8
no $ given	34	17	3	14	0	0

SOURCE: Compiled and computed from chattel mortgage records for Colusa, Contra Costa, Glenn, Kings, Madera, Merced, Monterey, Napa, Placer, Riverside, Sacramento, San Bernardino, San Joaquin, Santa Barbara, Santa Clara, Solano, Sonoma, Stanislaus, Sutter, and Yuba counties. For titles and dates of chattel mortgage books for each county, see the Bibliography.

There were six to seven hundred people from Heungshan district there, which made Lee Bing feel immediately at home. He decided to settle in Walnut Grove.

Working as a farm laborer in the daytime and studying English at night—paying several old white ladies a few dollars to teach him—Lee Bing soon found work as a dishwasher and janitor in the Walnut Grove Hotel, owned by Alex Brown, a prominent American merchant and banker. When the Chinese cook at the hotel returned to China, Lee Bing took his place, his salary jumping from $20 to $60 a month. The salary increase helped him to pay off various debts incurred by his older brother, who had become an alcoholic. He eventually saved up enough money to purchase a restaurant of his own, in partnership with a Mr. Chan from San Francisco. Soon, he also became a partner in a hardware store and an apothecary. His biggest investment was buying a share in the Shang Loy gambling house.

In 1900 Lee Bing returned to China to see his father and to marry. But he did not bring his wife, Lum Bo-ying, to the United States until ten years later. She bore him two sons, the older of whom, Ping Lee, owns a supermarket in Walnut Grove today and is an active community leader. During the first decade of the twentieth century, Lee Bing met Dr. Sun Yat-sen when the latter visited Chinese American communities to raise funds and to rally support for his revolutionary movement. Lee Bing contributed enthusiastically to the cause, and after the Manchu dynasty was deposed in China, he immediately cut off his queue.

In 1914 Chan [Chin] Tin San, the merchant who had a store on the land of P. J. van Löben Sels, had leased a lot measuring 49 by 100 feet from George W. Locke.[15] The plot was north of the boundary between the lands of Locke and Mary L. Wise and east of the railroad line and the Sacramento River. Rent was $10 a month the first year, going up by $5 a year until the end of the lease period in 1924.[16] Tin San built a store and a saloon on the lot he leased. Another merchant, Owyung Wing Chong, built a boardinghouse south of Tin San's store. Yet a third, Yuen Lai Sing, built a gambling parlor.[17] When fire destroyed the Chinatown in Walnut Grove the following year, Tin San, Lee Bing, and several other Chinese merchants from Heungshan (now renamed Chungshan) district decided to build their own town next to Tin San's store, while the Sze Yup people remained in Walnut

Grove to rebuild the Chinatown there. Lee Bing, Tin San, and their partners leased nine acres from the heirs of George Locke for the purpose. The Chungshan merchants chose plans for the buildings they wanted and hired white construction workers to put them up. This was the beginning of Locke, which the Chinese call "Lockee."

Locke was named after George Locke, a merchant who sold window shades, wallpaper, carpets, and other merchandise on J Street, between Third and Fourth Streets, in downtown Sacramento. Locke had invested in 490 acres of Delta land north of Walnut Grove, on which he raised cattle and planted a pear orchard. After Locke's death in 1909, his children continued to manage both his Sacramento business and his Delta land, which was becoming more and more valuable as asparagus became a boom crop in those years.[18]

The growing importance of the Walnut Grove area was recognized by the Southern Pacific Railroad when sometime during the first decade of the twentieth century, one of its subsidiaries, the Sacramento Southern Railroad Company, extended its line to Walnut Grove and built a packing shed on the levee across from the asparagus cannery on Andrus Island. Chinese worked in the Southern Pacific packing shed as well as in the cannery.

Lee Bing was one of the chief financiers of Locke's construction. Though he had suffered severe losses in the fire, his savings were sufficient for him to erect six buildings housing a restaurant, a boardinghouse, two gambling joints, a drygoods store, and a hardware store. Lee Bing also partly financed the construction of a community hall, which was soon used as a Chinese school. Finally, he contributed to the building of a Poor House to provide shelter for members of the Chinese community who were destitute.

After establishing his new business headquarters at Locke, Lee Bing began to branch out elsewhere. In 1919 he bought a meat market on Third Street in Sacramento. He then leased land near the city to raise poultry to supply the market. This venture was a success, and soon Lee Bing persuaded the owner of a vacant lot across the street to construct a large building on it to be leased to him. Lee Bing's new meat market had an icebox that measured 40 feet by 40 feet; to supply the needed amount of ice, he built his own ice plant on top of the market. Finding that he

Lee Bing, his wife, Lum Bo-ying, and their two sons, Ping Lee and On Lee. (Photograph Courtesy of Ping Lee, Walnut Grove, California)

had more chicken and meat than local customers could buy, Lee Bing began to truck his merchandise to three branch markets that he set up elsewhere in the city. With the money he made, he bought a ranch near Grass Valley in Nevada County and moved his family there. In 1925 he purchased 1,000 acres of the Brack Tract. To bypass the Alien Land Law, Lee Bing set up a corporation in whose name the property was held. He had the entire tract planted in asparagus. A few years later, however, unable to continue making mortgage payments on the tract, Lee Bing sold

it, with the aid of Pardow Hooper, to avoid foreclosure by the
Missouri State Life Company.[19]

Lee Bing lost a reported $80,000 during the 1929 crash, but by
1932 he had recovered sufficiently to open a restaurant in Grass
Valley. Retaining Locke as his headquarters, he eventually
established other businesses all over northern California and in
Oregon, including a restaurant and cardroom in Weed, a cardroom in Susanville, and a cardroom, bar, and dance hall in
Chester. Since few Chinese resided in these places, Lee Bing's
businesses apparently catered to white clients. At age seventy-
five, Lee Bing sold all his businesses, keeping only the two
gambling houses in Locke. Unable to visit China because it had
become Communist, Lee Bing died in America at age ninety-
seven in 1970.

Given the lack of biographical information about other Chinese merchants in agricultural California, it is difficult to assess
how representative Lee Bing was. Probably few others matched
him in the array of enterprises undertaken, because few lived as
long or took as many risks as he did. Men like Lee Bing and
Chin Lung became the ruling elite of Chinese communities in
agricultural California, less because they owned any means of
production than because they possessed the organizational
talents to link their fellow countrymen to white society at large.
Having access to those who did own means of production, they
themselves did not need to have any, so they were not members
of a bourgeoisie in the Marxist sense. Eventually, of course, they
bought property and accumulated capital through mercantile
operations, but they succeeded economically and became community leaders mainly on the basis of their managerial abilities.
Their social origins in China had little to do with the ultimate
positions they attained in California.

Chinese Commission Merchants

While men like Lee Bing operated businesses in fixed localities, other Chinese merchants made money by moving goods
from place to place. The most important of these mobile merchants were vegetable peddlers and fruit merchants. By the mid-

1870s, even before Chinese fruit-growers became numerous, some Chinese had already become fruit buyers or commission merchants. Available evidence suggests that they first entered this business in the foothills of the Sierra Nevada where some of the state's earliest orchards had been planted. One 1876 account tells of Chinese who "had bought the fruit of a great many orchards for purposes of drying" in Tuolumne County. The orchards around Shaw's Flat, Sonora, and Columbia were also said to be

> swarming with Celestials [Chinese], some drying on shares, others from purchase. This plan relieves the orchardists from risk and manual labor, and gives employment to the Chinese.... Whether the dried fruit is destined for the home market, or for China, is for the future to demonstrate, but it will be the means of stimulating an increased production. Much fruit will be saved which otherwise would have gone to waste.[20]

The narrator of this account was careful not to appear to be too enthusiastic about the entry of Chinese into the commission/ fruit-drying business. He added by way of apology:

> We do not advocate the employment of Chinese labor, or even the selling of the fruit crop in lump, but these things will regulate themselves. The present crop had to be gathered and attended to. No other means seemed so suitable to secure the fruit when ripe, for ripe fruit will not wait upon the convenience of anyone, but must be attended to; hence the sale of gardens to the Chinese, who seem to have enough laborers to take advantage of conditions when favorable.[21]

The 1880 manuscript population census listed scattered numbers of Chinese as "fruit merchants" or "fruit dealers." Since being commission merchants was probably a seasonal occupation, it is likely that more Chinese engaged in this business than the census (taken in June 1880) indicated. An account written in 1882 claimed that "the Chinese have become the leading manipulators of the fruit crop of California, and that they, by reason of the extent and boldness of their transactions, were driving the heavy fruit commission merchants of San Francisco out of the field." According to this writer,

their method of procedure is to visit the fruit districts as
soon as a reasonable estimate of the extent of the crops can
be made, and then contract for the fruit, paying so much per
tree, or acre, as the case may be—they taking all chances,
and picking and boxing the fruit themselves. In this manner
they buy up the products of entire districts. They peddle out
all they can throughout the surrounding country, and either
ship the balance to the city markets or dry it. Their action is
an advantage to many of the fruit growers, as they realize
as much as, or more than formerly for their fruit crops, and
are not subjected to the annoyance and trouble of picking
and shipping. And so John [a name commonly applied to
Chinese in the period] obtrudes himself into a new avenue
of trade.[22]

Though this account may be somewhat extravagant in its claims
in terms of the extent of Chinese operations, nonetheless it provides a glimpse of the manner in which Chinese commission
merchants operated.

From the above account, it appears once again that the Chinese managed to drive a wedge into the commission business by
combining three entrepreneurial functions: supplying the needed
labor to harvest the crops, distributing whatever portion they
could through retail peddling, and processing the rest for future
sale outside of the immediate area of production. Most important, the Chinese were willing to assume all risks.

In 1908 and 1909, agents of the Dillingham Commission found
that many Chinese were still buying fruit by the tree or by the
orchard. The Commission reported:

The usual method in buying fruit is for several Japanese
or several Chinese to associate together in making the contract for purchasing the fruit of an orchard. They share the
expenses of harvesting the crop and divide the profits realized. These groups require practically no capital, as they get
advances from the fruit-distributing companies with which
they carry on the work of the harvest, and all members of
the group work in gathering the crop. The shipping company obtains a mortgage on the crop to secure its advances,
so that the landowner and the shipping company both look
to the crop as their security.... All checks are made out to

the ranch owners and pass through the hands of the fruit company, so that both have a chance to take out what is due them before the proceeds are paid over to the Japanese or Chinese buyers. As a rule, ranchers and fruit companies dealing with Chinese, whose reputation for commercial honesty is above reproach, are not as rigid in their requirements as to security as they are when dealing with Japanese, who can not in all cases be relied upon.[23]

Chinese not only bought fruit but cured it. A lease signed in Monterey County in 1911 between Charley Fang and Marion T. Rowe indicated that on a plot of ground 200 feet square to be rented for five years at $850 a year, the landlord would build two brick kilns "similar to the dryer [he had erected for] Sing Wo and Company in Pajaro Township." The Pajaro Valley was a famous apple-producing area, and these kilns were to be used to dry the fruit. Charley Fang had to deposit $1,000 as security in the Bank of Watsonville, but he would be permitted to withdraw the sum if he should decide to build two additional kilns himself. On the lot rented there stood a bunkhouse, a kitchen, and toilets that the tenants and their employees could use. There was also a spur of the Southern Pacific Railroad running through the property.[24]

An intriguing question related to Chinese fruit-dealers is whether they developed this business skill in California or brought it with them from China. In his study of a village in the eastern corner of Kwangtung province in the early 1920s, Donald Kulp provided a two-page description of fruit-dealers whose operations sounded strikingly similar to the descriptions given above. Kulp found that in Phoenix Village, out of 167 persons whose occupations he could list, 10 were wholesale fruit-dealers, who

> buy the fruits before the harvest is even ripe; sometimes, when the trees are only in bloom. On the basis of the quality and abundance of the blossoms, the dealers estimate the probable quantity and quality of the crop and higgle with the gardener until a price per *picul* of harvested fruit is agreed upon. After the bargain has been struck the original owner is no longer responsible for the crop. Cultivation and care of the trees is then turned over to the fruit speculators.

In some instances, a second sale was made when the fruit was nearly ripe, when the original dealers sold their crop or part of it to another set of dealers.[25]

Phoenix Village was not in one of the districts where most of the emigrants to California hailed from, but it was only about two hundred miles (as the crow flies) east of the Pearl River delta and was itself an emigrant community whose inhabitants went primarily to Thailand, Malaysia, Vietnam, and Cambodia. It seems that cultural diffusion was taking place among the Chinese who traveled between their home communities and farflung places in Southeast Asia and the Americas, but it is hard to say in retrospect in which direction such diffusion spread.

Chinese Farm Cooks

Though one seldom thinks of cooks as performing managerial functions, Chinese who worked as farm cooks in fact ran the households and farms of many California landowners in the nineteenth and early twentieth centuries. Unlike the tenants, labor contractors, and merchants who often served as bargaining agents on behalf of large groups of Chinese, farm cooks interacted with the white community as individuals and were often treated as intimate members of their employers' households.

Because they were usually employed singly, Chinese farm cooks led a more isolated existence than almost any other group of Chinese in rural California. Frequently, a lone cook was the only Chinese in a locality. In San Joaquin County in 1860, for example, only nine Chinese lived outside of the Chinatown in Stockton; eight of them were cooks in the homes of white farmers dozens of miles away from town. Chinese farm cooks were especially prevalent in the Sacramento Valley where hundreds of them worked for farmers and town dwellers. Some of these cooks began their careers as teenage boys, and there are records of a number who adopted the last names of the white families they worked for. Because of their youth, isolation, and long term of service with the same families, these cooks probably became more "Americanized" than most of their compatriots. They had to learn to adapt to American mores even more quickly and thoroughly than their urban Chinese counterparts; the latter, at least, could visit Chinatown and interact with other Chinese from time to time. In the countryside, the nearest fellow country-men—often other Chinese farm cooks—were usually too far away for a convenient visit.

In some households, Chinese farm cooks assumed a role much like that of the English butler, managing the entire household. Two available accounts of Chinese cooks provide a glimpse into how they lived and worked. Yuen Yeck Bow, who eventually adopted the name Jack Ellis, worked as a cook for the Ellis family in Marysville.[26] (It is easy to see how "Yeck" became "Jack.") Yuen had come to California originally to mine for gold, but he soon turned to railroad construction. Like many other Chinese railroad workers, he performed the dangerous task of blasting rock with black powder. One day a boulder from a hillside he was blasting fell on his head, knocking him uncon-scious for two days. After recovering from this accident, he de-cided to abandon railroad building and found employment as a cook for a banker named H. Jewett in Marysville. (There is in-sufficient evidence to ascertain whether this H. Jewett was the same man who had employed Ah Sam, the irrigator, in Kern County.) When the Jewetts moved to Europe, Yuen Yeck Bow started working for the Ellis family in the late 1860s.

Yuen Yeck Bow went to market every day. Possibly, he bought some of the groceries from fellow Chinese vendors in Marysville. He was entrusted with responsibility for household finances, squaring away accounts with his employer only once a month. It is not clear if he had to advance the cash for the groceries or bought them on credit from the merchants. Yuen was very fond of children, but "like all Chinese, favored boys." He became friends with a large number of them as he made his daily shop-ping trip. He carried his purchases in a large basket, but some-times, because he passed out treats on his way home, his basket became empty before he reached home.

The narrator of Yuen's story was one of the boys in the Ellis family. He recalled that one of the most difficult decisions Yuen Yeck Bow ever had to make concerned cutting off his queue in 1911. Yuen wished to obey the order that the new republican government in China had issued to all Chinese men to cut off this symbol of Manchu oppression, but he "just hated to give up that

old pigtail of his, but he finally did and then became quite proud
of the change."

Yuen stayed with the Ellises until 1913 when the head of the
family died. Perhaps also feeling his own age, he decided to return to China. As a memento of his forty-seven years of faithful
service, he asked the son of the family for a family heirloom as
a souvenir—a Knights Templar Maltese Cross which had belonged to his deceased employer. Three years after his return to
China, the Ellis family received a letter from Jack, the old cook;
enclosed was a photograph of himself wearing a bowler hat and a
Western suit, proudly displaying the Maltese Cross and its gold
chain across his chest.

Another Chinese farm cook about whom something has been
recorded was Wong Gee.[27] In 1887 he came as a sixteen-year-old
boy to work as cook for the Moorehead family in West Chico,
Butte County. Since the Chinese Exclusion Law barring the entry of Chinese skilled and unskilled laborers was already in effect,
it is not known under what auspices he entered the country.

Wong Gee had a very busy schedule. He not only cooked but
cleaned house, washed clothes, watched over the Moorehead
children, processed farm products, and served as ranch foreman.
He got up at 6 A.M., made breakfast for the Moorehead family,
and packed lunches for the work crew on the ranch. After clean-
ing up the breakfash dishes, he separated the milk and cream,
churned butter, washed clothes, and cleaned the house. Then he
prepared lunch for the family and the farm laborers. He usually
had the early afternoon to himself, and during this time he cultivated Chinese vegetables for his own use in the yard around the
small house where he lived. During harvest season, he canned
and preserved fruit and vegetables for his employer's family, or
sorted nuts and other crops grown on the farm.

Wong Gee sometimes used his leisure time to make wooden
toys for the children in the family. One of the Moorehead daughters recalled that Wong Gee—who had by now adopted the name
Gee Moorehead—made many unique toys of his own design,
including airplanes with revolving propellers, for her and her
siblings. On Sunday afternoons, Wong Gee bicycled to Chico to
visit his friends. He was especially fond of visiting the Shanghai
Restaurant in Chico and of talking on the phone to two friends
who were cooks at the Gianella ranch and the Mary Rogers

ranch. At age sixty-nine, he bought a car and taught himself to drive. Wong Gee had learned English from Mrs. Moorehead, and eventually he bought an old typewriter, which he used to write letters (in English) to his friends.

Wong Gee worked for the Moorehead family for sixty years. He was a completely trusted member of the household, for whenever the Mooreheads went on vacation or went to live in their summer home in Alameda, Wong Gee was left in charge of the house and farm. He wrote them letters to let them know that he was taking care of everything and there was no need for them to worry.

During the sixty years Wong Gee lived and worked in Chico, he returned to China several times, and there he sired a family. When his two sons reached their early teens, he brought them to the United States. One, Wong Hing, lived with him at the Moorehead ranch for several years. When his sons grew older, Wong Gee sent them to Denver and gave them money to start a business. In the years before he retired, Wong Gee was earning $350 a month. Throughout his years in Chico, he asked his employer to send a large portion of his paycheck each month back to China to support his wife and daughter, who never came to the United States. Wong Gee returned to China in 1947 when he was seventy-six years old.

These two vignettes show that Chinese cooks on California farms were all-purpose servants who sometimes served as farm managers when the families for whom they worked were away. They not only fed the other hired hands but supervised them on occasion. Apparently, they were sufficiently respected by the other employees to be granted such status. Knowledgeable about the American way of doing things, they were entrusted with a great deal of responsibility. Like the English butler, they were at once servant and authority figure.

Farmers' wives were especially dependent on their Chinese cooks. A letter in the *Stockton Independent* of October 23, 1876, from a woman who signed herself simply as "Martha," expressed what many farm wives must have felt. Martha stated that she had been a homemaker since 1829 and had been in California since 1850. Her family consisted of herself, an invalid husband, two boys, and three hired hands.

For all these I had to cook, to wash for my own family, churn, prepare butter for market, with the many other employments that fall to the lot of the farmer's wife, besides supervise all the out-door work. I sought for help in vain, in Stockton and in San Francisco, offering as high as $50 per month, for I found myself wholly unable to accomplish all I had to do.[28]

Martha then recounted her trials and tribulations with a series of white female helpers, all of whom gave her endless trouble. Finally she tried a Chinese cook in 1866.

What a blessing he was. What bread and coffee and nice broiled steak he gave us, and no fuss nor noise. He staid nearly a year, then left for higher wages.

Since then I have hired Chinamen, and seldom one I could not trust. The one I have now has been with me nearly two years, is about 18 years of age, a good plain cook, washer and ironer, churns, takes care of pigs and poultry, harnesses my buggy horse, herds stock, is handy with carpenters' tools, or paint brush, and is in fact very quick to learn anything; can kill and dress a hog and take care of the meat and lard as well as any professional butcher. If I leave my home and there is any money in the house I give it into his charge. I pay him $20 per month. Will anyone tell me how or by whom I can replace him except by one of his countrymen? One of the objections raised against them is that they send their money to China. I know that this boy buys boots, hats, shirts, socks, etc., in Stockton,... I sell my wheat to Mr. Sperry; he grinds it into flour for shipment to China, and the money for which it is sold comes back to purchase my next crop.

In conclusion, I do not believe the Chinese usurp the place of white labor; they simply fill a want that cannot, at least at present, be otherwise supplied.... I think the prejudice against the Chinese and the warfare made upon them and against those who employ them [are] unjust and uncalled for.

I find my China boy honest and with principles that would do credit to a Christian.[29]

Chinese cook in a restaurant, Stockton, San Joaquin County, California.
(Photograph by Jack Sloan, courtesy of the California Historical Society,
San Francisco)

Chinese cooks were so much in demand that even after the number of Chinese in other agricultural work had declined greatly by the end of the nineteenth century, many Chinese continued to work as cooks on farms. Some data from the 1900 manuscript census showed how they enabled farm work to be done efficiently. On a ranch owned by a German-born stock raiser named John Schutz in Fresno County, 324 farm laborers of American, German, Italian, Irish, and Mexican nativity were employed. They were placed in fourteen camps, each of which contained men of only one nationality, or at the most, two. In each camp the cook was Chinese. Nine of them could speak English. They had been in the United States from thirteen to twenty-one years; seven of them had come after the passage of the 1882 Chinese Exclusion Act. Since half of them were in their late thirties, and the others were either forty or forty-one years old (except one who was fifty-five), most must have come to the United States as young men barely out of their teens. Schutz also had three Chinese laundrymen and two truck gardeners attached to the main house on the ranch. Other employees included three carpenters, born in England, Italy, and the United States, an American-born engineer, a German-born blacksmith, a California-born wheelwright, a California-born bookkeeper, and a ranch foreman from Italy.[30] On this cosmopolitan farm, which was a microcosm of agricultural California, Chinese cooks, laundrymen, and truck gardeners performed the housekeeping chores that kept everyone in fit working condition.

Cooks, like vegetable peddlers, tenant farmers, and merchants, served as individual links between Chinese communities and white society in agricultural California. Through the work they did, they made it possible for themselves and for other Chinese to live symbiotically with whites most of the time. However, the existence of the rural Chinese, like that of their urban brethren, was punctuated by sporadic outbreaks of violence against them. No account of the magnitude of what the Chinese achieved in California would be complete without an examination of how the Chinese managed to survive in the face of such violence and despite their failure to form complete communities in nineteenth-century America.

ELEVEN

Survival and Community

 CONTEMPORARY OBSERVERS charged the Chinese
with a refusal to assimilate to American ways, and
many scholars have stressed how the Chinese have
adamantly preserved their culture in the United
States. Many Chinese values, practices, and patterns
of social organization were indeed transferred to
American soil, but the fact remains that Chinese communities
that developed in America were by no means replicas of those in
China.

Chinese American communities differed in at least three fundamental ways from traditional Chinese society. First, Chinese
immigrant communities, though semiautonomous enclaves,
were profoundly shaped by forces emanating from the larger society around them. Not only did the Chinese quickly adapt to the
functioning of a capitalist economy, they persisted despite legal
exclusion and anti-Chinese violence. Second, since few members
of gentry families emigrated, no landowning-literati class established itself in America. Those scions of the ruling class who did
come to the United States were students who later returned to
China as bearers of Western civilization. Instead, as was described earlier, merchants who had a low status in China became
the elite in Chinese American communities by virtue of their
ability to deal with whites and to provide for the needs of other
Chinese. Finally, as few women emigrated to the United States,
Chinese America was virtually a womanless world, especially in
rural areas. Consequently, Chinese American communities were
socially incomplete.

Violence and Persistence

Several waves of anti-Chinese activities swept agricultural California in 1876–79, 1886, and 1893–94, all of which were years of economic hardship. A number of writers, particularly Carey McWilliams and Victor and Brett de Bary Nee, have concluded that these activities drove the Chinese out of rural areas, but such was in fact not the case. The Chinese not only persisted in agriculture, increasing both in numbers and in the amount of land they leased in several of the major agricultural regions in the decades just before and after the turn of the century, but also survived in one area—the Sacramento–San Joaquin Delta— where a small Chinese farming community was established and has remained to this day.

The economic dislocations of the 1870s enabled anti-Chinese groups in San Francisco to gather increasing support, as the Chinese were blamed for the economic problems all workers faced. Anti-Chinese sentiment spread to the countryside where the Order of Caucasians (sometimes known as the Caucasian League), a white supremacist group, was established in the late spring of 1876. The group first came to public notice in July 1876 when, in the aftermath of the shooting of a white man named Cheer by a Chinese worker at the American House in Truckee in Nevada County near Donner Summit, a mob had gathered menacingly in front of the Truckee jail where three Chinese witnesses were confined. On discovering that the man who had done the shooting was not in the jail, the mob dispersed, but someone apparently notified the engineer of the local fire brigade that he would be shot should he attempt to put out any fires in Chinatown. A reporter then sent telegrams to the leading newspapers along the Pacific Coast spreading the rumor that the Caucasian League would burn Chinatown. On July 14 two prominent millmen, George Schaffer and Elle Ellen, received similar threatening notices that said:

> Dear Sir: You are respectfully requested without further warning to discharge the Chinamen in your employ, and give your work to whites instead, whom you well know are suffering from the effects of all those heathens in our midst. Think well of the country of your adoption, and try to assist

the poor white man in making an honest living. Take heed
lest the course you are now pursuing shall fall upon your
own head with tenfold vengeance. [Signed]—Native
Americans.[1]

The public charged the Caucasian League with threats to
burn the town. In response, the organization's president, Hamlet
Davis, one of Nevada County's oldest pioneers, issued a denial.
A meeting was held, with some 150 of the 200 reported members
of the League in attendance. The group issued a statement
saying they deprecated mob violence and regretted that

> [a] certain bad element has sprung up in our midst, which
> has shown a tendency to riotous conduct by appearing upon
> our streets during the night, armed and masked, and by
> sending certain anonymous, threatening letters to some of
> our citizens and property-holders.[2]

The Chinese, meanwhile, left their wooden buildings and barri-
caded themselves in the few brick or stone houses they had. The
San Francisco Bulletin reported that the Chinese had bought six
hundred revolvers since the troubles began and could be heard
daily practicing their shooting. They sent for a detective, said to
be paid by the Chinese Six Companies in San Francisco, to help
them, while the Central Pacific Railroad Company sent its
own detectives to Truckee to guard the company's property. Insurance companies began canceling the fire insurance policies
they had issued.[3]

The Order of Caucasians also led rallies in Red Bluff in April
and in Marysville in May 1876. The group held its first state
convention in Sacramento the following September.[4]

In February and March, 1877, a series of violent anti-Chinese
acts occurred in Chico and its vicinity in Butte County. In mid-
February, the soap factory of John Bidwell, which had been
leased to some Chinese for use as a slaughter house, was burned.
A threatening note sent to Bidwell said, "Sir, you are given
notice to discharge your Mongolian help within ten days or suffer
the consequences." Then, on February 28, some men set fire to
the barn on the ranch of the widow, Mrs. Patrick, who had
leased her orchard and garden to Chinese. Six horses were
burned alive, and the dwelling house of the Chinese tenants

was consumed by flames as well. Shots were fired at the Chinese as they attempted to put the fire out. The Chinese suffered an estimated loss of $1,500.[5] A third incident took place on March 5 when the home of some Chinese in Nord was burned. The inhabitants escaped before fire engulfed the building, but the arsonists shot at the Chinese as they retreated. The day after, a washhouse operated by Ah Shu on Chico Creek was burned.[6]

Several unsuccessful attempts to burn both the "old" and the "new" Chinatown in Chico were made in March,[7] but the event that respectable citizens of Chico found most deplorable was the murder on the evening of March 14 of several Chinese who had been hired to grub and clear a piece of land on the Chris Lemm ranch, located about a mile outside of Chico on the Humboldt Road. Five armed men went to the ranch, shot four of the Chinese, doused their house with kerosene, and set it on fire. Two other Chinese were wounded but managed to escape. One died later, and the other managed to reach town to report the incident to the police. John Bidwell formed a Committee of Safety to patrol the town. Professional detectives were hired to search for suspects, and the citizens of Chico contributed $1,000 as a reward for their capture. The flood of threatening letters continued.[8]

Finally, on March 27 five men suspected of murdering the Chinese on the Lemm ranch, and seven others suspected of the other acts of arson, were captured. The confession of Ned Conway, who was caught first as he dropped a threatening letter in the mail, led to the arrest of the others. Confessions by Conway, Thomas Stainbrook, Henry C. Wright, Adam Holderbaum, H. T. Jones, and Charles Slaughter, which were published in the *Chico Enterprise* on March 30, 1877, revealed that members of the Laborers' Union—an off-shoot of the Order of Caucasians—and particularly those belonging to a group known as the Council of Nine, had planned to instigate a reign of terror in the Chico countryside in an attempt to oust the Chinese from the area. According to some of the confessions, Ned Conway, Charles and John Slaughter, Eugene Roberts, and Thomas Stainbrook, had gone to the Lemm ranch with the intention of robbing the Chinese, who they believed had plenty of money. They entered the cabin of the Chinese, held them at gunpoint, and searched for money. Only then did Roberts suggest that they each choose a Chinese to shoot. Some of the others objected, but four Chinese were killed anyway. Roberts had a bottle of kerosene in his pocket, threw it over the heads and clothes of the Chinese, and one of the Slaughter brothers lighted a match to set their victims on fire.[9]

The grand jury rushed through the hearings in three days. Five men were held for murder and arson, and six for arson. Pleasant Slaughter, brother of Charles and John, was sentenced to ten years for burning Bidwell's factory. Stainbrook, Roberts, Charles Slaughter, and John Slaughter each received a life sentence, while H. T. Jones, who refused to confess to anything, received twenty years. Writing in the Yuba City and Marysville *Independent Herald* three-quarters of a century later, a reporter commented:

> Had their victims been white men they would have been hanged. In fact, they would have been, undoubtedly, accorded that special type of hanging termed lynching.[10]

John Bidwell's factory was rebuilt, but on the night before it was to reopen, it was burned to the ground again.

Other acts of wanton violence against the Chinese continued after the Chico incidents. At the same time, employers elsewhere also received threatening notices. In August 1877 the San Francisco *Daily Morning Call* reported that white farmers in the Sacramento Delta had received letters dated July 1877, which said:

> Notice is heare given to all men who owns lands on the Sacramento River is heare ordered to dispense with Chinese labour or suffer sutch consiquinces as may follow within tenn days. We have heaved and puked over Chinese imposition long a nuff. Good-by John Long Taile.[11]

Sixty farmers, "in the main sturdy old pioneer settlers" who employed Chinese, met in Courtland under the leadership of J. V. Sims and O. R. Runyon and declared that "the Chinamen were of necessity employed and found to be sturdy, sober and industrious, the whites being the very reverse."[12] They bemoaned the fact that

> modern labor has degenerated.... It is no way for workmen to act, to demand work and then be resort to riot, arson

and murder when they cannot get it.... Good white labor,
willing to work for remunerative wages and to stand by his
employer, could get employment, but thieving, cowardly
tramps, who wander around, systematically lying, plunder-
ing, exciting and threatening, had no chance. Many were
city-bred and ignorant of manual labor, and think they can
turn their hands to any occupation. Farming is a special
trade, requiring special knowledge.[13]

Chinese were driven out of the Rocklin-Roseville area in Placer
County, and their quarters were burned, in September 1877.
In the same year, the Chinatowns of Grass Valley in Nevada
County, Colusa in Colusa County, and Lava Beds in Butte
County, were all burned, as were the fields and barns of farmers
who employed Chinese. From then through 1879, partly as a
result of the vehement anti-Chinese statements of Denis Kearney
and the Workingmen's Party before and during the second constitutional convention, acts of violence against the Chinese
occurred frequently but went unpunished.[14] In April 1882, when
some Chinese workers in Colusa County demanded wages equal
to those paid Americans, a new society called the American and
European Labor Association was established, whose goal was
to import girls from Europe and the East Coast to replace
Chinese servants in Colusa County. The organization also
encouraged whites to patronize only those who hired other
whites.[15]

The second phase of rural anti-Chinese activities erupted in
1886, when violence and boycotts were used. Not only were
white employers of Chinese laborers boycotted, but so were products made or cultivated by the Chinese. Soon after midnight on
February 25, 1886, about thirty masked men went to the ranch of
H. Roddan, a hop grower two miles east of Wheatland in Yuba
County, broke into a house where ten or eleven Chinese were
sleeping, roused them, and marched them across the fields to the
farm of a widow, Mrs. Fogg, where they gathered another half
dozen Chinese, before marching the entire group to the farm of
O. D. Woods, where the Chinese quarters were also broken into.
The last house was burned, along with the belongings of the
Chinese. One of the masked men hit an old Chinese on the head

with his pistol, cutting a big gash. The captured Chinese were
taken to Wheatland and turned loose. The following day, some
brave Chinese returned to Roddan's place to gather their belongings but were not allowed to take them. The reporter for the
Marysville Daily Appeal commented that "the object of the raid is
not apparent. It does not seem to have been to drive the Chinese
away, else they would not have been turned loose in Wheatland."[16] The writer then observed:

> Such work is most damaging to the cause. The sentiment of
> the people of the coast is against violence in any form, as is
> shown in the resolution of almost every anti-Chinese asso-
> ciation. The sentiment is to get rid of the Chinese by "all
> lawful and honorable means." Mob violence and incendiar-
> ism is neither lawful nor honorable and will not be counte-
> nanced by the people. Truckee is getting rid of her China-
> men without it, and so will every other place. Mob violence
> will not accomplish it. On the contrary, it tends to estrange
> the majority of the people from the cause, and makes them
> fight shy of the association. The eastern press is impressing
> the people with the idea that all the Western slope is a mob,
> and occurrences such as that at Wheatland only strengthens
> the assertions.[17]

The following day, the *Wheatland Graphic* reported that citizens of Nicolaus, another town in Yuba County, had asked the
Chinese to leave the week before, marching forty of them to the
river and putting them on a barge. The citizens, a number of
whom accompanied the Chinese to Sacramento, had raised $125
to pay the latter's fare. Three or four Chinese were allowed to
remain behind to settle the affairs of their community.[18] In
Wheatland itself, the Chinese posted four or five guards every
night even though the Anti-Chinese Association of Wheatland
had pledged that no violence would be used. Ching Ping, the
Chinese vice-consul, went to Wheatland accompanied by H. J.
Burns, a former deputy marshal, to investigate the situation
there. Meanwhile, the Anti-Chinese Club held its largest meeting to date, with a preacher and two professors as the featured
speakers.[19] Despite the anti-Chinese activities, the railroad company, which used Chinese maintenance crews, insisted that they

return to work on the bridge they were building in the section of the line between Wheatland and Sheridan. The Chinese, fearful for their lives, refused to comply with the orders.[20]

In March 1886 conventions were held concurrently in Sacramento by two anti-Chinese groups that were at odds with each other. A group based in Sacramento, which claimed to be broadly representative and advocated peaceful means to get rid of the Chinese in the state, met on March 10. Another group, based in San Jose, allegedly of limited representation, since "a large section of the State was debarred representation," had met earlier, but had adjourned in order to reassemble in Sacramento on the same day in competition with the first group. The *Wheatland Graphic* commented that the San Jose group's intention seemed to be

> a purposeful plan to bring about a collision of methods, that will furnish our opponents East precisely the opportunity they desire—reason to exclaim that this people do not know their own minds. It looks very much as if the promoters of the San Jose scheme mean to oppose any and all temperate methods, and to insist on foolish radicalism that threatens to do so much injury to industrial interests in California.[21]

One major outcome of these anti-Chinese conventions was the call for two kinds of boycott. First, members of the public were asked to sign a petition pledging "that they will not, knowingly, buy[,] sell[,] or eat any vegetables or fruit raised, cultivated, picked or packed by chinese, from the date of signing this mutual agreement until after the 1st day of January, 1887."[22] Second, people were asked not to trade with employers of Chinese. The *Pacific Rural Press* published a series of letters in April and May 1886 from influential fruit-growers criticizing the boycott. Judge W. C. Blackwood, a horticulturalist in Hayward and one of the most vocal advocates of the use of Chinese labor, in his letter printed on April 24, pointed out that

> the law of the "boycott" is a two-edged instrument and if it is to be enforced it can and will be made a boomerang, wounding and inflicting vital injuries on those who first resort to it.... No one industry can be seriously affected or disturbed without producing a general disturbance affecting injuriously other industries.[23]

In early May, Blackwood was joined by other prominent fruit-growers in Alameda County to pass a strongly worded resolution denouncing the boycott as

> an unlawful means of correcting the evil complained of;... as a bantling of foreign extractions,... to arbitrarily wrest from American citizens the liberties and rights guaranteed them by the Constitution of the United States,... to introduce among a peaceful and law-abiding people a system of anarchy and rapine.[24]

It was ironic that in defending the Chinese, these property owners showed another kind of nativism of a decidedly antiradical character. Their sentiments were most clearly expressed by W. W. McKaig of Oakland, who made a speech charging the boycott to be a

> species of warfare against property. The Chinese are entitled to the same protection granted by our laws as all foreigners. If allowed to go on and show its hideous proportions, it will antagonize all our efforts for the exclusion of the Chinese and will help the prejudice of the East. Wyoming, Tacoma and Seattle have injured us greatly already, and now comes "boycotting," a masked barbarism. I call it "masked" because it would not hesitate to rob, destroy, burn and murder, if it had the courage.... The boycott comes from the essence of German socialism, Russian nihilism and chattering French communism.[25]

Various farmers' associations, such as the Santa Clara Horticultural Society and the Mendocino Hop Growers' Association, also resolved that they "will not submit to any interference by any man or set of men in [their] business."[26] Ad hoc groups, such as the Fruit Growers and Citizens' Defensive Association, sprang up to fight the boycott.

Defenders and supporters of the boycott also had their say in the pages of the *Pacific Rural Press*. Their views echoed those expressed by urban anti-Chinese agitators, revealing once again that the anti-Chinese crusade was but a deflected manifestation of the struggle between capital and labor as American capitalism matured in the last decades of the nineteenth century. S. Harris Herring of Deer Ridge Farm (county unknown) attacked both the employers of Chinese and the Chinese themselves when he wrote:

> Labor [has been] materialized and reduced into a marketable or cash commodity and the humanity back of it lost sight of. It becomes in their vision mere machinery, subject to their will to increase their gains....
>
> A class supporting and degrading civilization in China has chained with the shackles of a relentless superstition and unresisting subservency an inbred race of miserable slaves, who are satisfied with a miserable fare the merest pittance can procure. What more natural than that the services of such a race is desirable by men disposed to regard labor as a mere article or commodity to be purchased low that the proceeds resulting, may, in competition with his fellows, bring him greater profit?...
>
> Our philanthropic invitation to the oppressed of all nations to seek our shores ... was never meant either in spirit or in fact to invite the owners of slaves to bring their hordes here to be used by conscienceless corporations and hard scrupled men to increase their profits....
>
> The Chinese slave invasion of our State has been imposed upon us against our constant expostulation as a people. It has been encouraged only by scheming companies and equally selfish individuals.... This slave system must not only stop its invasion, but it must leave our shores.[27]

A third wave of anti-Chinese activities broke out in rural California in 1893–94 when attempts were made to drive the Chinese out of Fresno, the Napa Valley, Compton, Redlands, Tulare, Visalia, Ukiah, and the Vaca Valley.[28] Carey McWilliams, using Bertram Schrieke as an authority, stated that over half a million acres were put out of cultivation as a result of the alleged exodus of the Chinese.[29] Other writers, relying on McWilliams, have since then reinforced the supposition that the Chinese were indeed driven out altogether from agricultural California in those years. However, a careful compilation of acreages leased by the Chinese—based on information in the extant lease and crop mortgage documents of over forty agricultural counties— indicates that this assumption is simply not true.[30] Figure 1, which shows the recorded acreages leased by Chinese from 1870 to 1920 in California's six major agricultural regions, reveals that in the Sacramento Valley, the mid-1880s, the mid-1890s, and the last years of the second decade of the twentieth century were the peak years of Chinese tenancy in terms of the total number of acres leased. In the Sacramento–San Joaquin Delta, Chinese leased the largest acreages from the mid-1910s till the end of World War I. Chinese tenancy was at its height during the 1890s and between 1906 to 1912 in the northern and central coastal valleys. In both the San Joaquin Valley and the southern coastal plains, the peak occurred in the early 1890s. Thus, despite the flurry of anti-Chinese activities, Chinese tenant farmers and farm laborers did not depart from the land.

Table 31, which shows the acres leased by Chinese as a percentage of total improved acres in each county, reveals that although the amount of land leased by the Chinese never constituted more than a small fraction of the cultivated area in each county, still, compared with the Japanese, who were repeatedly charged with "monopolizing" and taking over California agriculture, the Chinese share was not insubstantial. As late as 1920, Chinese leased thousands of acres in the rice-growing areas of Colusa, Yolo, and Yuba counties, in the fruit-growing Vaca Valley of Solano County, and in the central Delta in San Joaquin and Contra Costa counties.

The aggregate number of Chinese truck gardeners, farmers, and farm laborers in the major agricultural counties listed in the manuscript schedules of the 1880 and 1900 censuses also underscores Chinese persistence in agriculture in the last decades of the nineteenth century. As table 32 shows, although the number of Chinese farmers decreased in the Sacramento–San Joaquin Delta between 1880 and 1900, more Chinese tenants could be found at the later date in other major agricultural regions that had developed later, including the Vaca Valley of Solano County, the Santa Clara Valley, and the fruit belt in the "Uplands" of Placer County. Moreover, the number of Chinese farmworkers had increased, despite a sharp decline in the overall Chinese population. The fact that the number of Chinese agriculturalists increased in the same period that the overall Chinese population decreased meant that the proportion of Chinese agriculturalists within the total Chinese population rose significantly. Chinese who earned a living through farming in 1900 constituted over

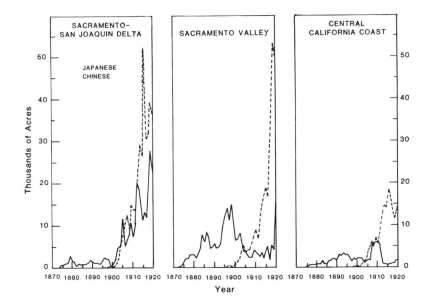

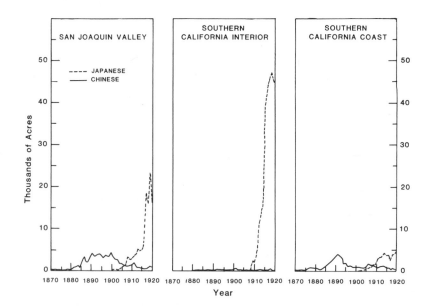

two-fifths of the Chinese population in at least nine counties: Tehama, Placer, Sutter, Yolo, Solano, Sacramento, San Joaquin, Contra Costa, and Monterey. If only the agricultural townships in these counties are considered, the Chinese farming population made up an even larger proportion of the total Chinese population.

That the Chinese managed to stay in agriculture long after they had allegedly been driven out was the result of three factors: their own tenacity, their utility to landowners, and the ineffectiveness of their opponents. The first two factors have been documented in the preceding account, but some comments must be made about the third. Unlike urban areas where workingmen who feared Chinese competition in the labor market swelled the ranks of anti-Chinese groups, many of the anti-Chinese activities in the rural areas seemed to have been led either by men espousing white supremacist values or by hoodlums out to enjoy themselves by tormenting "Chinamen." An organized white supremacist group such as the Order of Caucasians undermined its own effectiveness through its use of terrorist tactics. On the other hand, it was not easy to congregate less-violence-prone groups, such as small farmers who opposed the Chinese because they saw the latter as tools of "land monopolists," for common action because they operated farms in scattered locations. Finally, white farmworkers, who might have been most vehemently opposed to the Chinese, were not an easily organizable lot either. Despised as tramps, stereotyped as drunken, single, shiftless men, it was said in 1877 that the group's

Figure 1   *Total Acres Leased by Chinese and Japanese Tenants as Recorded in California County Archival Documents*
SOURCE: *Computed from data in "Leases" and "Chattel/ Crop/Personal Property Mortgages" of 40 California counties. For a complete listing, see the section on California County Archival Materials in the Bibliography.*
NOTE: *The acres in the following three regions are underrepresented: Central California Coast, San Joaquin Valley, and Southern California Coast. The mortgage records have been disposed of in most of the counties in those regions.*

Table 31 Total Improved Acres and Acres Leased by Chinese and Japanese Tenants by County and Decade, 1890–1920

	1890			1900		
Counties Grouped according to Regions	Total Improved Acres	Chinese Leased Acres	% Chinese	Total Improved Acres	Chinese Leased Acres	% Chinese
SACRAMENTO VALLEY						
Butte	360,273	1,467	0.41	302,029	844	0.28
Colusa	772,355	NS	NS	358,227	594	0.17
Glenn	NYE	NYE	NYE	355,781	180	0.05
Placer	140,023	411	0.29	121,063	289	0.24
Sacramento	341,601	648	0.19	327,159	1,232	0.38
Shasta	88,547	495	0.56	86,540	158	0.18
Solano	366,419	0	0	344,058	90	0.03
Sutter	225,997	1,795	0.79	206,877	2,060	1.00
Tehama	395,661	1,765	0.45	269,693	2,347	0.87
Yolo	308,923	0	0	351,213	0	0
Yuba	165,916	356	0.21	154,013	936	0.61
SAN JOAQUIN VALLEY						
Fresno	951,490	858	0.09	786,337	1,672	0.21
Kern	139,659	617	0.44	324,031	720	0.22
Kings	NYE	NYE	NYE	262,148	1,637	0.62
Madera	NYE	NYE	NYE	277,721	240	0.09
Merced	626,596	945	0.15	613,376	NS	NS
San Joaquin	559,784	515	0.09	652,923	0	0
Stanislaus	654,041	30	0	622,700	0	0
Tulare	588,118	791	0.13	546,289	267	0.05
CENTRAL COAST						
Alameda	222,885	319	0.14	226,118	0	0
Contra Costa	260,236	600	0.23	262,617	13	0.01
Monterey	479,350	421	0.09	373,605	952	0.25
Napa	147,311	0	0	111,966	18	0.02
San Luis Obispo	273,179	30	0.01	412,356	40	0.01
San Mateo	117,581	0	0	72,429	0	0
Santa Barbara	237,178	37	0.02	202,982	483	0.24
Santa Clara	281,351	383	0.14	290,285	813	0.28
Santa Cruz	76,125	25	0.03	62,849	149	0.24
Sonoma	308,080	576	0.19	221,374	187	0.08
SOUTHERN COAST						
Los Angeles	436,792	3,377	0.77	518,744	1,086	0.21
Orange	192,559	40	0.02	236,847	5	0
San Diego	238,942	55	0.02	229,791	10	0
SOUTHERN INTERIOR DESERT						
Imperial	NYE	NYE	NYE	NYE	NYE	NYE
Riverside	NYE	NYE	NYE	216,033	0	0
San Bernardino	307,845	153	0.05	96,920	523	0.54

SOURCE: Compiled and computed from U.S. Bureau of the Census, *Eleventh Census of the U.S.: Statistics of Agriculture; Twelfth Census of the U.S.: Statistics of Agriculture; Thirteenth Census of the U.S.: Statistics of Agriculture*; and *Fourteenth Census of the U.S.: Statistics of Agriculture*; and records of leases and chattel mortgages for the counties listed above. (For a full listing, see the Bibliography.)
NOTE: NYE = not yet established; NS = not stated.

		1910					1920		
Total Improved Acres	Chinese Leased Acres	% Chinese	Japanese Leased Acres	% Japanese	Total Improved Acres	Chinese Leased Acres	% Chinese	Japanese Leased Acres	% Japanese
247,097	307	0.12	60	0.02	253,745	1,980	0.78	7,100	2.80
336,509	640	0.19	203	0.06	302,429	11,062	3.66	17,529	5.80
309,765	0	0	0	0	336,482	0	0	3,081	0.92
98,608	512	0.52	2,684	2.72	136,455	293	0.21	1,504	1.10
275,682	715	0.26	8,204	2.98	399,024	1,136	0.28	11,434	2.87
96,217	0	0	0	0	103,470	0	0	0	0
310,452	1,066	0.34	1,883	0.61	299,264	6,182	2.07	2,307	0.77
199,510	572	0.29	843	0.42	232,070	1,349	0.58	10,963	4.72
186,642	633	0.34	120	0.06	232,722	0	0	140	0.06
317,268	0	0	1,785	0.56	300,094	3,765	1.25	7,490	2.50
94,250	30	0.03	0	0	98,997	1,405	1.42	7,116	7.19
590,205	271	0.05	480	0.08	672,591	40	0.01	1,432	0.21
315,387	0	0	0	0	390,932	0	0	0	0
196,569	525	0.27	125	0.06	259,639	0	0	2,745	1.06
391,086	160	0.04	0	0	262,971	220	0.08	0	0
607,742	NS	NS	1,045	0.17	506,582	0	0	80	0.02
611,762	5,381	0.88	8,991	1.47	599,403	13,407	2.24	27,023	4.51
512,189	60	0.02	0	0	477,871	0	0	1,776	0.37
507,024	264	0.05	210	0.04	544,598	168	0.03	2,173	0.40
177,314	200	0.11	513	0.29	185,324	0	0	8	0
262,152	2,221	0.85	908	0.35	238,369	2,538	1.06	824	0.35
371,509	886	0.24	2,191	0.59	398,320	1,099	0.28	1,766	0.44
101,114	11	0.01	131	0.13	116,723	0	0	0	0
326,928	523	0.16	0	0	402,269	0	0	348	0.09
100,800	7	0.01	3	0	77,736	NS	NS	14	0.02
215,552	4,414	2.05	2,154	1.00	210,353	338	0.16	12,093	5.75
237,170	334	0.14	829	0.35	206,890	0	0	844	0.41
66,875	9	0.01	224	0.33	67,838	175	0.26	289	0.43
248,271	85	0.03	81	0.03	251,730	0	0	106	0.04
418,998	889	0.21	939	0.22	483,096	660	0.14	2,729	0.56
189,463	0	0	785	0.41	200,945	0	0	259	0.13
234,045	293	0.13	0	0	262,646	63	0.02	2,348	0.89
176,069	0	0	1,913	1.09	310,708	0	0	0	0
278,151	0	0	0	0	348,538	125	0.04	2,123	0.61
136,625	180	0.13	2	0	175,272	NS	NS	1,660	0.95

Table 32 The Chinese Agricultural Population as a Percentage of the Total Chinese Population, 1880 and 1900, Selected Counties

	1880					1900				
	Truck Gardeners	Farm Laborers[a]	Farmers	Total Agric. Popul.	% of Total Popul.	Truck Gardeners	Farm Laborers	Farmers	Total Agric. Popul.	% of Total Popul.
SACRAMENTO VALLEY										
Butte	82	17	21	120	3.2	31	223	8	262	36.8
Colusa	27	169	0	196	20.2	8	43	3	54	19.7
Placer	75	150	0	225	10.3	8	334	55	397	37.8
Sacramento	184	537	428	1,149	23.5	58	1,165	231	1,454	44.7
Solano	26	240	0	266	26.8	2	417	29	448	49.6
Sutter	43	33	2	78	29.3	50	52	5	107	39.1
Tehama	183	53	0	236	30.5	7	302	74	383	52.5
Yolo	46	30	0	76	12.5	7	93	42	142	41.0
Yuba	87	124	0	211	9.8	83	25	1	109	15.2
SAN JOAQUIN VALLEY										
Fresno	8	23	4	35	4.6	71	403	17	491	27.7
Kern	20	78	15	113	16.1	8	473	36	517	57.1
San Joaquin	86	471	99	684	34.3	5	917	156	1,078	57.5

COASTAL COUNTIES

Alameda	21	455	76	552	12.6	54	360	44	458	20.7
Contra Costa	0	168	52	220	30.1	5	180	24	209	44.9
Los Angeles	208	140[b]	0	348	29.8	95	690[b]	298	1,083	33.7
Monterey	27	62	1	90	24.2	20	411	26	457	53.3
Santa Clara	18	763	2	783	29.1	41	670	40	751	43.2
Sonoma	12	202	0	214	23.3	9	148	14	171	28.5

SOURCE: Compiled and computed from U.S. National Archives, Record Group 29, "Census of U.S. Population" (manuscript), 1880 and 1900.

[a] Includes those labeled as farm laborers, as farm hands, and as working on farm.

[b] Includes ranch hands and vaqueros.

demoralization ... [was] the result of a long train of circumstances running through the course of many years. Some of them [were] fugitives from the law. Others [were] men who came to the State long ago to dig for gold, and they became dissipated because the lack of titles in the mining districts did not encourage matrimony and permanent residence. Wages have been so high that farmers could not afford to employ them, except for a few months of seeding and harvesting, and the other months were spent in intoxication as the most agreeable pastime.[31]

Some four decades later, when Carleton Parker investigated the social background of white farm laborers in the aftermath of the bloody hop-pickers' uprising in Wheatland in 1913, he found that three-quarters to nine-tenths of the men were unmarried, over 40 percent were under thirty years of age, and more than half had no trade whatsoever. As many as two-thirds admitted that they intended to continue "floating," with no desire to obtain steady employment.[32] Men such as these might engage in sporadic violence against the Chinese, but they hardly constituted a sustained force that could succeed in driving the latter out.

Survival and Community

The ability of the Chinese to persist in farming did not necessarily lead to the establishment of permanent Chinese communities in rural California. After about 1920, of all the places where Chinese had farmed in California, a Chinese farming community survived only in the Sacramento–San Joaquin Delta. Many temporary communities did become established elsewhere, but since they were not normal in either a demographic or a social sense, they failed to continue in existence beyond the natural life-spans of their members. These communities were peculiar in that they consisted almost entirely of men, many of whom were married, but few of whom had their wives with them. Few Chinese women emigrated, because Chinese culture dictated that they should remain home to perpetuate the family line and to serve their mothers-in-law. If their husbands succeeded in making some money abroad, the expectation was that they would sooner or later return. But equally important, even if no cultural constraints had existed, poor men going abroad to earn money could ill afford to take wives along. And of the married women who did join their husbands in America, some did not enjoy life there and returned to China. Among unmarried women, only those who could bring economic returns—such as prostitutes, laundresses, or seamstresses—were considered valuable enough to ship overseas. Finally, those men who wished to send for wives from China after the passage of the Chinese Exclusion Acts could no longer do so; consequently, the very small number of American-born Chinese nubile females were highly sought after. Their birth and existence enabled Chinese American communities to survive in the harsh years of exclusion.

Womanless households, then, rather than families, formed the essential units of Chinese communities in both urban and rural California. Like the households of extended families in traditional China, Chinese households in rural California functioned as production units. In agricultural areas, households frequently consisted of tenants and their associated laborers. The controlling factor in the composition of households was the number of men needed to operate a farm. In the early years, roles seemed to be interchangeable, as an inverse relationship existed between the number of partners and farm laborers in a household: the more partners, the fewer laborers, and vice versa; but as time passed a more hierarchical household structure developed. Table 33 depicts the household composition of Chinese farmers in the Sacramento Delta. The left half of the table shows the composition of households in 1880; the right half shows household composition in 1900. (One more row was added for 1900 because the number of households with fifteen or more Chinese farm laborers had become significant by 1900.) In 1880 thirty-six of the eighty-five households contained no farm laborers. In these, Chinese tenant farmers lived and worked with other Chinese, who were listed by census takers as partners of the heads of the households, indicating that household members were of equal status. Farm laborers living with Chinese farmers no doubt hoped to save up enough money eventually to become partners of the farmers. In 1880, among those households with farm laborers, most had no more than five or six. The modal size of

Table 33   The Sacramento–San Joaquin Delta: The Household
Composition of Chinese Farmers, 1880 and 1900

Number of Laborers and Farm Laborers Living in Household	1880							
	Number of Additional Farmers Living in Household							
	0	1–2	3–4	5–6	7–8	9–10	11+	Total
0	1	6	8	11	4	4	2	36
1–2	3	7	2	0	0	0	0	12
3–4	8	5	3	1	0	0	0	17
5–6	3	1	1	1	0	0	0	6
7–8	3	2	1	0	0	0	0	6
9–10	2	1	0	0	0	0	0	3
11+	4	0	1	0	0	0	0	5
Total	24	22	16	13	4	4	2	85

SOURCE: Compiled and computed from U.S. National Archives, Record Group 29, "Census of U.S. Population" (manuscript), 1880 and 1900.

households among the Delta's Chinese farmers in 1880 was six
persons.

Twenty years later, two changes had appeared: there were
fewer households without farm laborers, and the average number of farm laborers within those households containing them
had increased. Not only had farming become larger in scale, requiring more hands per production unit, but social relations
appeared to be less egalitarian. This did not mean that the partnership system was no longer available to Chinese farmworkers
as an avenue of upward mobility; it simply indicated that by
1900 the gap had widened between Chinese tenants near the
top of the status hierarchy and the men who worked for them.
Whereas in the early years there seemed to be no rigid class
barriers among different groups in the Chinese population in
rural California—movement in and out of various occupations
being quite fluid—such divisions had become more apparent by
the turn of the century as upward social mobility became more
difficult to achieve. On the other hand, it is important to realize
that whatever class divisions that might have existed were miti-
gated by mutual dependence, kinship and village ties, and ethnic
solidarity in the face of hostility from society at large.

Few real Chinese families existed in nineteenth-century Cali-

| Number of Laborers and Farm Laborers Living in Household | 1900 | | | | | | | |
| | Number of Additional Farmers Living in Household | | | | | | | |
	0	1–2	3–4	5–6	7–8	9–10	11+	Total
0	2	0	3	2	2	1	0	10
1–2	6	2	2	2	4	0	0	16
3–4	9	2	1	3	2	0	0	17
5–6	14	2	0	1	0	0	0	17
7–8	9	2	1	1	1	0	0	14
9–10	5	1	1	0	0	0	0	7
11–14	8	1	1	0	0	0	0	10
15+	13	1	0	0	0	0	0	14
Total	66	11	9	9	9	1	0	105

fornia because, even though there were Chinese women present, the great majority of them were women who had been brought to America to work as prostitutes.[33] Most of the women were in their twenties, though there were also teen-aged girls and women in their thirties and forties. From the 1850s through the 1870s, prostitutes could be found widely scattered in the larger mining camps and towns and usually made up 3 to 7 percent of the Chinese population in each county. The largest concentrations, as might be expected, were in the larger towns of the more prosperous mining counties.

To get an accurate count of the total number of prostitutes, it is necessary to include not only those who were explicitly listed as prostitutes but also those who can be deduced to have been. That is to say, it is almost entirely certain that single women living together in all-female households whose members had no listed occupation were prostitutes. Virtually all the Chinese women in northern rural California in 1860 were prostitutes. Ten years later, even though the number of wives had increased, 82 percent of the 557 Chinese women in the Northern Mines, 79.1 percent of the 440 women in the Southern Mines, and fully 88 percent of the 276 women in Sacramento County were prostitutes. (No attempt was made to count the number of Chinese

Chinese servant of the William Keith family, Berkeley, Alameda County, California, ca. 1905. (Photograph courtesy of the California Historical Society, San Francisco)

women in the other counties of the Sacramento Valley, because the number of Chinese there was small.)

Most of the women lived in groups of two to six, but there were also one-person households consisting of lone prostitutes. Not all prostitutes lived in brothels, and few brothels had owners in residence. Most brothel owners listed in the census were men, but a small number of madams could also be found. The 1900 manuscript census showed that some brothels in San Francisco's Chinatown were run by couples, among whom a few wives were apparently continuing in their profession. Some prostitutes boarded in households with various men. In a number of cases, the composition of households as listed in the census suggests that groups of miners or laborers probably shared one woman.

Married women became numerous only around 1880. The 1880 manuscript census provided information on the relationship of individuals to the heads of households, and quite a number of Chinese women were listed as wives. As table 34 reveals, 29 percent of the 317 women in the Southern Mines, 30.8 percent of the 439 women in the Northern Mines, 41.8 percent of the 225 women in the Sacramento Valley, and 35.7 percent of the 157 women in the San Francisco Bay Area (excluding the city and county of San Francisco) were listed by census takers as wives. The great majority of them did not work, although a handful did. Sewing and washing clothes were the main occupations open to Chinese women in that period. There were even a few married women who still earned a living as prostitutes. Among the women who were not specifically listed as wives, by far the largest proportion were prostitutes. (In table 34, two figures are given: one for "prostitutes [stated]," the other for "prostitutes [probable].") Perhaps as a reflection of the money that miners had for various forms of recreation, there were more prostitutes in the two mining regions than there were in the agricultural areas.

Twenty years later, as a result of the strict enforcement of redlight district laws and of Chinese exclusion, the number of Chinese women, as well as the number and proportion of prostitutes among them, had declined. Now, prostitutes represented only 16.0, 11.6, and 14.5 percent of the Chinese women in the Northern Mines, the Sacramento Valley, and the Bay Area, respectively. Conversely, the number of wives had increased to 44.6, 49.5, and 46.5 percent. The number of children (as reflected in the number of persons without occupations among the "non-wives," most of whom were girls under age fifteen) had also gone up. Resident families, though few and far between, were being established in the last two decades of the nineteenth century.

Relatively few Chinese agriculturalists had resident wives.[34] In 1880 in the counties of northern California which were surveyed for table 34, only seven farmers, eight gardeners, and two farm laborers had wives living with them. In 1900, not counting the Southern Mines, there were eighteen farmers, six gardeners, and eighteen farm laborers in northern California with resident wives. About one-third of these women were found in the Sacramento–San Joaquin Delta. Since the Delta among all the agricultural areas where Chinese farmed had the clearest beginnings of a family society, the situation there will be described in greater detail.

In the Sacramento–San Joaquin Delta in 1900, there were fifty-nine married Chinese women in Sutter, Franklin, Georgiana, O'Neal, and Union townships. By 1910, however, only

Table 34   Distribution of Chinese Women by Region, Marital Status, and Occupations, 1880 and 1900

	1880		
	Southern Mines[e]	Northern Mines[f]	Sacramento Valley[g]
LISTED WIVES:[a]	92	135	94
Housewives[b]	82	121	84
Seamstresses	0	4	7
Laundresses	2	1	1
Prostitutes (stated)	4	6	0
Miners	1	1	0
Laborers	0	0	0
Others	3	2	2
NON-WIVES:[c]	225	304	131
Prostitutes (stated)	81	89	5
Prostitutes (probable)[d]	83	115	66
Seamstresses	3	0	4
Laundresses	3	11	0
Miners	2	0	0
Laborers	2	10	4
"Keeping house"	7	37	21
No occupation stated	42	41	22
Others	2	1	9
TOTAL NO. OF WOMEN	317	439	225
% who were wives	29.0	30.8	41.8
% who were prostitutes (stated and probable)	53.0	47.8	31.6

SOURCE: My tally from U.S. National Archives, Record Group 29, "Census of U.S. Population" (manuscript), 1880 and 1900.

[a] This number is not the same as the number of married women. Only women specifically listed as "wife" are included.

[b] Housewives include wives listed as "keeping house" and those with no stated occupation.

[c] Non-wives include women who were single, widowed, married but not listed as wives, and those for whom no information was given in the column "Relationship to Head of Household."

[d] Women living in all-female households with no stated occupations were counted as "probable" prostitutes.

[e] Includes El Dorado, Amador, Calaveras, Tuolumne, and Mariposa counties.

[f] Includes Plumas, Butte, Sierra, Yuba, Nevada, and Placer counties.

[g] Includes Tehama, Colusa (for 1880), Glenn (for 1900), Sutter, Sacramento, Yolo, and Solano counties.

[h] Includes Sonoma, Napa, Marin, Contra Costa, Alameda, Santa Clara, and San Mateo counties; *excludes* San Francisco.

[i] Information for Colusa County for 1900 is not available, so the numbers in this column are undercounts.

	1900		
S.F. Bay Area[h]	Northern Mines	Sacramento Valley[i]	S.F. Bay Area
56	78	98	74
46	75	94	69
4	0	0	0
2	0	0	0
1	0	0	0
0	0	0	0
0	2	3	1
3	1	1	4
101	97	100	85
13	6	14	0
22	22	9	23
8	0	6	1
4	1	0	1
0	0	0	0
1	2	3	3
14	3	3	1
29	58	50	50
10	5	15	6
157	175	198	159
35.7	44.6	49.5	46.5
22.9	16.0	11.6	14.5

Table 35   *The Sacramento–San Joaquin Delta: Occupations of Men with Resident Wives, 1900 and 1910*

Occupation of Husbands	Number in 1900	Number in 1910
Business Entrepreneurs	(32)	(12)
Merchant/grocer	21	10
Restaurant keeper	4	0
Butcher	1	0
Boardinghouse keeper	1	1
Broom factory owner	1	0
Cigar stand owner	1	1
Vegetable peddler	1	0
Laundryman	2	0
Artisans and Professionals	(5)	(2)
Clerk	1	1
Doctor	1	0
Dentist	1	0
Printer	1	0
Shoemaker	0	1
Tailor	1	0
Farmers and Farm Laborers	(16)	(10)
Fruit grower	1	1
Farmer	5	4
Rancher	0	2
Wine maker	0	1
Farm foreman	0	1
Farm laborer	10	1
Nonagricultural Laborers and Personal Service	(6)	(2)
Cook	2	0
Common laborer	3	2
Porter	1	0
TOTAL	59	26

SOURCES: U.S. National Archives, Record Group 29, "Census of U.S. Population," 1900 (manuscript), and "Census of U.S. Population," 1910 (manuscript), for Sutter, Franklin, and Georgiana townships in Sacramento County, and O'Neal and Union townships in San Joaquin County.

twenty-six men in the same townships had resident wives. In both years, there were some married women whose husbands must have been away when the census takers came, because the latter were not listed. At the same time, a number of women were shown as heads of households. Even more unexpectedly, a few women lived alone, including Ah Sen, a seventy-year-old poultry

raiser in Cosumnes Township, Sacramento County, who had
been in the United States since 1860 and owned her own farm.
The occupations of the husbands of resident wives in the Delta in
1900 and 1910 are given in table 35, which shows that the largest
number of resident wives were married to merchants and other
entrepreneurs, while agriculturalists ranked second. That ten
farm laborers had wives with them in 1900 meant that perhaps
not all farm laborers led a completely poverty-stricken or migratory existence. By 1910, however, only one farm laborer in the
Delta had a wife with him.

Many of the wives were considerably younger than their husbands. The age difference was especially notable among farm
laborers, some of whom were twice or even almost three times
as old as their wives. The age difference resulted from the long
years men took to accumulate sufficient means to marry and
from the fact that since women from China could no longer come
easily after 1882, the only nubile females available were the
handful of second-generation girls who were coming of age at the
turn of the century. Twenty-one of the wives in the Delta in 1900
had been born in California, and a 25-year-old woman, May,
who was married to a printer, had been born in Idaho. This
couple had given their children American names: Pauline, Eva,
and Edmund, and lived as a nuclear family in Sutter Township,
Sacramento County, this fact perhaps reflecting the mother's
more Americanized orientation.

There were other children with American names. Also in
Sutter Township, the children of a cook were named Mamie,
George, and Lillie and attended a mission school. The largest
group of American-named children was found in the family of
Ah Sing, a 49-year-old farmer in Center Township, Sacramento
County, and his wife Lillia, who had named their seven children,
ranging in age from 17 to 1, Charley, Johnnie, Sadie, Lottie,
Bessie, Fanny, and Hattie.

In the Delta, only six Chinese farmers out of over four hundred in the area in 1900 had wives living with them. Chun
Lou-see, a 39-year-old woman born in California, was married
to Chun Lum, a 60-year-old fruit grower in Franklin Township,
who could read and write (presumably Chinese), and speak
English. Lou-see, who was illiterate, had three sons, Chun Lum,
Jr., 17, Chun Lum Sing, 15, Chun Lum Ching, 8, and a 1-year-

Mrs. Hing Owyang, Sr., and her children, Runyon-Dorsey Ranch, near Courtland, Sacramento Delta, California. (Photograph courtesy of the Sacramento River Delta Historical Society, Locke, California)

old daughter, Chun Yoke. With this family lived four farm laborers, ranging in age from 22 to 52, and four fruit packers, aged 49 to 63. Though the census taker did not state how long Chun Lum had been in the United States, six of his employees had arrived before 1882, one in 1884, while the eighth and youngest was a native Californian. All of Chun Lum's employees could read and write, and half of them supposedly could speak some English. Four of them were married and four single.

Four of the six farming families lived in Georgiana Township in the central Delta. One consisted of Yen Lee, aged 47, who had been in the United States for thirty years and could read and write, as well as speak English, his 28-year-old wife (for whom no first name was shown), their two sons, aged 9 and 5, a daughter, aged 7, and a baby girl of 5 months. All the children had been born in California. No one else resided with this rare nuclear family. Another family consisted of a 32-year-old literate farmer named Ah Yen who had come to the United States in 1889, his 20-year-old wife, Low See, and two farm laborers, aged 52 and 60. The older man, who was single, was an old California hand, having been in the country since 1864. A third family was headed by Gee Chom, aged 37, a twenty-year resident of the United States who could not read or write but could speak English, his wife Kee, aged 23, who had been in the United States since she was 5 years old and was one of the few women working as a farm laborer, and eight farm laborers, ranging in age from 15 to 62. All had been in the United States for at least twenty years except the 15-year-old boy, who had immigrated only in 1897. None could read or write, but all claimed to be able to speak English. Two of these men were married and one was a widower. The fourth family consisted of Jack Wong, aged 50, in the United States since 1874, unable to read or write but able to speak English, his 16-year-old wife, Quoy Get, who came in 1893, and fourteen farm laborers, aged 22 to 61, three married, one a widower, all except five of whom had been American residents for at least twenty years.

The lone Chinese farmer in the San Joaquin part of the Delta with a resident wife was 51-year-old Ah Bin of O'Neal Township who came to the United States in 1880. Able to read and write and to speak English, he lived alone with his wife, Bin Wa-fa, who was 60. She could also read and write, but could not speak English. She, too, had come to the United States in 1880. This couple was unusual in that she was older than he, was literate, and probably immigrated together with him; moreover, they had no one else living with them.

Farm laborers with wives in the Delta in 1900 included forty-one-year-old Chang Bha, married to 27-year-old California native Chang Young-su, who had borne him two sons and two daughters; 58-year-old Hoo La, married to 48-year-old Jess Lo, who had given him one daughter and two sons; 50-year-old Ah Hat, who had two wives, 30-year-old Ah Loke and 31-year-old

Choy Won, both of whom had come from China ten years earlier; 34-year-old Yet Mein, who also had two wives, 19-year-old
Chuck Ee and 39-year-old Young Ou, both born in California;
49-year-old Cong Low, married to 28-year-old Seen Yoke; 45-
year-old Poon Leon, married to 30-year-old Ah Hoe, who had
borne him a son and a daughter; 51-year-old Lim Ku, married to
26-year-old Choy Yoke; 63-year-old Cho Wong, married to 53-
year-old Ah Gun; 51-year-old Ah Gu, married to 45-year-old Ah
Toy; and 60-year-old Ah Jim, married to 20-year-old Sim Choy,
who lived alone with him. Most of these farm laborers' households contained additional members, largely other farmworkers.

Only seven Chinese farmers in the Delta had wives in residence in 1910. Owyang Ah Cho of Franklin Township, aged 51,
in the United States since 1885, had been married to 38-year-old
California-born Sun Lau for twenty-three years. They had three
boys and three girls, ranging in age from 4 to 19. Tai-kou, the
oldest son, worked as a farm laborer. Living with them were
seven other farm laborers, most of whom had come to the United
States in the preceding decade. Also in Franklin Township was
Jan Ting Sun, a 31-year-old farmer, himself born in California,
who was married to Jan Ting Wong Su, 25, a fellow Californian.
They had a baby boy, Jan Ting Sing. Seven farm laborers lived
in this household also. In Georgiana Township, an orchardist,
Wong Gong, 40, had a wife, Wong Yet Gum, 25, two sons and a
daughter, and two farm laborers in his household. Another farmer, 36-year-old Chung Jow, a native of California, was married
to Chung Eck Your, 25, also born in California, by whom he had
two daughters; one farm laborer completed their household. A
third family in Georgiana Township consisted of 54-year-old Son
Quay, in the United States since 1876, and his wife, 45-year-old
Son Sing Lan, who had borne him two daughters in California.
The other members of their household were four farm laborers.
In O'Neal Township, a truck farmer, Wong Ting, 38, was married to 34-year-old Wong Gow, by whom he had three sons. He
must have owned a large truck farm, because his household contained thirty-four farm laborers and a bookkeeper, the latter a
relative. Finally, Wa Chinn, 63, an asparagus grower, had a 23-
year-old wife, She Cheung, and a daughter. A total of thirty-two
laborers lived with and worked for him, twenty-seven of whom
were Chinese and five Japanese.

Though far fewer farmers than merchants brought their China-born wives to live in America or married American-born women, it would be simplistic to argue that this was because farmers were less well-to-do. The evidence presented in this study indicates that many Chinese farmers in the Sacramento–San Joaquin Delta were quite well off. A more likely reason was that fewer of them could live in permanent homes than merchants, artisans, or professionals, moving as they did from one tract to another as their leases expired. Unlike merchants who lived in the small Chinatowns, where at least some amenities of civilization were available, farmers could not provide secure or pleasant surroundings for their women, who more often than not had to share their homes with large numbers of rough, single men. The resident wives of Chinese farmers and farm laborers were thus another group of unsung Chinese American pioneers.

Though few in number, the very fact that families did become established in the Delta is historically significant because what happened there might have been replicated elsewhere had immigration exclusion not been imposed. Chinese families sprang up in the Delta because there was a sufficiently large concentration of Chinese in that area for over half a century. In 1880, Chinese farmers made up 69 and 49 percent of all farmers in Georgiana and Union townships, respectively; twenty years later, the proportion of Chinese farmers still remained high at 54 and 49 percent. Elsewhere in agricultural California, Chinese failed to establish permanent communities, because their numbers did not constitute a critical mass and they farmed in each area for shorter spans of time.

Other factors contributed to the persistence of the Chinese presence in the Delta, and identifying these factors enables us to conjecture in what other localities the Chinese might have established a firm foothold had they remained long enough to do so. The geography of the Delta had a lot to do with the ability of the Chinese to stay. Farming on the peat islands of the central Delta was unpleasant but financially rewarding. Chinese tenants, who were among the earliest "tule farmers," established a stronghold there, and were not joined by others till three decades or more later when Japanese, Italians, Koreans, and Asian Indians appeared on the scene. In contrast, around the Delta's periphery, most of the tenants were white Americans from the South and Midwest, or Italians and Portuguese. Owner-operators, most of whom also farmed in the periphery of the Delta, were mainly Americans from the northeastern and north central regions and northern European immigrants.

Another aspect of the Delta's physical geography that helped Chinese communities to survive there to this day was the Delta's perceived remoteness. Though in fact quite close to San Francisco, until recent times it was accessible only by water. With its sunken fields, ribbons of levees, and hundreds of waterways often shrouded in fog, it was a world apart. Few people outside of the region knew of its existence, and anti-Chinese agitators, even if they had wanted to, would have had difficulty in reaching it.

Furthermore, a relatively stable ethnic division of labor enabled the Chinese to find a niche there. An analysis of data in the 1860, 1870, and 1880 manuscript agricultural censuses indicates that there was a relationship between the nativity of Delta farmers and the kinds of crops they grew.[35] In Franklin Township, farmers born in New England and the Middle Atlantic states, as well as British immigrants, dominated grain farming. Farmers from the north central and south central regions of the country were mainly engaged in diversified farming in 1860, but many members of this group had also become grain and hay growers by 1870. Until the Chinese came along, relatively few farmers grew potatoes. Between 1860 and 1880 an increasing number of European immigrants, particularly Germans, entered the northern Delta, where they grew grain, hay, vegetables, fruit, potatoes, and beans, and raised livestock. The Irish farm-operators cultivated mostly grain.

In Georgiana Township there were few farms that produced only grain and hay. As in Franklin Township, it was again farmers from the north central and south central states who were most prominent as operators of diversified farms in the early period. Potato cultivation was more important in the central Delta, and Americans and European immigrants as well as Chinese engaged in it. Unlike the Chinese, however, few white farmers specialized in potato cultivation.

Another way in which the ethnic division of labor can be demonstrated is the relationship between the nativity of farmers and their tenure status. A larger percentage of Americans born in New England and the Middle Atlantic states than those from other regions of the country were owner-operators—79 versus 59 percent. Northern European immigrants ranked second in terms of the percentage who owned farms. Eighty percent of those from Britain (excluding Ireland) were owner-operators. Among the Irish, 67 percent were owner-operators. Of the Germans, 65 percent owned the land they farmed, while 69 percent of the immigrants from other parts of Europe also owned their farms. Share-cropping was more than twice as common as cash rental among American tenants from the Midwest and the southeastern seaboard. It is not known which form of tenancy Italians and Portuguese preferred, since they did not become numerous until after 1880, and the manuscript schedules of the post-1880 agricultural censuses (which would provide such information) have not been released. Chinese tenants preferred to pay cash rent, although some also gave their landlords shares of the crop. And as was shown earlier, during the 1870s and 1880s Chinese dominated the farm labor force in the Delta. The niche that Chinese fitted into in Delta farming was thus quite clear: many specialized in cash tenancy and grew potatoes in the backswamps of the peat islands in the central Delta, while an even larger number engaged in all kinds of farm labor all over the region.

Finally, the diversity of Delta farming also made it possible for the Chinese to establish real communities in the area. Within any twenty-mile radius one could find potato and bean fields, hay and grain farms, orchards, vineyards, and truck gardens. The proximity of fields growing crops that ripened at different times allowed farm laborers headquartered in the Delta to find work for at least nine months of the year, as harvests began in March and continued until December. Since Chinese were valued as pruners, some were even employed in the slack winter months. So, instead of having to travel up and down the state, Chinese farmworkers found work almost throughout the year in the area and consequently were able to form relatively stable social groupings that became the nuclei for permanent communities.

Other areas possessing similar physical and social features where Chinese farming communities might also have been established were the Sutter Basin, the rice-lands of Colusa, Glenn, Butte, Yuba, and Sutter counties in the upper Sacra-

mento Valley, and the Imperial Valley. Each of these areas
was characterized by a harsh landscape, remoteness from
population centers, an ethnically mixed rural population
conducive to a clear division of labor along racial lines, and
the ability to support diversified farming within a reasonable
commuting distance. The Chinese did farm in considerable
numbers in the first two areas but not the third. However, they
formed no permanent communities in either the Sutter Basin,
which they entered in the late 1880s, or the adobe rice-lands
which they began to farm in the late 1910s, because appearing in
these areas after the various Chinese exclusion laws had cut off
immigration—the major source of replenishment for the Chinese
population in America—the Chinese found it difficult to sustain
their numbers. Finally, almost no Chinese farmed in the
Imperial Valley, because it too did not become arable until the
end of the first decade of the twentieth century when irrigation
works were built. Except for the Delta, by the late 1920s there
were few Chinese farmers or farm laborers left anywhere because
the population that remained grew old, their children did not
wish to continue in farming, and other groups—particularly the
Japanese, the Italians, and various European immigrants—
quickly displaced them from California's fields and orchards.
Their departure was a quiet one: having persisted through
decades of anti-Chinese violence and having survived in one
small pocket despite the virtual absence of women, the Chinese
left behind few visible marks on the landscape. Only thousands
upon thousands of faded records in dusty basements remain as
monuments to their pioneering spirit and, indeed, their valor.

Conclusion

THEY departed, but they did not disappear. After the 1920s, some Chinese continued to eke out a living in the small towns of agricultural California, others sought solace and safety in the Chinatowns of larger urban centers, many returned to China, others passed away. The real significance of the six or seven decades of active Chinese participation in California's agricultural development, however, lies not in their continued physical presence on the land but in the legacy of the diverse and integral role they played in the development of California agriculture. Working as truck gardeners, vegetable peddlers, commission merchants, farm cooks, tenant farmers, and owner-operators of farms, thousands of Chinese brought new land under cultivation, experimented with various crops, and provided much of the labor needed to plant, harvest, pack, preserve, and sell the crops in almost every major agricultural region of California.

In addition to being a significant part of California's agricultural history, the achievements of the Chinese in California agriculture are also an important part of Chinese American history for the simple reason that before the turn of the century, the majority of the Chinese in America resided in rural areas. Even if this had not been so, the rural Chinese experience would still have been notable because it differed in many ways from the urban one, which most scholars have erroneously assumed to be representative for all Chinese in America.

Three differences stand out. While the same classes—an entrepreneurial elite, a group of independent professionals and artisans, and a working class of laborers, cooks, and servants—

emerged in Chinese communities in both urban and rural areas, the relative proportions of the three social strata in each setting differed. In nineteenth-century California, the more urban the setting, the more Chinese belonged to the first two classes. Forty percent or more of the Chinese in San Francisco and Sacramento from about 1870 to the turn of the century earned a living as entrepreneurs, 5 to 12 percent were professionals and artisans, while the working class made up less than half of the urban Chinese population. In agricultural California, on the other hand, tenant farmers, merchants, and labor contractors, who constituted the rural elite, seldom exceeded 15 percent of the Chinese population; professionals and artisans of all sorts added up to only 1 to 3 percent; laborers and providers of personal service, who worked for both Chinese and whites, made up over 80 percent of the Chinese in agricultural regions.

Since the Chinese in the countryside, more than their urban peers, depended on white employers for their livelihood, the degree of interaction between Chinese and whites also differed in the two kinds of localities. Because of their greater interaction with whites, the rural Chinese, particularly the leaders, were less insulated from the larger society than the dwellers of urban Chinatowns, who could exist for years without interacting with whites at all. Chinese tenant farmers and rural merchants were an integral part of the economy—if not always the society—of agricultural California.

The urban and rural community leaders also had different bases of power. Chinatowns in urban America were ethnic enclaves in which the community leaders' status and power came mainly from their contacts with China. The largest merchants owned import-export businesses, and the most important political leaders headed organizations established on the basis of homeland ties. Moreover, since almost all the imported goods were sold to Chinese—who favored food, clothing, and utensils from China—it was in the interests of these merchants and directors of community associations to help their customers and clients to maintain strong ties to China. After all, provisioning their fellow countrymen was the chief source of the merchants' profit, and helping Chinese retain an orientation to the homeland was the main basis of the association leaders' power. The power and status of the rural community leaders, by contrast, were based on more diverse factors. One was the role they played as agents of city-based firms in distributing imported goods to Chinese scattered throughout North America. But perhaps more important than the connection to big-city merchants was their ability to facilitate the interaction between employers (Chinese and white) and Chinese laborers in farming as well as construction projects. By enabling large numbers of their fellow countrymen to earn a living and to survive in an alien, non-Chinese-speaking environment, on the one hand, while helping white landowners to find the labor and managerial expertise to bring new land under cultivation, on the other, they were depended upon by both groups. Since it was difficult for Chinese workers to survive without the knowledge and contacts that the tenant farmers, rural merchants, and labor contractors possessed, the latter were in a position to exploit as well as to aid the former. At the same time, since white landowners found their services so convenient, these members of the rural Chinese elite were also in a position to strike relatively good bargains in land leasing and other business transactions.

It is interesting, too, to compare the experience of the Chinese in rural California with that of the Chinese in Southeast Asia. Though in both areas of the world Chinese worked in mining and agriculture, there were differences that should be noted. With the exception of Chinese in Malaysia and Singapore who grew gambier, pepper, and rubber, some in Java who cultivated sugarcane, and those in Cambodia who produced pepper, Chinese in Southeast Asia engaged in agricultural production as food processors and distributors and not as direct producers. This was because in several of the Southeast Asian countries Chinese were expressly prohibited from cultivating rice and other crops. Chinese in California, as this study has demonstrated, participated in the direct cultivation of crops.

A more important difference is that for some three centuries, the Chinese served as a racially distinct minority who provided an economic link between the European colonial powers and the non-Chinese, nonwhite, peoples in Southeast Asia. Because the middleman role they played was so prominent, the Chinese in that region of the world—as a group—have been dubbed a "middleman minority." The Chinese in America, however—with the exception of a small number who settled in the Mississippi Delta, where they served as a buffer between white and black Americans—cannot be so characterized. There are two reasons for this: only a relatively small percentage of them performed middleman functions; and like the compradors in China, but unlike the Chinese in Southeast Asia, the Chinese American elite came from the same race and culture as the laborers whom they recruited and the consumers to whom they sold merchandise.

What the Chinese in nineteenth-century rural California did have in common with the Chinese in Southeast Asia was their great entrepreneurial drive—a drive that proved remunerative in capitalist economies where the production of raw materials was more important than the manufacture of finished goods. In such a context, it was possible for men with little capital, but a great deal of energy and a willingness to take risks, to achieve a measure of economic success. The length of time that Chinese were able to perform entrepreneurial functions in Southeast Asia and in the United States differed, however. In Southeast Asia, Chinese entrepreneurs gained a firm foothold in the colonial economies over the course of several centuries; in California, where industrialization did take place and where whites dominated all sectors of the economy from the American conquest onward, Chinese were found to be useful only for several decades after the gold rush when labor was so scarce and high-priced that white Americans had to call on Chinese middlemen to help them tap the vast reservoir of Chinese labor available.

Another feature common to the Chinese in Southeast Asia and those who came to California is that being niche-fillers—doing work that others either did not wish to or could not do—has been costly. The history of the Chinese diaspora is replete with examples of Chinese being persecuted as conditions changed: functions that they performed which were deemed valuable at one time were no longer considered desirable later on. Compared with other areas of the world to which they have emigrated, the Chinese in agricultural California encountered relatively less hostility than elsewhere because in the decades when they farmed most actively they tended to facilitate rather than dominate agricultural production. Just at the point when they might have gained sufficient control over certain areas of California agriculture to have posed a threat to white farmers, the Chinese agriculturalists gave up farming. They did so because passage of laws that excluded Chinese laborers from entering the United States robbed them of the needed number of hired hands that had previously enabled them to be competitive. Thus, had they not left agriculture when they did, the Chinese probably would have encountered the same fate as the Japanese who began arriving in the late 1880s and who were eventually barred from buying and leasing land altogether. The brevity of their venture, however, in no way reduces the measure of their achievement in the bitter-sweet soil of California.

Essay on Sources

 THIS study relied heavily on two kinds of sources: the
manuscript schedules of the U.S. Census of Popula-
tion for 1860, 1870, 1880, 1900, and 1910, and the
manuscript schedules of the Census of Agriculture for
1860, 1870, and 1880; and legal records of property
transactions in the official archives of over forty California counties. The 1890 census was not used, because the schedules were lost in fire in the U.S. Department of Commerce
building in 1921; the 1900 and 1910 manuscript schedules of the
agriculture census have not been released to the public. (The
Bureau of the Census could not tell me where the documents
have been stored.)

The Manuscript Census

Current legislation requires the original census schedules to
be kept confidential for seventy-two years before being released.
The U.S. National Archives has undertaken the responsibility
for microfilming the schedules and selling the microfilms. For
that reason, when citing the manuscript population census,
the U.S. National Archives, instead of the Bureau of the Census,
is given as the author. On the other hand, the manuscript schedules of the agriculture census were returned to each state. I used
a microfilm copy of these schedules—filmed by the California
State Library—on deposit at the Bancroft Library of the University of California, Berkeley.

It is important to know something about the history of the

census in order to assess the validity of the information the manuscript schedules contain. The first United States decennial census was taken in 1790. Before 1850 the primary purpose of the census was to fulfill the constitutional requirement that the number of representatives for each state in the House of Representatives be apportioned according to population. The Three-Fifths Compromise called for a separate count of the free population and of slaves. There was no federal office to direct the census effort, and reports were made by local marshals.[1]

In 1850 the census law was revised and a central Census Office was established. An effort was henceforth made to gather important social and economic data in addition to vital statistics. Also, questions for the schedule were revised to remove as much discretionary judgment from the heads of the household and the census takers as possible. Thus, instead of asking—as was previously done—a head of household how many persons in a certain age range lived in his household, the enumerator merely took down the ages of each household member. Or, instead of having the head of the household decide what category his occupation fell under—"manufacturing," "farming," and so forth—his occupation was simply written down without categorization. Decisions on how to classify the raw data were then made by census administrators and their staff in Washington, D.C.[2]

One of the major problems the Census Office encountered and took years to overcome was the occupational classification scheme. Only in the Ninth census taken in 1870 was a systematic scheme adopted. Francis Amasa Walker, an economist and statistician, who directed the 1870 and 1880 censuses, tried to collect occupational statistics in a more scientific way: instructions to the enumerators became more precise, a new classification scheme was devised, and the first calculating machines were introduced in the census office. Walker followed two criteria in selecting the occupations to be included in the published census: the number of persons engaged in a particular occupation, and occupations which merited attention by their prestigious nature. Under the first criterion, lower- and working-class occupations which had a large number of practitioners were included, whereas under the second criterion, most of the professions were included even though some had relatively few practitioners.[3]

Table 36   *Information Contained in Manuscript Schedules of Census of*
Population for 1860–1900

	1860	1870	1880	1900
Address			x	x
Name	x	x	x	x
Relation to head of household			x	x
Age	x	x	x	x
Month of birth if born in census year		x	x	
Sex	x		x	x
Race	x	x	x	x
Value of real estate owned	x	x		
Value of personal estate owned	x	x		
Occupation	x	x	x	x
Number of months in year unemployed			x	x
Birthplace	x	x	x	x
Whether parents are foreign born		x		
Father's birthplace			x	x
Mother's birthplace			x	x
Married within the year	x	x		
Marital status			x	x
Women: Number of children born				x
Women: Number of children living				x
School attendance	x	x	x	x
Literacy	x	x	x	x
Ability to speak English				x
Sickness/temporary disability			x	
Deaf/dumb/blind/insane/idiot/pauper/convict	x	x	x	
Number of slaves owned	x			
Citizen if 21 years or older		x		
Year of immigration				x
Whether naturalized				x
Whether on farm				x
Home owned or mortgaged				x

NOTE: The 1890 census schedules were lost in a fire in the Department of Commerce building in Washington, D.C. in 1921.

Table 36 shows the questions contained in the 1860, 1870, 1880, and 1900 population census schedules. The questions asked reflect some of the major concerns of the time. Immigration and ethnicity had become important issues by 1880, so the birthplaces of a person's parents were listed; in 1900 questions on the ability to speak English, year of immigration, and whether the person was naturalized were included because these had become issues.

Using the manuscript schedules from the population census for studying the demographic history of the Chinese population in America requires certain precautions. Since reliability of interpreters varied, so the quality of the information also varied from one locality to another. Owing to the language barrier, in some enumeration districts census takers left blank many items on the Chinese.

Information on age presents a problem, since the Chinese count a person's age from conception rather than birth. Moreover, all individuals automatically become one year older at Chinese New Year. So the age stated on the census schedules could be as much as two years more than the Western equivalent. This becomes a problem when the size of age cohorts is determined, skewing the age of the Chinese population upward.

Information on relationship to the head of household may also be inaccurate because it may reflect Chinese concepts of kinship and fictive kinship. For example, a fellow villager who, according to American usage, would not be considered a cousin, may be listed as such. Nonetheless, many census takers seemed quite careful in taking down information under this heading. For example, a clear distinction was made between "partners" and "servants" in households with members who were unrelated to the heads of the household.

Another problem was the undercount of the Chinese population. With the prevalence of anti-Chinese sentiment, and the fact that many Chinese frequently were harassed by all manner of whites, census takers were in all likelihood not perceived to be friendly or trustworthy. It is impossible to guess how many Chinese may have hidden from the census takers. Finally, after 1882, when laborers were barred from entry into the United States, some persons who were in fact laborers may have told the census enumerators that they worked in nonmenial occupations. The undercount in the post-1880 censuses may have been especially great after Chinese exclusion went into effect. The data on Chinese culled from the census for this study, therefore, should be considered as minimum numbers.

The census can be used to study the Chinese only as cohorts or groups. It is next to impossible to trace individuals from one census to the next, because the majority of the Chinese only gave their nicknames to the enumerators, using the diminutive "Ah" for their names (which some enumerators mistook for family names). Only an occasional Chinese listed his family name. Moreover, census takers seldom put down the names of cities or districts and simply listed "China" as their birthplace. So birthplace could not be used as corroborating evidence to trace individuals from one census period to the next.

The careful reader who is familiar with Ping Chiu, *Chinese Labor in California, 1850–1880: An Economic Study*, may have noticed that some of my figures on Chinese truck gardeners, farmers, and farm laborers differ considerably from his. There are two reasons for the discrepancies. Ping Chiu and I used different criteria for inclusion and exclusion. He included all members in a truck gardener's or farmer's household as gardeners or farmers, whereas I did not. For truck gardeners, I included only individuals specifically listed as "gardener," "garden laborer," and "works in garden." For farmers, I included those listed as "farmer," "orchardist," "fruit grower," and "vineyardist." For farm laborers, I included those listed as "farm laborer," "farm hand," "works on farm," and any "laborer" living in any household where the head of the household was listed as a "farmer," "orchardist," "fruit grower," or "vineyardist." I tallied "garden laborers" as "gardeners," but did not tally "farm laborers" or "laborers" living in the households of farmers as "farmers," because very few Chinese were listed as garden laborers, whereas a large number were shown to be farm laborers. Thus, there seemed little point in creating a separate category for the first instance, but there was every reason to count farm laborers separately.

The second reason for discrepancies between Ping Chiu's data and mine is that many of the microfilmed rolls were photographed from books with badly faded ink. In some rolls, the handwriting is so faint as to be almost illegible; in other rolls, ink spills and other smudges make it impossible to decipher certain lines. Whenever I found a large discrepancy between Ping Chiu's figures and mine, I made a re-tally. In some instances, I tallied my figures three times before deciding to use my count instead of his, even though the latter has been published.

There are also discrepancies between my data and those in the published census. The manuscript schedules contained a considerable number of arithmetical errors made by the staff of the

Census Bureau. Mistakes frequently occurred when numbers were transposed from one column to another or from one page to another. The magnitude of some of the errors made may be seen in the following example. In the 1870 manuscript agriculture census, schedules showing individual farms were given, as well as summary figures for each township and county. In Township Number One in San Mateo County, on a certain page of the schedules, there were three farmers who produced $1,600, $1,600, and $400, respectively, in market garden crops. This total of $3,600 was copied as $36,000 on the summary page, causing the total value of market garden crops for the township to be given as $96,860 instead of $64,460, as it should have been. This created an error of 50.3 percent. The figure in the published census was based on the erroneous figure on the summary page. Other similar errors were found in the pages for other counties. For that reason, I have used my own additions throughout. The only exception is table 21, where the figures given are from the published census. However, I used my own figure for Santa Clara County for 1870 because its ranking would have been out of order owing to an arithmetical error made by the Census Bureau. This fact is noted in the footnote to the table.

In contrast to the population census, very few Chinese were listed in the agriculture census. All those listed have been reproduced in several of the tables in the text of this study. The number of Chinese farmers listed in the census of population was far larger than the number of Chinese farmers found in the census of agriculture, for unknown reasons. An analysis of the information in the latter, therefore, was used mainly to assess the Varden . Fuller thesis rather than to shed additional light on Chinese agriculturalists.

California County Archival Records

County archives contain documents which certain units of county government are required by law to keep. Archival materials may be found in the offices of the County Clerk, the County Recorder, the County Auditor, the County Assessor, the County Board of Supervisors, and the County Board of Education.[4]

Though these documents are generated and kept for legal purposes, they may also be used fruitfully by researchers to study
many aspects of local social, economic, and political history. The
documents that were used extensively in this study were property records which may be found in several offices: the County
Recorder, the County Auditor, and the County Assessor.

There are three kinds of records pertaining to titles to property: those dealing with the original acquisition of titles, those
dealing with the transfer and encumbrance of property, and
miscellaneous records such as maps and surveys. In California,
original titles derived from two sources: Spanish and Mexican
land grants confirmed by the U.S. federal government, and
land in the public domain that was disposed of. Since the federal
government made various grants of land to the state, individuals
can receive patents to land in the public domain either directly
from the federal government or from the state.

In California, the greater part of the public domain containing
arable land was disposed of by the 1870s. Records of the acquisition of original titles may be found at several locations. Duplicate
copies of patents to land acquired from the federal public
domain—the originals of which are in Washington, D.C.—are
available in the Sacramento regional office of the Bureau of Land
Management. Patent files to state lands are kept at the office of
the State Lands Commission, also in Sacramento. County offices
have documents on the sale of swamp and overflowed lands and
homestead claims.

After the 1870s the majority of the property records were for
the transfer of titles rather than the original acquisition of titles.
The office of the County Recorder contains records of grant
deeds, leases, mortgages, releases and satisfaction of leases and
mortgages, tax sales, attachments and liens, certificates of sales
under execution, transcript of judgments, mining claims, and
water rights. The offices of the County Auditor and County
Assessor have assessment rolls and plat maps.

In general, the County Recorder has kept many more kinds of
records and with longer "runs" of them than the County Auditor
or County Assessor. The most complete property records are the
grant deeds because these constitute proof of ownership to real
property. Some counties have records going back to the Spanish
period. Different counties have preserved varying runs of other

kinds of property records. County Recorders with ample space have usually preserved all or most of their records, but those counties that lack space or which have moved the County Recorder's office from one building to another often either removed such documents from public access or disposed of records that are no longer legally in force. For example, because they are so voluminous, many chattel mortgage records were thrown away because once such mortgages were satisfied or released, the County Recorder was no longer required by law to keep them.

Assessment rolls in the office of the County Assessor usually go back only five years, so while it is possible to do research on historical trends in the ownership of property in a particular locality by using the deed records, it is virtually impossible to do research on long-term changes in the value of property, as the latter requires information from assessment rolls. Among the exceptions are Sierra County, where old assessment rolls from the 1850s to the present are available in the court house, and Yuba County, where one set of old assessment rolls is in the basement of the court house, with an incomplete duplicate set at the county library.

The different records—called "instruments"—were entered in separate books prior to the early 1920s. Each type of instrument was indexed separately, with a dual indexing system. One index provides the names of grantors (e.g., sellers, landlords, mortgagors); the second gives the names of grantees (e.g., buyers, tenants, mortgagees). Between 1920 and 1927, counties changed their system of record keeping, combining all instruments into the same set of books, known as the Official Records of the county. Only one pair of grantor-grantee indexes was then prepared. In the early 1970s, many County Recorder's offices began to computerize their record keeping. The present indexing system consists of a single index which mixes together the names of grantors and grantees. From a researcher's point of view, the pre-1920s records are much less time-consuming to use, because if one is interested in only one kind of instrument, one can simply look at the index (where available) for that particular set of records, without having to wade through a far larger index for all the different kinds of documents combined.

Although relatively few Chinese bought real property, grant

deeds were nevertheless useful for this study. Grant deeds provide information on the names of buyers and sellers, the legal description of the property, and the date of the transaction. A considerable number state the amount of cash involved, although there is no certainty whether such amounts represented the full sale price. During those periods when revenue stamps were used, the amount of these stamps provides a clue to the sale price, since a ratio of $1.10 per $1,000 was used.[5] For example, if a grant deed bears a revenue stamp of $16.50, then one knows that the property sold for $15,000.

The bulk of the information on Chinese tenancy came from lease documents, which usually provide far more information than grant deeds. In addition to the names of the landlords and tenants—known respectively as "party of the first part" and "party of the second part"—most leases gave the legal description, the amount of the cash rent or the share of the crops to be handed over to the landlord, the length of the lease period, and the uses to which the leased land could be put. In many leases, it was stated what improvements were already on the land, such as buildings, wells, fences, various equipment and machinery, farming implements, orchards or vineyards, irrigation works, and so forth. Such information provides insight into the changing state of agricultural technology in an area. For example, in Sacramento County, by piecing together information from many leases, it is possible to tell when farmers changed from using windmills to gasoline engines and electrical pumps as a source of power for drawing water for irrigation.

Clauses in some leases also offer glimpses into the social relations between landlords and tenants. Some landlords were fussy about the moral conduct of their tenants. The sale and consumption of alcohol was sometimes prohibited. (Such clauses appeared prior to Prohibition.) A number of leases forbade Chinese and Japanese tenants to work on Sundays, others prohibited Chinese women on the leased premises. Evidence of local feuds sometimes also surfaced in leases when landlords enjoined tenants from employing members of certain families.

In both deeds and leases, one of the most important items of information is the legal description, which refers to the exact location and size of the land under transaction. In this study, it has been possible to pin down the locations of land leased by the

Chinese and the size of these tracts because of the legal descriptions available in the lease records. So it is important to understand how these legal descriptions are given in the documents.

In towns, cities, and other incorporated areas, the legal description was often given as sections of named streets in a town plat or in terms of numbered and alphabetized blocks and lots. Without the relevant plats and subdivision maps, therefore, the exact locations cannot be determined. The size of town lots or city lots is frequently given in feet, but when a parcel is shaped irregularly, a far more complex system consisting of units of measurement used by surveyors—chains, links, and rods, together with compass points—is employed. Unfortunately, it is impossible for laymen to figure out the exact sizes of parcels so identified.

In the rural areas, legal descriptions of real property are given in several ways. The most convenient form is the United States rectangular survey system. Land is surveyed according to township and range coordinates emanating from given basepoints, of which there are three in California: the Humboldt, Mount Diablo, and San Bernardino Basepoints and Meridians. Township coordinates run north and south, while range coordinates run east and west. Each township and range is divided into thirty-six sections, each of which normally contains 640 acres, or one square mile. When the natural topography interferes with division into exact squares, however, sections are marked off as fractional sections. Sections are further subdivided into halves, quarters, and smaller units. A quarter section, or 160 acres, is considered in many parts of the United States to be a reasonable size for a family farm. (The famous "a mule and 40 acres" is based on the supposition that a quarter of a quarter section is sufficient for subsistence.) When legal descriptions are given in township and range coordinates, it is possible to be precise in determining the size of a purchased or rented parcel even when it is oddly shaped.

Not all real property is described in township and range coordinates, however. In those areas where Spanish and Mexican land grants existed, legal descriptions are given as subdivisions or lots in particular ranchos. In these cases, unless a subdivision map of the rancho can be found, or unless the lease or deed document specifically stated the number of acres contained in a lot, it is not possible to find either the exact location or the number of acres in a parcel of land.

In some areas, farms were simply referred to by their owners' names. If the legal description was given as "the northwest 40 acres of Mrs. Smith's ranch five miles southeast of Auburn," unless an ownership map of the area for that period can be found, it would not be possible to locate the ranch, though one might know approximately where it was. If a legal description simply stated that so and so had rented "the garden plot in the south half of the Parkinson Ranch," then one would not even know on which section of a county map to begin to look for the Parkinson Ranch. Since ownership maps do not have indexes, whether the researcher is able to locate a parcel of land depends on sheer diligence, keen eyesight, and a good deal of luck. Since many of the tracts leased by Chinese could not be located, the maps in this study understate the extent of Chinese leasing.

In almost all counties, the great majority of the Asian tenants rented their land on a cash basis. In most cases, no initial rental payment was required, but in others, rather large sums had to be deposited when the leases were signed. Tenants operating very large farms also needed working capital. Mortgage records provide information on the sources of credit available to tenants who needed it.

During the nineteenth century, credit was usually extended in the form of a two-party mortgage between a mortgagor (borrower) and a mortgagee (lender). Filing such mortgage documents with the County Recorder was a way to insure that the mortgagor complied with the terms of the loan, particularly its repayment. By the early years of the twentieth century, deeds of trust began to be used with increasing frequency. A deed of trust involves three parties: the trustor (borrower), the trustee (person or company to whom the property of a debtor is attached), and a beneficiary (lender). Because of the confusion between a deed and a deed of trust, many deeds of trust were mistakenly entered into the deed books rather than the mortgage books. The researcher who wishes to find a full run of the mortgage documents for a locality should look through both sets of books.

Mortgage records in the offices of California's County Recorders were divided according to the type of collateral used. Mortgages secured by real property were usually separated from mortgages secured by personal property. The latter included crops, farm implements and machinery, automobiles, furniture, appliances, furnishings, and other personal property. When horses were still widely used for farming, they were frequently used as collateral also. Individual horses were described not only by size, age, and markings, but often by name also.

Documents for leases of agricultural land were entered in several kinds of books. Clerks in the County Recorder's offices were rather inconsistent in which set of books they chose to use. Most leases which did not involve a concomitant crop mortgage —to guarantee the payment of the rent—were usually entered in the Book of Leases. Mortgages made *after* leases went into effect were usually entered in the Book of Crop Mortgages, or the Book of Chattel Mortgages, or the Book of Personal Property Mortgages. However, if a lease and a chattel mortgage were signed simultaneously, then, depending on the whim of the clerk, either set of books could have been used. To confuse matters further, in some counties lease and chattel mortgage documents were sometimes found in books entitled Miscellaneous Records, and other times in Agreements and Contracts. Researchers who have gone through one set of books cannot assume that they have found all the pertinent documents they are trying to locate.

Property records in county archives are most valuable for research because they provide systematic, time-series data not available from other sources. However, those wishing to use these records should be aware that different sets of property records are complete to varying degrees. Also, different kinds of records serve to answer different research questions, so potential researchers should be fully aware of what information each kind of instrument contains before completing their research designs.

It is also useful to know that for many Asian immigrants, after the passage of the 1913 Alien Land Act in California, many deeds to property they purchased were not in their own names, so deed records are useful for studying the purchase of land by Asian immigrants only before that date. After 1913 a considerable number of Japanese and a handful of Chinese purchased land in the names of their American-born children or in the names of corporations. While the children can be identified, the shareholders in corporations cannot, unless incorporation papers were filed. Incorporation papers for some "dead" corporations are available in the California State Archives in Sacramento.

Table 37   County Archival Records Used

County	Mining Records	Deeds	Leases	Chattel Mortgages	Misc. Records	Other Records
Alameda			x			
Amador	x	x			x	
Butte		x	x			
Calaveras	x	x	x			
Colusa			x	x		
Contra Costa			x	x		
El Dorado	x	x	x		x	
Fresno		x	x			
Glenn			x	x		
Imperial			x	x	x	
Kern		x	x			
Kings			x	x		
Los Angeles		x	x			
Madera			x	x		
Merced		x	x	x		
Monterey			x	x		
Napa			x	x		
Nevada	x	x	x			
Orange			x			
Placer		x	x	x		
Riverside			x	x		
Sacramento		x	x	x		
San Bernardino			x	x		
San Diego			x			
San Joaquin			x	x		x
San Luis Obispo			x			
San Mateo			x			
Santa Barbara			x	x		
Santa Clara			x	x	x	
Santa Cruz			x			
Shasta		x	x		x	x
Sierra	x	x	x			x
Solano			x	x		
Sonoma			x	x		
Stanislaus			x		x	
Sutter			x	x	x	
Tehama		x	x			
Tulare		x	x			
Tuolumne		x	x			
Yolo			x			
Yuba	x	x	x	x		x

In the case of leased land, it is extremely difficult to estimate what percentage of the total land leased was actually recorded. Mr. James Johnstone, County Recorder of San Joaquin County, guessed that leases with attendant chattel mortgages were probably recorded much more frequently than plain leases. The reason is that legal claim to collateral was involved. He thought "most" chattel mortgages were recorded, but estimated that only "10 to 50 percent" of the leases were.[6] Other County Recorders queried did not even wish to venture a guess. The number of acres leased by Asian tenants shown in table 31 is therefore an undercount. Table 37 indicates that in only seventeen of the counties studied were mortgage books still available or accessible—the figures for the number of acres leased by Chinese in those counties were thus more accurate. While the figures in the table and in the graphs in figure 1 may not show the actual total acreages leased by Chinese and Japanese tenants, nonetheless, they are useful representations of *trends*. If we look at the availability of records by regions, pre-1920s chattel mortgage records are still in existence for most of the counties in the Sacramento–San Joaquin Delta, the Sacramento Valley, and the southern interior desert, while chattel mortgage records are available for about half of the counties in the northern and central California coast, so the acreages shown for these regions are more accurate than the acreages shown for the San Joaquin Valley and the southern coastal plains, where the old chattel mortgage books are no longer available.

County archival records are extremely time-consuming and tedious to use, but without them and the manuscript schedules of the censuses of population and agriculture, this study could not have been made. Up to this point, only data sporadically collected by the California Bureau of Labor Statistics in the late 1880s and by the Immigration Commission of the U.S. Senate between 1907 and 1909 have been available. These contained only a few inaccurate figures on the Chinese. Moreover, since they were for just one or two particular years, they provided only "snapshots" of the Chinese at best. Perhaps that is another reason that Chinese contributions to the development of California agriculture have been neglected.

Notes

PREFACE

1. Varden Fuller, "The Supply of Agricultural Labor as a Factor in the Evolution of Farm Organization in California"; Carey McWilliams, *Factories in the Fields: The Story of Migratory Farm Labor in California*; Lloyd H. Fisher, *The Harvest Labor Market in California*.

2. Information from the manuscript schedules of the census of population was used by John W. Stephens, "A Quantitative History of Chinatown, San Francisco, 1870 and 1880," In *The Life, Influence and the Role of the Chinese in the United States, 1776–1960* (San Francisco: Chinese Historical Society of America, 1976), pp. 71–88; Ralph Mann, "The Decade After the Gold Rush: Social Structure in Grass Valley and Nevada City, California, 1850–1860," *Pacific Historical Review* 41 (1972): 484–504; and Ralph Mann, *After the Gold Rush: Society in Grass Valley and Nevada City, California, 1849–1870*. Studies that used the archival records of California counties are William S. Hallagan, "Labor Contracting in Turn-of-the-Century California Agriculture," *Journal of Economic History* 40 (1980): 757–776; Karen Leonard, "Marriage and Family Life Among Early Asian Indian Immigrants," *Population Review* 25 (1981): 67–75; Karen Leonard and Bruce La Brack, "Conflict and Compatibility in Punjabi-Mexican Immigrant Families in Rural California, 1915–1965," *Journal of Marriage and the Family* 46 (1984): 527–37; and Sucheng Chan, "Chinese Livelihood in Rural California: The Impact of Economic Change, 1860–1880," *Pacific Historical Review* 53 (1984): 273–307.

3. Paul W. Gates, *Landlords and Tenants on the Prairie Frontier: Studies in American Land Policy* (Ithaca: Cornell University Press, 1973); Margaret Beattie Bogue, *Patterns from the Sod: Land Use and Tenure in the Grand Prairie, 1850–1900* (Springfield, Ill.: Illinois State Historical Library, 1959); and Yasuo Okada, *Public Lands and Pioneer Farmers, Gage County, Nebraska, 1850–1900* (Tokyo: Keio University Press, 1971).

4. Gunther Barth, *Bitter Strength: A History of the Chinese in the United States, 1850–1870*; and Stanford M. Lyman, "The Chinese Diaspora in America, 1850–1943."

5. Most writers have relied on documents and reports issued by the United

States Congress and the California State Legislature and on contemporary newspapers. Three of the most commonly used sources are California, State Legislature, "Report of Joint Select Committee Relative to the Chinese Population of the State of California," *Journal of the Senate and Assembly, Appendix*, vol. 3 (Sacramento: State Printing Office, 1862); U.S. Congress, Joint Special Committee, *Report*, 44th Cong., 2nd sess., 1877; and California, Senate, Special Committee on Chinese Immigration, *Chinese Immigration: Its Social, Moral, and Political Effect, Report to the California State Senate of its Special Committee on Chinese Immigration* (Sacramento: State Printing Office, 1878).

6. Readers interested in these topics may consult the bibliographies at the beginning of the section "Selected Bibliography on the History of the Chinese in California, 1850s to 1920s."

INTRODUCTION

1. Hubert Howe Bancroft, *History of California, 1860–1890*, vol. 7; Lucile Eaves, *A History of California Labor Legislation*; Ira B. Cross, *A History of the Labor Movement in California*; Elmer Clarence Sandmeyer, *The Anti-Chinese Movement in California*; and Alexander Saxton, *The Indispensable Enemy: Labor and the Anti-Chinese Movement in California*.

2. Mary Roberts Coolidge, *Chinese Immigration* (New York: Henry Holt and Co., 1909).

3. Barth, *Bitter Strength;* Stuart Creighton Miller, *The Unwelcome Immigrant: The American Image of the Chinese, 1785–1882*; Robert McClellan, *The Heathen Chinee: A Study of American Attitudes toward China, 1890–1905*; Rose Hum Lee: *The Growth and Decline of Chinese Communities in the Rocky Mountain Region* (New York: Arno Press and the New York Times, 1979), which is in fact her unedited Ph.D. dissertation completed at the University of Chicago in 1947, and *The Chinese in the United States of America*; Stanford M. Lyman: "The Structure of Chinese Society in Nineteenth-Century America," *The Asian in the West*, and *Chinese Americans*.

4. S .W. Kung, *Chinese in American Life: Some Aspects of Their History, Status, Problems, and Contributions*; Betty Lee Sung, *Mountain of Gold: The Story of the Chinese in America*; Francis L. K. Hsu, *The Challenge of the American Dream: The Chinese in the United States* (Belmont, Calif.: Wadsworth Publishing Co., Inc., 1971); and Jack Chen, *The Chinese of America*.

5. Clarence E. Glick, *Sojourners and Settlers: Chinese Migrants in Hawaii* (Honolulu: Hawaii Chinese History Center and University Press of Hawaii, 1980); Victor G. and Brett de Bary Nee, *Longtime Californ': A Documentary Study of an American Chinatown*; James W. Loewen, *The Mississippi Chinese: Between Black and White* (Cambridge: Harvard University Press, 1971); and Lucy M. Cohen, *Chinese in the Post-Civil War South: A People without a History* (Baton Rouge: Louisiana State University Press, 1984).

6. Shih-shan Henry Tsai, *China and the Overseas Chinese in the United States, 1868–1911*.

7. A large proportion of the pre-1970 writings on Chinese Americans focused on their "failure" to assimilate. See, for example, Kit King Louis: "Prob-

lems of Second Generation Chinese," *Sociology and Social Research* 16 (1932): 250–58, and "Program for Second Generation Chinese," *Sociology and Social Research* 16 (1932): 455–62; Francis Y. Chang, "An Accommodation Program for Second-Generation Chinese," *Sociology and Social Research* 18 (1933): 541–53; David T. Cheng, "Acculturation of Chinese in the United States (Ph.D. diss., University of Pennsylvania, 1946); Robert Lee, "Acculturation of Chinese Americans," *Sociology and Social Research* 36 (1952): 319–21; Kian M. Kwan, "Assimilation of the Chinese in the United States: An Exploratory Study in California" (Ph.D. diss., University of California, Berkeley, 1958); James W. Chin, "Problems of Assimilation and Cultural Pluralism Among Chinese-Americans in San Francisco" (M.A. thesis, University of the Pacific, 1965); Stanley L. M. Fong: "Assimilation of Chinese in America: Changes and Social Perception," *American Journal of Sociology* 71 (1965): 265–73, and *The Assimilation of Chinese in America: Changes in Orientation and Social Perception* (San Francisco: R and E Research Associates, 1974); and Mely Giok-lan Tan, *The Chinese in the United States: Social Mobility and Assimilation* (Taipei: The Chinese Association for Folklore, 1973). Post-1970 writings cover more diverse topics. See the annual bibliography in *Amerasia Journal* for complete listings.

8. George F. Seward, *Chinese Immigration, in Its Social and Economic Aspects*.

9. Ping Chiu, *Chinese Labor in California, 1850–1880: An Economic Study*.

10. Fuller, "Supply of Agricultural Labor"; Paul S. Taylor, "Foundations of California Rural Society"; Lloyd Fisher, *The Harvest Labor Market in California*; Carey McWilliams, *Factories in the Field*; Cletus E. Daniel, *Bitter Harvest: A History of California Farmworkers, 1870–1941*; and Linda C. Majka and Theo J. Majka, *Farm Workers, Agribusiness, and the State*.

CHAPTER ONE

1. C. P. FitzGerald, *The Southern Expansion of the Chinese People: "Southern Fields and Southern Ocean"* (London: Barrie and Jenkins, 1972); and Harold J. Wiens, *China's March toward the Tropics* (Hamden, Conn.: The Shoe String Press, 1954).

2. Lo Jung-pang, "China as a Sea Power, 1127–1368" (Ph.D. diss., University of California, Berkeley, 1957), pp. 244–47. *Cohong* was the Anglicized name for the Chinese foreign trade guild which had been granted a state monopoly in foreign trade in exchange for guaranteeing that foreign merchants would behave properly while in China.

3. G. William Skinner, *Chinese Society in Thailand: An Analytical History* (Ithaca: Cornell University Press, 1957).

4. Ibid., p. 44.

5. There is a considerable literature on the Chinese in the Philippines, the best of which is Edgar Wickberg, *The Chinese in Philippine Life, 1850–1890* (New Haven: Yale University Press, 1965). A number of interesting essays on the first period of Chinese settlement may be found in Felix Alfonso, ed., *The Chinese in the Philippines, 1570–1770*, vol. 1 (Manila: Solidaridad Publishing House, 1966).

6. The most useful history of the Chinese in Thailand is Skinner, *Chinese*

Society.

7. Chen Ta, *Chinese Migrations, with Special Reference to Labor Conditions* (1923; repr. ed., Taipei: Ch'eng-wen Publishing Co., 1967), pp. 37–50; and Victor Purcell, *The Chinese in Southeast Asia*, 2nd ed., enl. (London: Oxford University Press, 1965), p. 24.

8. Yen Ching-hwang, "Ch'ing Changing Images of the Overseas Chinese (1644–1912)," *Modern Asian Studies* 15 (1981): 261–85.

9. Shih-shan H. Tsai, "Preserving the Dragon Seeds: The Evolution of Ch'ing Emigration Policy," *Asian Profile* 7 (1979): 497–506; and Robert Irick, *Ch'ing Policy toward the Coolie Trade, 1847–1878* (San Francisco: Chinese Materials Center, 1982).

10. Sing-wu Wang, *The Organization of Chinese Emigration, 1848–1888* (San Francisco: Chinese Materials Center, 1978), pp. 8–9.

11. Thomas W. Chinn, Him Mark Lai, and Philip C. Choy, eds., *A History of the Chinese in California: A Syllabus*, p. 2.

12. Ibid., p. 4; and Zo Kil Young, "Chinese Emigration into the United States, 1850–1880" (Ph.D. diss., Columbia University, 1971), p. 56.

13. Chinn et al., *History*, pp. 2–4.

14. Interview with Chin Shou and John Chin (Milpitas, California, October 12, 1979).

15. Theodore Shabad, *China's Changing Map: National and Regional Development, 1949–71* (New York: Praeger Publishers, 1972), pp. 184–86; T. R. Tregear, *A Geography of China* (Chicago: Aldine Publishing Company, 1956), pp. 256–59; and Barth, *Bitter Strength*, p. 20.

16. Frederic E. Wakeman, Jr., *Strangers at the Gate: Social Disorder in South China, 1839–1861* (Berkeley and Los Angeles: University of California Press, 1966), pp. 109–11.

17. Zo, "Chinese Emigration," pp. 64 and 67–68.

18. Wakeman, *Strangers at the Gate*, pp. 139–48.

19. Chinn et al., *History*, pp. 11–12; and Zo, "Chinese Emigration," pp. 121–23.

20. Official reports on the American involvement in the coolie trade may be found in U.S. Congress, House of Representatives, "Chinese Coolie Trade," *Message of the President of the United States, communicating, in compliance with a resolution of the House of Representatives, information recently received in reference to the coolie trade*, 36th Cong., 1st sess., 1860; and Jules Davids, ed., *American Diplomatic and Public Papers: The United States and China, Series I: The Treaty System and the Taiping Rebellion, 1842–1860*, vol. 17, *The Coolie Trade and Chinese Emigration* (Wilmington: Scholarly Resources, Inc., 1973). American involvement has been discussed in Hosea Ballou Morse, *The International Relations of the Chinese Empire*, vol. 2, *The Period of Submission, 1861–1893* (London: Longmans, Green, and Co., 1918), pp. 166–81; M. F[oster] Farley, "John E. Ward and the Chinese Coolie Trade," *American Neptune* 20 (1960): 209–16; M. Foster Farley, "The Chinese Coolie Trade, 1845–75," *Journal of Asian and African Studies* 3 (1968): 257–70; Shih-shan H. Tsai, "American Involvement in the Coolie Trade," *American Studies* 6 (1976): 49–66; and Robert J. Schwendinger, "Coolie Trade: The American Connection," *Oceans* 14 (1981): 38–44.

21. Davids, *American Diplomatic Papers*, pp. 15, 113, and 192.

22. Great Britain ended its participation in the slave trade in 1807, and slavery was abolished in the British colonies in 1833, to take effect in 1834. Initially, planters were permitted to keep their ex-slaves as apprentices for twelve years. The period was later reduced to seven years, and the apprentice system was abolished altogether in 1838.

23. Hugh Tinker, *A New System of Slavery: The Export of Indian Labour Overseas, 1830–1920* (London: Oxford University Press, 1974), pp. 104 and 108–9.

24. On the importation of Chinese coolies into Latin America, see Watt Stewart, *Chinese Bondage in Peru, 1849–1874* (1951; repr. ed., Westport, Conn.: Greenwood Press, 1970); Ching Chieh Chang, "The Chinese in Latin America: A Preliminary Geographical Survey with Special Reference to Cuba and Jamaica" (Ph.D. diss., University of Maryland, 1956); Eugenio Chang-Rodriguez, "Chinese Labor Migration into Latin America in the Nineteenth Century," *Revista de Historia de America* 46 (1958): 375–97; Arnold J. Meagher, *The Introduction of Chinese Laborers to Latin America: The Coolie Trade, 1847–1874* (San Francisco: Chinese Materials Center, 1975); and Duvon Clough Corbitt, *A Study of the Chinese in Cuba, 1847–1947* (Wilmore, Ky.: Asbury College, 1971).

25. I. D. DuPleissis, *The Cape Malays* (Cape Town: Miller, 1947); and Craig A. Lockard, "The Javanese as Emigrant: Observations on the Development of Javanese Settlements Overseas," *Indonesia* 11 (1971): 41–62.

26. Irick, *Ch'ing Policy*, p. 7.

27. Tinker, *New System*, pp. 76–77.

28. Any good modern history of China will have an account of the Opium War and the terms of the Treaty of Nanking. One of the best recent studies of the Opium War is Peter Ward Fay, *The Opium War, 1840–1842* (Chapel Hill: The University of North Carolina Press, 1975).

29. Wakeman, *Strangers at the Gate*, contains a detailed description of the impact of the events of this period on the Canton area. The influence that joint Western and Chinese administration of Canton had on Chinese emigration has been covered at length by Irick in *Ch'ing Policy*.

30. Although the "Act to Prohibit the Coolie Trade by American Citizens in American Vessels," *U.S. Statutes at Large*, 37th Cong., 2nd sess., 1862, vol. 12, pp. 340–41, forbade American ships to take coolies not only to the United States but to other foreign countries as well, it was easy for ship owners and captains to bypass the law when they did not land at American ports.

31. Wang, *Organization*, pp. 86–87.

32. Ibid., p. 123.

33. Irick, *Ch'ing Policy*, pp. 184–96.

34. Wang, *Organization*, pp. 68–69; and Irick, *Ch'ing Policy*, pp. 257–72.

35. Wang, *Organization*, p. 123.

36. For a discussion of the Burlingame-Seward Treaty, see chapter 2 below.

37. Persia Crawford Campbell, *Coolie Emigration to Countries within the British Empire* (London: P. S. King and Son, Ltd., 1923), p. 29. (Campbell gave no source for her figures.)

38. Wang, *Organization*, p. 105.

39. Tinker, *New System*, pp. 162–63 and 165; and Wang, *Organization*,

pp. 209–13 and 307–8.

40. The Huntington Library has several volumes of inventories of merchandise carried by ships of the Pacific Mail Steamship Company that plied back and forth between China and the west coast of the United States.

41. Robert J. Schwendinger, "Investigating Chinese Immigrant Ships and Sailors," p. 18.

42. Dorothy J. Perkins, "Coming to San Francisco by Steamship," p. 29.

43. Schwendinger, "Immigrant Ships," pp. 20 and 22–23.

44. Before the immigration detention barracks on Angel Island in San Francisco Bay were built, the sheds of the Pacific Mail Steamship Company were used to hold arriving Chinese passengers as they awaited questioning and clearance by customs officials.

45. Alexander McLeod, *Pigtails and Golddust: A Panorama of Chinese Life in Early California*, pp. 86–92.

46. Ibid.

47. Chinese immigrants formed several kinds of mutual aid associations: family associations, clan associations, and district associations. The last were for people who came from the same provincial subdivisions in China.

48. Saxton, *Indispensable Enemy*.

49. For the anti-Chinese movement in Canada, see W. Peter Ward, *White Canada Forever* (Montreal: McGill-Queens University Press, 1978), pp. 3–78; Edgar Wickberg, ed., *From Canton to Canada* (Toronto: McClelland and Stewart, Ltd., 1982), pp. 42–72 and 135–47; Patricia E. Roy, "The Preservation of the Peace in Vancouver: The Aftermath of the Anti-Chinese Riot of 1887," *BC Studies* 31 (1976): 44–59; and Donald Avery and Peter Neary, "Laurie, Borden and a White British Columbia," *Journal of Canadian Studies* 12 (1977): 24–34. For the anti-Chinese movement in Australia, see A. T. Yarwood, *Asian Migration to Australia: The Background to Exclusion, 1896–1923*, rev. ed. (Melbourne: Melbourne University Press, 1967), pp. 104–24; H. I. London, *Non-White Immigration and the "White Australia" Policy* (New York: New York University Press, 1970), pp. 3–25; A. C. Palfreeman, *The Administration of the White Australia Policy* (Melbourne: Melbourne University Press, 1967); and D. Gibb, ed., *The Making of "White Australia"* (Melbourne: Melbourne University Press, 1973). For comparative studies of attempts to bar Asian immigrants from "white" countries, see Charles A. Price, *The Great White Walls Are Built: Restrictive Immigration to North America and Australia, 1836–1888* (Canberra: Australian National University Press, 1974); Robert Huttenback, *Racism and Empire: White Settlers and Colored Immigrants in the British Self-Governing Colonies, 1830–1910* (Ithaca: Cornell University Press, 1976); and Andrew Markus, *Fear and Hatred: Purifying Australia and California, 1850–1901* (Sydney: Hall and Iremonger, 1979).

CHAPTER TWO

1. Oscar Handlin, *The Uprooted* (Boston: Little, Brown and Co., 1951).

2. For eyewitness accounts of the California Indians, see *As the Padres Saw Them: Californian Indian Life and Customs as Reported by the Franciscan Missionaries. Historical Introduction, Notes and Translation*, by Maynard Geiger; *Anthropological*

Commentary, Notes and Appendices, by Clement W. Meighan (Santa Barbara: Santa Barbara Mission Archives Library, dist. by A. H. Clark Co., 1976); and Stephen Powers, *Tribes of California*, vol. 3, ed. J. W. Powell, *Contributions to North American Ethnology* (Washington, D.C.: Government Printing Office, 1877). Of the older scholarly works on the subject, the writings of Sherburne Friend Cook, Alfred Louis Kroeber, and Robert Fleming Heizer are the most important. Of special interest to historians are Sherburne F. Cook, *Population Trends among the California Mission Indians* (Berkeley and Los Angeles: University of California Press, 1940); Sherburne F. Cook, *The Conflict between the California Indian and White Civilization* (Berkeley and Los Angeles: University of California Press, 1943); Alfred L. Kroeber, *Handbook of the Indians of California* (Washington, D.C.: Government Printing Office, 1925); Robert F. Heizer, *Languages, Territories and Names of California Indian Tribes* (Berkeley and Los Angeles: University of California Press, 1966); Robert F. Heizer and M. A. Whipple, comps. and eds., *The California Indians: A Source Book*, 2nd. rev. enl. ed. (Berkeley and Los Angeles: University of California Press, 1971); Robert F. Heizer, *The Destruction of California Indians: A Collection of Documents from the Period 1847 to 1865 in which are Described Some of the Things that Happened to Some of the Indians of California* (Santa Barbara: Smith and Co. 1974); and Robert F. Heizer and Robert B. Elsasser, *The Natural World of the California Indians* (Berkeley and Los Angeles: University of California Press, 1980). A newer work that reflects a Native American perspective is Jack D. Forbes, *Native Americans of California and Nevada* (1969; rev. ed., Happy Camp, Calif.: Naturegraph Publishers, 1982). George Harwood Phillips, *Chiefs and Challengers: Indian Resistance and Cooperation in Southern California* (Berkeley and Los Angeles: University of California Press, 1975), also offers a fresh interpretation of Indian-white relations.

   3. For a list of Spanish and Mexican land grants, see Robert G. Cowan, *Ranchos of California* (Fresno: Academy Library Guild, 1956). Contemporary views may be found in Reuben L. Underhill, *From Cowhides to Golden Fleece; A Narrative of California, 1832–1858, Based Upon Unpublished Correspondence of Thomas Oliver Larkin of Monterey, Trader, Developer, Promoter, and Only American Consul* (Stanford: Stanford University Press, 1946). Several works of synthesis include Charles C. Baker, "Mexican Land Grants in California," *Historical Society of Southern California Annual Publications* 9 (1914): 236–43; R. H. Allen, "The Spanish Land Grant System as an Influence in the Agricultural Development of California," *Agricultural History* 9 (1935): 127–42; Paul Wallace Gates, "Adjudication of Spanish-Mexican Land Claims in California," *Huntington Library Quarterly* 21 (1958): 213–36; and Robert Glass Cleland, *The Cattle on a Thousand Hills: Southern California, 1850–1880*.

   4. In the voluminous literature on California history, Hubert Howe Bancroft, *History of California*, in 7 vols., and Theodore H. Hittell, *History of California*, in 4 vols., remain monumental though flawed classics. A recent work that focuses on the American conquest is Neal Harlow, *California Conquered: War and Peace on the Pacific, 1846–1850* (Berkeley and Los Angeles: University of California Press, 1982).

   5. Leonard Pitt, *The Decline of the Californios: A Social History of the Spanish-Speaking Californians, 1846–1890* (Berkeley and Los Angeles: University of Cali-

fornia Press, 1966), is the classic treatment on the subject.

6. For accounts of the use of Chinese workers in the building of the western end of the first transcontinental railroad, see Ping Chiu, *Chinese Labor*, pp. 40–51; Barth, *Bitter Strength*, pp. 117–20; and Alexander Saxton, "The Army of Canton in the High Sierra," *Pacific Historical Review* 35 (1966): 141–52.

7. The text of the Burlingame-Seward Treaty may be found in W. M. Malloy (ed.), *Treaties, Conventions, International Acts, Protocols and Agreements between the United States of America and Other Powers, 1776–1909*, 2 vols. (Washington, D.C.: Government Printing Office, 1910), vol. 1, pp. 234–36. Analyses of the Burlingame Mission are available in Knight Biggerstaff, "The Official Chinese Attitude toward the Burlingame Mission," *American Historical Review* 41 (1936): 682–701; Frederick Wells Williams, *Anson Burlingame and the First Chinese Mission to Foreign Powers* (New York: Russell and Russell, 1912); and Tsai, *China and the Overseas Chinese*, pp. 24–29.

8. The anti-Chinese movement has been analyzed more than any other aspect of Chinese American history. Major works on the topic include Coolidge, *Chinese Immigration*; Eaves, *Labor Legislation*; Cross, *Labor Movement*; Sandmeyer, *Anti-Chinese Movement*; Barth, *Bitter Strength*; Miller, *Unwelcome Immigrant*; and Saxton, *Indispensable Enemy*. The following account is synthesized from the above works.

9. Coolidge, *Chinese Immigration*, p. 504.

10. The following discussion is based on an analysis of data on the number of Chinese in the United States by state and county. U.S. Bureau of the Census, *Eighth Census of the United States: Population, 1860* (Washington, D.C.: Government Printing Office, 1864), pp. 25–26; *Ninth Census of the United States: Population, 1870* (Washington, D.C.: Government Printing Office, 1872), pp. 89–93 and 338; *Tenth Census of the United States: Population, 1880* (Washington, D.C.: Government Printing Office, 1883), pp. 498–99; *Eleventh Census of the United States: Population, 1890* (Washington, D.C.: Government Printing Office, 1895), pp. 401 and 437–41; and *Twelfth Census of the United States: Population, 1900* (Washington, D.C.: Government Printing Office, 1901–2), pp. cxxiii and 565–70.

11. Rodman Wilson Paul, *Mining Frontiers of the Far West, 1848–1880* (New York: Holt, Rinehart and Winston, 1963), pp. 39–40.

12. In these tables, I have grouped individual occupations into categories differently from the way the U.S. Bureau of the Census did. The Bureau of the Census used the following broad categories: Agriculture; Professional and Personal Services; Trade and Transportation; Manufacturing and Mining. Thus, the Bureau of the Census lumped together barbers and servants with physicians and surgeons under "Professionals and Personal Services," and placed miners, fishermen, blacksmiths, clerks, factory owners, and tailors under "Manufacturing and Mining." I have separated the professionals (defined as persons with skills acquired through higher education) from the providers of more menial personal services (such as barbers, cooks, waiters, dishwashers, and servants). Instead, I have placed skilled artisans (defined as persons who acquire specialized skills through apprenticeships to become journeymen) together with professionals. Furthermore, instead of separating those in

agriculture from those in fishing and mining, I have grouped farmers, truck gardeners, fishermen, and miners together as primary producers and extractors (i.e., persons who earn a living by working with the natural resources of the earth).

13. The *San Francisco Bulletin* reprinted an item from the *Auburn Placer Press* of May 19, 1857, which described how Chinese miners had been robbed by men with double barreled guns, who then hanged three of the Chinese witnesses.

14. Coolidge, *Chinese Immigration*, p. 32.

15. This decision was rendered by the California Supreme Court in the case of *The People v. George W. Hall*, 4 *Cal.* 399 (1854).

16. The article from the *Shasta Republican* was reprinted in the *San Francisco Bulletin* of December 18, 1856, and stated that "hundreds of Chinese" had been "slaughtered in cold blood during the last five years by 'desperados.'" Though the papers were hardly pro-Chinese, the reporter sounded as though he thought it was proper and just to hang a white man for murdering Chinese.

17. Sucheng Chan, "Chinese Livelihood in Rural California: The Impact of Economic Change, 1860–1880," pp. 282–83.

18. Chiu, *Chinese Labor*, p. 23.

19. The best analysis of the Chinese in the San Francisco labor market is Paul M. Ong, "Chinese Labor in Early San Francisco: Racial Segmentation and Industrial Expansion."

20. For the Chinese in fishing, see Arthur F. McEvoy, "In Places Men Reject: Chinese Fishermen at San Diego, 1870–1893"; and Eve Armentrout-Ma, "Chinese in California's Fishing Industry, 1850–1914."

21. The following account is based on my tally, computation, and analysis of data on Chinese individuals listed in the manuscript schedules of the 1870 U.S. census of population as summarized in table 4.

22. The figures for San Francisco are not from my own tally but have been culled from U.S. Bureau of the Census, *Ninth Census of the United States: Population, 1870* (Washington, D.C.: Government Printing Office, 1872), p. 799, table XXXII. The figures in the published table refer only to the city of San Francisco. Certain occupations that were important for the Chinese population were not listed, so these have been estimated in table 4.

23. Chiu, *Chinese Labor*, pp. 55–61 and 89–128.

24. The following account is based on my tally, computation, and analysis of data on Chinese individuals listed in the manuscript schedules of the 1880 U.S. census of population as summarized in table 5.

25. See table 6.

26. U.S. Bureau of the Census, *Fourteenth Census of the United States: Population, 1920* (Washington, D.C.: Government Printing Office, 1921–22), vol. 4, pp. 342–59.

CHAPTER THREE

1. Whether there really is an "agricultural ladder" has been discussed in W. J. Stillman, "The Agricultural Ladder," *American Economic Review, Supplement* 9 (1919): 170–79; George H. von Tungeln, "Some Observations on the

So-called Agricultural Ladder," *Journal of Farm Economics*, 9 (1927): 94–106; Carl F. Wehrwein: "The Pre-Ownership Steps on the 'Agricultural Ladder' in a Low Tenancy Region," *Journal of Land and Public Utility Economics* 4 (1928): 416–25, "The Post-Ownership Steps on the 'Agricultural Ladder' in a Low Tenancy Region," *Journal of Land and Public Utility Economics* 6 (1930): 65–73, and "The 'Agricultural Ladder' in a High Tenancy Region," *Journal of Land and Public Utility Economics* 7 (1931): 67–77; E. D. Tetreau, "The 'Agricultural Ladder' in the Careers of 610 Ohio Farmers," *Journal of Land and Public Utility Economics* 7 (1931): 237–48; Lawanda Fenlason Cox, "Tenancy in the United States, 1865–1900: A Consideration of the Validity of the Agricultural Ladder Hypothesis," *Agricultural History* 18 (1944): 97–105; Shu-ching Lee, "The Theory of the Agricultural Ladder," *Agricultural History* 21 (1947): 53–61; Marshall Harris, "A New Agricultural Ladder," *Land Economics: A Quarterly Journal of Planning, Housing and Public Utilities* 26 (1950): 258–67; and Erven J. Long, "The Agricultural Ladder: Its Adequacy as a Model for Farm Tenure Research," *Land Economics: A Quarterly Journal of Planning, Housing and Public Utilities* 26 (1950): 269–73.

2. There is a considerable literature on farm tenancy, almost all of it focusing on the midwestern United States. Some scholars, particularly Paul W. Gates, consider tenancy to be an unhealthy result of land speculation made possible by a faulty public-land-disposal policy; others see it as a useful step in the farmer's advancement up the agricultural ladder; yet others view it as a normal part of the maturation of an agricultural economy. Book-length studies on the phenomenon of tenancy include Donald L. Winters, *Farmers Without Farms: Agricultural Tenancy in Nineteenth-Century Iowa* (Westport, Conn.: Greenwood Press, 1978); Allan G. Bogue, *From Prairie to Corn Belt: Farming on the Illinois and Iowa Prairies in the Nineteenth Century* (Chicago: University of Chicago Press, 1963); Margaret Beattie Bogue, *Patterns from the Sod* (n. 3 in Preface); and Gates, *Landlords and Tenants* (n. 3 in Preface). See also Richard T. Ely and Charles J. Galpin, "Tenancy in an Ideal System of Landownership," *American Economic Review, Supplement* 9 (1919): 180–212; Leon E. Truesdell, "Farm Tenancy Moves West," *Journal of Farm Economics* 8 (1926): 443–50; George S. Wehrwein, "Place of Tenancy in a System of Farm Land Tenure," *Journal of Land and Public Utility Economics* 1 (1925): 71–82; F. A. Buechel, "Relationships of Landlords to Farm Tenants," *Journal of Land and Public Utility Economics* 1 (1925): 336–42; George S. Wehrwein, "The Problem of Inheritance in American Land Tenure," *Journal of Farm Economics* 9 (1927): 163–75; Gustav W. Kuhlman, "A Study of Tenancy in Central Illinois," *Journal of Land and Public Utility Economics* 3 (1927): 290–97; William ten Haken, "Land Tenure in Walnut Grove Township, Knox County, Illinois," *Journal of Land and Public Utility Economics* 4 (1928): 13–24 and 189–98; M. M. Kelso, "A Critique of Land Tenure Research," *Journal of Land and Public Utility Economics* 10 (1934): 391–402; John D. Black and R. H. Allen, "The Growth of Farm Tenancy in the United States," *The Quarterly Journal of Economics* 51 (1937): 393–426; David M. Ellis, "Land Tenure and Tenancy in the Hudson Valley, 1790–1860," *Agricultural History* 18 (1944): 75–82; Max M. Tharp, "A Reappraisal of Farm Tenure Research," *Land Economics: A Quarterly Journal of Planning, Housing and*

Public Utilities 24 (1948): 315–30; Paul W. Gates, "Research in the History of American Land Tenure: A Review Article," *Agricultural History* 28 (1954): 121–26; Allan G. Bogue, "Foreclosure Tenancy in the Northern Plains," *Agricultural History* (1965): 3–16; and Donald L. Winters, "Tenant Farming in Iowa, 1860–1900: A Study of the Terms of Rental Leases," *Agricultural History* 48 (1974): 130–50, and "Tenancy as an Economic Institution: The Growth and Distribution of Agricultural Tenancy in Iowa, 1850–1900," *Journal of Economic History* 37 (1977): 382–408.

3. The most important recent studies of Southern sharecropping have been done by economic historians. Roger L. Ransom and Richard Sutch, in *One Kind of Freedom: The Economic Consequences of Emancipation* (Cambridge: Cambridge University Press, 1977), argue that the status of the black freedmen was largely determined by institutional arrangements, particularly debt peonage, that not only discriminated against them in particular but retarded Southern agricultural development after the Civil War in general. Stephen J. DeCanio, *Agriculture in the Postbellum South: The Economics of Production and Supply* (Cambridge: The MIT Press, 1974), and Robert Higgs, *Competition and Coercion: Blacks in the American Economy, 1865–1914* (Cambridge: Cambridge University Press, 1977), conclude that market forces played the most important role in determining the relative well-being of blacks.

4. R. L. Adams and Richard H. Smith, Jr., *Farm Tenancy in California and Methods of Leasing*, pp. 20–21.

5. William Stephen Richards, "Georgraphical Aspects of Rice Cultivation in California," chapter 3.

6. Chinese listed in the 1870 and 1880 censuses of agriculture were shown to be growing considerable amounts of sweet potatoes.

7. Tregear, *Geography of China* (n. 15, ch. 1), pp. 127–28.

8. Robert G. Spier, "Food Habits of Nineteenth-Century California Chinese."

9. Armentrout-Ma, "Chinese in Fishing," p. 152; and McEvoy, "In Places Men Reject," p. 14.

10. Charles Nordhoff, *California for Health, Pleasure, and Residence* (New York: Harper and Brothers, 1872), p. 190.

11. Business records of Chung Tai and Company (Berkeley: Bancroft Library, University of California, Berkeley).

12. Paul L. Langenwalter II, "The Archaeology of 19th Century Chinese Subsistence at the Lower China Store, Madera County, California," in *Archaeological Perspectives on Ethnicity in America: Afro-American and Asian American Culture History*, ed. Robert L. Schuyler (Farmingdale, N.Y.: Baywood Publishing Co., Inc., 1980), pp. 105–6.

13. U.S. National Archives, Record Group 29, "Census of the United States Population" (manuscript), 1900, Alameda County, California.

14. Business records of Chung Tai and Company and Kwong Tai Wo Company (Berkeley: Bancroft Library, University of California, Berkeley); William S. Evans, "Food and Fantasy: Material Culture of the Chinese in California and the West, circa 1850–1900"; and Patricia A. Etter, "The West Coast Chinese and Opium Smoking."

15. Harold P. Anderson, "Wells Fargo and Chinese Customers in Nineteenth Century California," p. 742.

16. My tally from National Archives, "Census of the United States Population" (manuscript), 1860, El Dorado, Amador, Calaveras, and Tuolumne counties, California.

17. Business Records of Wing On Wo and Company (Berkeley: Bancroft Library, University of California, Berkeley).

18. Interview with James J. Sinnott (Downieville, California, July 22, 1980); and James J. Sinnott, *Downieville: Gold Town on the Yuba* (Fresno: Mid-Cal Publishers, 1977), p. 108.

19. Chiu, *Chinese Labor*, p. 77.

20. My tally from National Archives, "Census of the United States Population" (manuscript), 1880, Los Angeles County, California.

21. See the discussion in the section "Truck Gardening in Sacramento Valley Towns" below.

22. Sinnott, *Downieville*, p. 107.

23. Calaveras County, "Deeds," Book G, p. 1.

24. Calaveras County, "Deeds," Book F, p. 238.

25. Calaveras County, "Leases and Bills of Sale," Book C, p. 136.

26. Amador County, "Deeds," Book D, p. 709.

27. Amador County, "Deeds," Book F, p. 177.

28. Amador County, "Deeds," Book F, p. 220.

29. Amador County, "Deeds," Book H, p. 521.

30. Amador County, "Deeds," Book A, pp. 338 and 456; Book C, p. 264; Book D, pp. 202, 620, 786, and 788; Book E, pp. 129, 392, and 491; Book F, pp. 51 and 184; Book G, pp. 310 and 727; Book H, p. 273; Book I, p. 383; Book L, p. 45; Book M, p. 16; Book O, pp. 70, 91, and 700; and Book S, pp. 410 and 412.

31. Amador County, "Deeds," Book I, p. 381.

32. Amador County, "Deeds," Book K, p. 430.

33. Amador County, "Deeds," Book K, p. 536.

34. Amador County, "Deeds," Book R, p. 94; Book W, p. 285; Book 2, p. 575; and Book 17, p. 430.

35. Amador County, "Miscellaneous Records" (leases), Book C, pp. 300 and 339.

36. El Dorado County, "Leases," Book A, p. 59.

37. El Dorado County, "Leases," Book A, p. 121.

38. El Dorado County, "Deeds," Book L, p. 265.

39. El Dorado County, "Deeds," Book M, p. 273.

40. El Dorado County, "Deeds," Book 31, p. 367.

41. El Dorado County, "Deeds," Book 35, p. 434.

42. El Dorado County, "Deeds," Book S, p. 95; Book T, p. 233; Book W, p. 38; and Book Y, p. 167.

43. My computation from U.S. Bureau of the Census, "Census of Productions of Agriculture" (manuscript), 1860, Amador County, California. (The manuscript schedules of the 1860, 1870, and 1880 agricultural censuses for California have been deposited in the California State Library, Sacramento; a

microfilmed copy is available at the Bancroft Library, University of California, Berkeley.)

44. Amador County, "Miscellaneous Records" (leases), Book C, p.283.

45. Nevada County, "Deeds," Book 4, p. 229.

46. Nevada County, "Deeds," Book 9, p. 169.

47. Nevada County, "Leases," Book 1, pp. 158, 159, and 412; Book 2, pp. 52, 123, 155, 170, 253, 257, 310, 359, 383, 430, and 479; and Book 3, p. 161.

48. Nevada County, "Leases," Book 1, p. 159; and Book 2, pp. 155, 318, and 430.

49. Nevada County, "Leases," Book 1, p. 412.

50. Nevada County, "Leases," Book 2, p. 257.

51. Susan W. Book, *The Chinese in Butte County, California, 1860–1920*, p. 6.

52. Butte County, "Deeds," Book A, p. 235.

53. Book, *Chinese in Butte County*, p. 9.

54. Joseph A. McGowan, *History of the Sacramento Valley*, vol. 1, p. 322.

55. Book, *Chinese in Butte County*, p. 20.

56. My computation from U.S. Bureau of the Census, "Census of Productions of Agriculture" (manuscript), 1860, Yuba County, California.

57. Ibid.

58. Ibid.

59. Yuba County, "Leases," Book 1, p. 131.

60. Interview with Joe Lung Kim (Marysville, August 9, 1980).

61. "Minutes of the City Council Meeting" (Marysville, April 14, 1876).

62. "Minutes of the City Council Meeting" (Marysville, October 2, 1876).

63. Auburn *Stars and Stripes*, 1866 (in *Bancroft Scraps*, vol. 6, p. 28).

64. *Wheatland Graphic*, January 30, 1886.

65. Ibid.

66. "Minutes of the City Council Meeting" (Marysville, January 6, 1890, and March 3, 1890).

67. "Minutes of the City Council Meeting" (Marysville, November 2, 1890).

68. "Minutes of the City Council Meeting" (Marysville, January 14, 1893).

69. My tally from U.S. National Archives, Record Group 29, "Census of the United States Population" (manuscript), 1880, Sacramento County, California.

70. U.S. National Archives, Record Group 29, "Census of the United States Population" (manuscript), 1860, Sacramento County, California.

71. [Thomas H.] Thompson and [Albert A.] West, *History of Sacramento County*, pp. 73–74.

72. My tally from U.S. National Archives, Record Group 29, "Census of the United States Population" (manuscript), 1880, Sacramento County, California.

73. U.S. National Archives, Record Group 29, "Census of the United States Population" (manuscript), 1880, Sacramento County, California.

74. U.S. Bureau of the Census, "Census of Productions of Agriculture" (manuscript), 1860, Sacramento County, California.

75. My tally from U.S. National Archives, Record Group 29, "Census of

the United States Population" (manuscript), 1880, Sacramento County, California.

76. Charles Morley, trans., "The Chinese in California As Reported by Henryk Sienkiewics," p. 308.

CHAPTER FOUR

1. G. William Skinner, "Marketing and Social Structure in Rural China: Part I," *Journal of Asian Studies* 24 (1964): 3–43; "Part II," *Journal of Asian Studies* 24 (1965): 195–228; and "Part III," *Journal of Asian Studies* 24 (1965): 363–99.

2. *Newark Courier* (New Jersey), 1869 (in *Bancroft Scraps*, vol. 6, pp. 117–18).

3. *Wheatland Graphic*, June 5, 1886.

4. Santa Clara County, "Deeds," Book L, p. 675; Book M, p. 125; Book G, pp. 547 and 434; Book P, p. 620; and Book Q, pp. 131, 210, and 503.

5. Santa Clara County, "Leases," Book B, p. 116.

6. Santa Clara County, "Leases," Book B, p. 118.

7. Santa Clara County, "Leases," Book B, p. 75.

8. Santa Clara County, "Deeds," Book U, p. 263.

9. Santa Clara County, "Leases," Book B, p. 106.

10. Santa Clara County, "Leases," Book A, p. 648; Book B, p. 353; and "Miscellaneous Records," Book D, pp. 14 and 43.

11. Santa Clara County, "Miscellaneous Records," Book C, p. 220.

12. Santa Clara County, "Miscellaneous Records," Book E, pp. 307, 452, and 455.

13. Santa Clara County, "Miscellaneous Records," Book C, p. 271.

14. *Pacific Rural Press*, April 14, 1877.

15. Santa Clara County, "Leases," Book D, pp. 35 and 59.

16. Santa Clara County, "Miscellaneous Records," Book E, p. 307.

17. Santa Clara County, "Miscellaneous Records," Book E, p. 452.

18. Santa Clara County, "Leases," Book D, p. 66.

19. Santa Clara County, "Leases," Book D, p. 35.

20. Santa Clara County, "Leases," Book F, p.519.

21. *Pacific Rural Press*, April 14, 1877.

22. Ibid.

23. Ibid.

24. Santa Clara County, "Leases," Book D, p. 66.

25. *Pacific Rural Press*, April 14, 1877.

26. *Pacific Rural Press*, August 28, 1886.

27. The use of the term "yuen" was especially prevalent in Tehama County: Quong Sun Yuen, Shing Shang Yuen, Wing Mow Yuen, Hop Lee Yuen, and other partnerships even recorded their names in Chinese characters in that county's lease books in the 1880s.

28. *Pacific Rural Press*, July 3, 1886.

CHAPTER FIVE

1. Colusa County, "Leases," Book A, p. 82.

2. Colusa County, "Leases," Book B, p. 162.

3. Sacramento County, "Leases," Book B, p. 95.

4. Yuba County, "Leases," Book 1, p. 303.

5. Much of the literature on the Sacramento–San Joaquin Delta is of a technical nature. The most detailed historical account is John Thompson, "The Settlement Geography of the Sacramento–San Joaquin Delta, California." For short introductions, see John Thompson, "How the Sacramento–San Joaquin Delta was Settled," *The Pacific Historian* 3 (1959): 49–58; and Harold Gilliam, "San Joaquin County—The Conquest of the Delta." For some historical accounts of Delta reclamation, see U.S. Department of Agriculture, "Reclamation of Swamp and Overflowed Lands in California," *Report of the Commissioner of Agriculture for the Year 1872* (Washington, D.C.: Government Printing Office, 1874), pp. 179–87; "Reclamation of Marsh Lands in California," *Pacific Rural Press*, May 30, 1885; and George A. Atherton, "Reclamation and Development in the Sacramento–San Joaquin Delta." A contemporary account of the difficulties and rewards of farming Delta land may be found in A. J. Wells, "Tilling the 'Tules' of California."

6. W. A. Orton, "Potato Diseases in San Joaquin County, California," U.S. Department of Agriculture, Circular no. 23, n.d.; and Thompson, "Settlement Geography," p. 332.

7. Marcia Hall McClain, "The Distribution of Asparagus Production in the Sacramento–San Joaquin Delta," pp. 29–31.

8. Thompson, "Settlement Geography," pp. 359–60.

9. Paul W. Gates, for the United States Public Land Law Review Commission, *History of Public Land Law*, pp. 301–4 and 319–39. For the exact wording of these laws, see United States, *Statutes at Large*, V: 455, IX: 519, X: 244, and XII: 503.

10. Different authors give conflicting figures for the number of acres granted to railroads in California because they use different computations and do not always include the same railroad companies. Figures vary from over 8,000,000 acres to over 11,000,000 acres. For a general analysis of railroad grants, see Gates, *History*, pp. 362–68 and 373–78. For a detailed study based on county deed books of what happened to railroad grants, see Walter A. McAllister, "A Study of Railroad Land-Grant Disposal with Reference to the Western Pacific, the Central Pacific, and the Southern Pacific Railroad Companies" (Ph.D. diss., University of Southern California, 1939). Also useful is E. A. Kincaid, "The Federal Land Grants of the Central Pacific Railroad" (Ph.D. diss., University of California, Berkeley, 1922).

11. Paul W. Gates, "California's Embattled Settlers."

12. These were "An Act to Provide for the Sale of the Swamp and Overflowed Lands Belonging to this State," 151 *Cal. Stat.* (1855), p. 189; "An Act Creating a State Land Office for the State of California," 176 *Cal. Stat.* (1858), p. 127; "An Act for the Relief of Purchasers of Land from the State of California," 177 *Cal. Stat.* (1859), p. 180; "An Act Amendatory of An Act Entitled 'An Act to Provide for the Sale and Reclamation of the Swamp and Overflowed Lands of this State' Approved April 21, 1858," 314 *Cal. Stat.* (1859), pp. 340–42; "An Act to Amend an Act Entitled An Act Supplemental to An Act Entitled

An Act to Provide for the Reclamation and Segregation of Swamp and Over-flowed and Salt Marsh and Tide Lands Donated to the State of California, by Act of Congress, Approved May 13, 1861, Approved April 11, 1862," 232 *Cal. Stat.* (1864), p. 230; "An Act to Provide for the Management and Sale of Lands Belonging to this State," 215 *Cal. Stat.* (1868), pp. 507-30; and *California Political Code*, Title VIII: Property of the State (1872).

13. There is a story, perhaps apocryphal, that Henry Miller used a team of horses to drag a boat across thousands of nonswampy acres in the San Joaquin Valley so that he could testify it was wet land.

14. Thompson, "Settlement Geography," p. 202.

15. Thompson, "Settlement Geography," p. 301.

16. "An Act to Provide for Funding the Indebtedness of the Reclamation and Levee Districts of the State," 570 *Cal. Stat.* (1871-72), p. 835.

17. Thompson, "Settlement Geography," p. 199.

18. Atherton, "Reclamation and Development," pp. 129-30.

19. The land companies formed in the twentieth century differed from those of the nineteenth century in that they tended to hold on to the land they had reclaimed, leasing out most of it to tenants, many of whom were Asians. Nineteenth-century companies more often than not sold their reclaimed land.

20. Thompson, "Settlement Geography," pp. 307 and 309-10.

21. The following account is based on an analysis of data in the manuscript schedules of the 1860, 1870, and 1880 censuses of agriculture. The statistical computations on which this account is based may be found in Sucheng Chan, "Bittersweet Harvest: Chinese Immigrants and the Transformation of California Agriculture, 1860-1920" (available in the Asian American Studies Library, University of California, Berkeley, and in the College of Agriculture Library, Cornell University, Ithaca, New York), pp. 170-74 and 199-204.

22. Tide Land Reclamation Company, *Fresh Water Tide Lands of California* (San Francisco: M. D. Carr and Co., 1869).

23. Thompson, "Settlement Geography," p. 490.

24. This map, for which no author is given, is in the California State Archives, Sacramento.

25. San Joaquin County, "Book G of Miscellaneous", Book 11, p. 602.

26. Gilliam, "San Joaquin County," p. 3.

27. Julian Dana, *The Sacramento: River of Gold*, p. 169.

28. U.S. Congress, Joint Special Committee to Investigate Chinese Immigration, *Report*, 41st. Cong., 2nd. sess., Report No. 689 (Washington, D.C., 1877), p. 441.

29. Dana, *The Sacramento*, p. 164.

30. Interview with Cared van Löben Sels (Courtland, May 26, 1979).

31. Pieter Justus van Löben Sels, untitled manuscript, no pagination. This manuscript was shown to me through the courtesy of Mr. Cared van Löben Sels, grandson of P. J. van Löben Sels.

32. Van Löben Sels manuscript, n.p.

33. Van Löben Sels manuscript, n.p.

34. Van Löben Sels manuscript, n.p.

35. Rules of the Vorden Ranch, unpublished manuscript in the family

papers of P. J. van Löben Sels.

36. Thompson, "Settlement Geography," p. 262.

37. Van Löben Sels manuscript, n.p.

38. Sacramento County, "Leases," Book K, p. 256; and interview with Cared van Löben Sels.

39. Tide Land Reclamation Company, incorporation papers, 1869 (Sacramento: California State Archives).

40. Ibid.

41. For purchases made in Roberts's own name, see San Joaquin County, "Book A of Deeds," Book 23, p. 494; Book 24, pp. 40, 43, 47, 49, 51, 53, 55, 57, 63, 67, 80, 201, 506, 508, 509, and 511; Book 25, pp. 419 and 482; and Book 27, p. 83. For purchases made in the name of the Tide Land Reclamation Company, see San Joaquin County, "Book A of Deeds," Book 24, pp. 59, 61, 64, 66, and 69; and Book 27, p. 84.

42. Sacramento County, "Deeds," Book 62, pp. 384, 386, and 388; Book 63, pp. 443, 444, 446, and 448; Book 64, p. 444; Book 77, pp. 404 and 407; and Book 87, p. 363.

43. San Joaquin County, "Book A of Deeds," Book 39, p. 355.

44. U.S. Congress, Joint Special Committee, *Report*, p. 441.

45. Ibid., pp. 436 and 441.

46. Thompson, "Settlement Geography," pp. 226 and 229.

47. Ibid., p. 230.

48. Ibid., p. 227.

49. San Joaquin County, "Book A of Deeds," Book 39, p. 355.

50. U.S. Congress, Joint Special Committee, *Report*, p. 441.

51. Ibid., p. 437.

52. Ibid., pp. 436–37.

53. Ibid., pp. 441; and Thompson, "Settlement Geography," p. 229.

54. U.S. Congress, Joint Special Committee, *Report*, p. 442.

55. Ibid., p. 443.

56. Ibid.

57. Ibid., p. 437.

58. Ibid., pp. 437 and 440.

59. Sacramento County, "Leases," Book B, p. 95.

60. U.S. Congress, Joint Special Committee, *Report*, p. 440.

61. George Chu, "Chinatowns in the Delta: The Chinese in the Sacramento–San Joaquin Delta, 1870–1960."

CHAPTER SIX

1. For a description of the backbreaking work in preparing prairie land for cultivation, see David E. Schob, *Hired Hands and Plowboys: Farm Labor in the Midwest, 1815–1869* (Urbana: University of Illinois Press, 1975), pp. 21–42.

2. Thompson, "Settlement Georgraphy," p. 295.

3. A. J. Wells, "Tilling the 'Tules'," pp. 313–14.

4. Unlike most parts of the country, where cutting down trees and grubbing their roots were the most labor-consuming part of the set-up work in

preparing land for cultivation, land reclamation and the building of irrigation works demanded far more labor and expense in California. Most of the Central Valley is devoid of trees, so California farmers did not have to contend with removing them. However, the absence of timber also meant that until barbed wire was invented, fences were extremely expensive to build.

5. Gregory A. Stiverson, *Poverty in a Land of Plenty: Tenancy in Eighteenth–Century Maryland* (Baltimore: Johns Hopkins University Press, 1977), p. 10.

6. *Sacramento Bee*, November 11, 1869.

7. U.S. Congress, Senate, Reports of the Immigration Commission, "Immigrants in Industries," Part 25: *Japanese and Other Immigrant Races in the Pacific Coast and Rocky Mountain States*, vol. 24 (Washington, D.C.: Government Printing Office, 1911), p. 325.

8. Sacramento County, "Leases," Book B, p. 95.

9. Sacramento County, "Leases," Book B, p. 68.

10. Edward F. Treadwell, *The Cattle King: A Dramatized Biography* (New York: Macmillan Co., 1931).

11. Sacramento County, "Deeds," Book 76, p. 411, and Book 77, p. 2.

12. Sacramento County, "Deeds," Book 105, p. 545.

13. Sacramento County, "Deeds," Book 135, p. 613.

14. Sacramento County, "Deeds," Book 153, p. 476.

15. Sacramento County, "Leases," Book B, p. 280.

16. Sacramento County, "Leases," Book B, p. 322.

17. Thompson and West, *History of Sacramento County*, p. 263.

18. Sacramento County, "Leases," Book B, p. 542.

19. Sacramento County, "Chattel Mortgages," Book 27, p. 284.

20. Sacramento County, "Chattel Mortgages," Book 27, p. 363.

21. U.S. Bureau of the Census, "Census of Productions of Agriculture" (manuscript), 1880, Sacramento County, California.

22. James J. Parsons, "The California Hop Industry: Its Eighty Years of Development and Expansion" (M.A. thesis, University of California, Berkeley, 1939).

23. Sacramento County, "Leases," Book C, p. 150.

24. Thompson and West, *Sacramento County*, p. 261; and U.S. Bureau of the Census, "Census of Productions of Agriculture" (manuscript), 1880, Sacramento County, California.

25. Sacramento County, "Leases," Book D, p. 43.

26. Sacramento County, "Leases," Book D, p. 644.

27. Interview with Chin Shou and John Chin (Milpitas, October 12, 1979).

28. Thompson, "Settlement Geography," pp. 322–23.

29. *An Illustrated History of San Joaquin County, California*, p. 515.

30. *An Illustrated History*, pp. 221–23.

31. San Joaquin County, "Book G of Miscellaneous," Book 11, pp. 393, 582, 585, 588, 592, 598, 602, 605, and 608; Book 14, pp. 348, 389, 397, 401, 405, 409, 413, 417, 421, 425, 432, and 436; Book 15, pp. 316, 320, 323, 327, 330, 337, 341, 344, 351, and 362.

32. *An Illustrated History*, p. 223.

33. San Joaquin County, "Book G of Miscellaneous," Book 23, pp. 272, 220,

and 414; Book 36, p. 152; Book 37, p. 269; and Book 39, p. 65; "Book I of Miscellaneous," Book 14, pp. 400, 404, and 407; Book 22, p. 197; Book 31, p. 373; Book 39, p. 588; Book 40, p. 513; and Book 59, p. 210; and Sacramento County, "Leases," Book L, p. 344. See also: California State Archives, incorporation papers for English-Wallace Company, 1903; Middle River Farming Company, 1901; Middle River Navigation and Canal Company, 1902; Roberts Island Improvement Company, 1903; Venice Island Land Company, 1906; Victoria Island Company, 1898; Western Company, 1908; and Rindge Land and Navigation Company, 1905.

34. *Los Angeles City Directory*, 1907 (Los Angeles: Los Angeles City Directory Co., Inc., 1907).

35. Incorporation papers files for Middle River Farming Company; Rindge Land and Navigation Company; Empire Navigation Company; Holland Land and Water Company; and Middle River Navigation and Canal Company.

36. Rindge file; Holland file; Middle River Navigation file; and Empire file.

37. Interview with Rindge Shima (Oakland, March 17, 1979).

38. Middle River Navigation file.

39. San Joaquin County, "Book A of Deeds," Book 126, pp. 173, 175, and 177; Book 132, pp. 447 and 625; and Book 136, p. 453.

40. San Joaquin County, "Book A of Deeds," Book 144, p. 100.

41. San Joaquin County, "Book G of Miscellaneous," Book 23, p. 220.

42. Chin Lung's fourth and fifth sons, Chin Gway and Chin Shou, do not agree on the date of their father's birth: they date it sometime between 1860 and 1864. Interview with Chin Gway (conducted by Judy Yung, San Francisco, July 29, 1979); and interview with Chin Shou and John Chin (Milpitas, California, October 12, 1979).

43. Chin Shou interview.

44. Sacramento County, "Leases," Book F, p. 598.

45. San Joaquin County, "Book I of Miscellaneous," Book 15, p. 429.

46. San Joaquin County, "Book G of Miscellaneous," Book 20, pp. 167 and 399.

47. San Joaquin County, "Book I of Miscellaneous," Book 14, p. 330.

48. San Joaquin County, "Book I of Miscellaneous," Book 20, p. 585.

49. Chin Gway interview.

50. San Joaquin County, "Book G of Miscellaneous," Book 20, p. 272.

51. San Joaquin County, "Book G of Miscellaneous," Book 20, p. 473.

52. San Joaquin County, "Book I of Miscellaneous," Book 14, p. 330.

53. San Joaquin County, "Book G of Miscellaneous," Book 21, pp. 210 and 395; and Book 23, p. 146.

54. San Joaquin County, "Book I of Miscellaneous," Book 37, p. 489; and Book 35, p. 113; Sacramento County, "Leases," Book L, p. 344; San Joaquin County, "Book I of Miscellaneous," Book 52, p. 383; Book 53, p. 165; Book 57, p. 64; and Book 58, p. 370.

55. Chin Shou interview.

56. Chin Gway interview.

57. Chin Gway interview. The 1913 California Alien Land Law prohibited aliens "ineligible to citizenship" from buying agricultural land or leasing it for

more than three years. (206 *Cal. Stat.* [1913].) Eligibility for naturalization had been previously defined by an 1870 federal statute that granted it only to "free white persons, and aliens of African nativity, and persons of African descent." (*U.S. Rev. Stat.* 2169, *U.S. Comp. Stat.* [1918], 4358.) The California Initiative Act of 1920 closed some of the "loopholes" in the 1913 Act by forbidding aliens ineligible to citizenship from leasing land at all and from buying shares in corporations which owned land. (lxxxiii *Cal. Stat.* [1921].) Oregon passed an Alien Land Law in 1923 (145 *Ore. Laws* [1923] and 61 *Ore. Comp. Laws* [1940] 101–111).

58. Chin Shou interview.

59. Chin Shou interview.

60. Chin Shou and Chin Gway interviews.

61. Chinese American Farms, incorporation papers, 1919.

62. The following account is based on computations from the manuscript agricultural censuses of 1860, 1870, and 1880. For statistical tables, see Chan, "Bittersweet Harvest," pp. 228–44.

63. W. A. Orton, "Potato Diseases," pp. 4–5.

CHAPTER SEVEN

1. Harold E. Jackson, "A Geography of the Sacramento Peach Belt," p. 58.

2. The best account of the struggle by farmers against mining debris is Robert L. Kelley, *Gold versus Grain: The Hydraulic Mining Controversy in California's Sacramento Valley.*

3. For a study of Chinese contributions to the growth of viticulture and viniculture in Sonoma County, see William F. Heintz, "The Role of Chinese Labor in Viticulture and Wine Making in Nineteenth-Century California."

4. Parsons, "The California Hop Industry" (n. 22, ch. 6); and James J. Parsons, "Hops in Early California Agriculture," pp. 110–16.

5. Heintz, "Chinese in Viticulture," p. 23.

6. *Daily Alta California,* July 23, 1863, quoted in Heintz, p. 23.

7. Heintz, "Chinese in Viticulture," p. 24.

8. "Napa Valley," *San Francisco Chronicle,* October 14, 1883, quoted in Heintz, p. 46.

9. William F. Heintz, lecture to Chinese American History class, University of California, Berkeley, winter quarter, 1982.

10. "Wine Making in California," *Western Broker* (n.d.), repr. in *San Francisco Merchant* 17 (1887): 53, quoted in Heintz, "Chinese in Viticulture," p. 35.

11. Heintz, "Chinese in Viticulture," pp. 62–65.

12. Ibid., p. 65.

13. Ibid., p. 66.

14. Sonoma County, "Leases," Book D, p. 284.

15. Sonoma County, "Leases," Book F, pp. 74, 321, and 579; Book H, pp. 13 and 236; and Book I, p. 129; and Sonoma County, "Chattel Mortgages," Book F, p. 535; and Book P, p. 335.

16. Interview with Riichi Satow [*sic*] (Sacramento, May 30, 1979).

17. Immigration Commission, *Reports,* vol. 24, p. 200.

18. E. Gregg, "History of the Famous Stanford Ranch at Vina," *Overland*

Monthly 52 (1908): 334–48. Heintz, "Chinese in Viticulture," p. 31, stated that Stanford employed "hundreds" of Chinese to plant, prune, and harvest the almost three million vines on this property.

19. Tehama County, "Leases," Book A, p. 297.

20. Tehama County, "Leases," Book A, pp. 297, 366, 379, and 409.

21. Tehama County, "Leases," Book A, pp. 366, 370, 410, 476, and 495.

22. Tehama County, "Leases," Book B, pp. 240, 315, and 346.

23. U.S. Bureau of the Census, "Census of Productions of Agriculture" (manuscript), 1870, Tehama County, California.

24. Private communication from Paul W. Gates, March 28, 1983.

25. Dana, *The Sacramento*, p. 181.

26. Tehama County, "Leases," Book B, p. 29.

27. Tehama County, "Leases," Book B, p. 254.

28. Tehama County, "Leases," Book B, pp. 133, 179, 251, 294, 298, 378, and 384.

29. Almost all the Chinese tenants in Tehama County leased land in the names of copartnerships called "companies" in English.

30. Tehama County, "Leases," Book B, p. 29.

31. Tehama County, "Leases," Book A, pp. 290, 294, 300, 394, 414, 418, 443, and 471.

32. Tehama County, "Deeds," Book I, p. 736.

33. Tehama County, "Deeds," Book J, p. 145.

34. Tehama County, "Deeds," Book P, p. 384.

35. Tehama County, "Deeds," Book R, p. 455.

36. Tehama County, "Deeds," Book T, p. 272; Book X, p. 543; and Book 2, p. 55.

37. Tehama County, "Deeds," Book 9, p. 91.

38. Immigration Commission, *Reports*, vol. 24, p. 24.

39. Sutter County, "Leases," Book A, p. 428, and Book B, p. 37.

40. Yuba County, "Leases," Book 2, pp. 321, 542, and 575.

41. Sutter County, "Leases," Book A, pp. 302, 451, and 467; Book B, pp. 340 and 453; Yuba County, "Leases," Book 2, pp. 97, 279, 312, 427, 517, and 542; Book 3, pp. 104, 141, 207, and 343.

42. Yuba County, "Leases," Book 2, p. 109; and *Edward Woodruff vs. The North Bloomfield Gravel Mining Company et al.*, Circuit Court of the United States for the Ninth Circuit and State of California, 1883 [Briefs], vol. 7, pp. 2495, 2503, 2518, 2562, 2563, 2683–84, 2719, and 2828; vol. 8, pp. 2855, 2875, and 2877; vol. 9, pp. 3189–92, 3202, 3209, 3245, 3265, and 3282; vol. 12, pp. 4323, and 4338; vol. 13, pp. 4687, 4689–91 and 4719; vol. 15, p. 5711; vol. 19, pp. 7462 and 7475; and vol. 21, p. 8472.

43. Placer County, "Leases," Book A, p. 87.

44. Placer County, "Leases," Book F, p. 285.

45. Placer County, "Leases," Book F, p. 374.

46. (The) Producers' Fruit Co., incorporation papers file, 1896 (Sacramento: California State Archives); Placer County, "Personal Mortgages," Book 5, pp. 264 and 298; Book 6, p. 497; Book 7, pp. 10, 96, and 122; Book 9, pp. 34, 36, 38, 46, 56, 58, 82, 84, 100, 102, 116, 132, 138, 140, 142, 212, 214,

220, 238, and 278; Book 11, pp. 14, 15, 22, 27, 38, 48, 80, 111, 120, 122, and 123; Book 16, pp. 7, 72, and 144; Book 19, p. 78; Book 20, pp. 206, 253, and 283.

47. For mortgages extended by Schnabel Brothers and Co., see Placer County, "Personal Mortgages," Book E, p. 374; Book 5, p. 88; Book 6, pp. 6 and 376; for mortgages extended by George D. Kellogg, see Placer County, "Personal Mortgages," Book E, p. 169, and Book 5, pp. 118, 126, and 168; for mortgages extended by Penryn Fruit Co., see Placer County, "Personal Mortgages," Book 7, pp. 74, 130, 160, and 166; Book 9, pp. 80 and 152; and Book 13, p. 323; for mortgages extended by the Loomis Fruit Growers Association, see Placer County, "Personal Mortgages," Book 7, pp. 210, 212, 218, 220, 236, and 238; for mortgages extended by the Pioneer Fruit Co., see Placer County, "Personal Mortgages," Book 9, pp. 202, 240, and 288; Book 11, p. 116; Book 16, pp. 37, 38, and 60; Book 20, p. 14; and Book 22, p. 282; and for mortgages extended by the Porter Brothers' Co., see Placer County, "Personal Mortgages," Book 5, p. 332; Book 7, pp. 84 and 152; and Book 9, pp. 5, 52, and 60.

48. Immigration Commission, *Reports*, vol. 24, pp. 428–29.

49. Immigration Commission, *Reports*, vol. 24, p. 421.

50. El Dorado County, "Deeds," Book 44, p. 327.

51. El Dorado County, "Deeds," Book 47, p. 151.

52. El Dorado County, "Deeds," Book 47, p. 152.

53. El Dorado County, "Deeds," Book 47, p. 456.

54. El Dorado County, "Deeds," Book 66, p. 114.

55. El Dorado County, "Deeds," Book 53, p. 391.

56. El Dorado County, "Bills of Sale and Miscellaneous," Book A, pp. 51 and 55.

57. U.S. Bureau of the Census, "Census of Productions of Agriculture" (manuscript), 1880, Solano County, California.

58. Norris N. Bleyhl, "A History of the Production and Marketing of Rice in California," p. 88.

59. Ibid., p. 90.

60. Ibid., pp. 91–95.

61. Ibid., p. 106.

62. Butte County, "Leases," Book G. pp. 121, 124, and 390; Colusa County, "Leases," Book 2, p. 234; Book 3, pp. 16, 32, 49, 121, 140, 281, 283, 286, 301, 321, 324, 326, 389, and 391; and Book 5, p. 59; Colusa County, "Personal Property Mortgages," Book Z, pp. 232, 244, 246, and 307; Book 1, pp. 102, 216, and 226; Book 2, p. 434; Book 3, pp. 196, 223, 313, 436, 445, 457, 460, 463, 469, 472, and 475; Book 5, p. 297; Book 6, pp. 185, 252, 305, and 379; Book 7, pp. 42, 66, 78, 178, 186, 190, 198, 218, and 254; Glenn County, "Leases," Book 2, p. 399; Sutter County, "Leases," Book C, p. 498; Sutter County, "Crop and Chattel Mortgages," Book K, p. 227; and Yuba County, "Chattel Mortgages," Book 7, p. 6; Book 8, p. 562; Book 9, pp. 574 and 627; Book 10, pp. 41, 59, and 584.

63. The leases for land on the west side of the Sacramento River and lying some distance from the river, in particular, all carried the safety clause stating

that all payments would be refunded should the landlords be unable to supply sufficient water for the rice to grow properly.

CHAPTER EIGHT

1. Other studies of farmworkers which touch superficially on the Chinese are Lamar B. Jones, "Labor and Management in California Agriculture, 1864–1964"; Laurence I. Hewes, "Some Migratory Labor Problems in California's Industrialized Agriculture"; J. Donald Fisher, "A Historical Study of the Migrant in California"; Jerard D. Wagers, "History of Agricultural Labor in California Prior to 1880"; and James R. Hatfield, "California's Migrant Farm Labor Problem and Some Efforts to Deal with it, 1930–1940" (M.A. thesis, California State University, Sacramento, 1968).

2. In addition to relevant chapters in Daniel, *Bitter Harvest*; Majka and Majka, *Farm Workers*; Hewes, "Some Migratory Labor Problems"; Fisher, "A Historical Study of the Migrant"; and Hatfield, "Migrant Farm Labor Problem," studies on white migrant farm laborers include John Steinbeck's classic novel *The Grapes of Wrath* (New York: The Viking Press, 1939); Walter J. Stein, *California and the Dust Bowl Migration* (Westport, Conn.: Greenwood Press, Inc., 1973); State Relief Administration of California, *Migratory Labor in California* (San Francisco: State Relief Administration of California, 1936); California Employment Commission, *Handbook on Farm Labor Placement in California* (Sacramento: California Department of Employment, 1941); Tyr V. Johnson and Frederick Arpke," Interstate Migration and County Finance in California"; and U.S. Department of Agriculture, Bureau of Agricultural Economics, *Migration and Settlement on the Pacific Coast*, Report no. 10 (Washington, D.C.: U.S. Department of Agriculture, 1942). Besides the detailed accounts in Daniel, *Bitter Harvest*, and Majka and Majka, *Farm Workers*, studies of farmworkers' strikes and unionization efforts in the 1930s may be found in Stuart Jamieson, *Labor Unionism in American Agriculture*; and James Gray, *The American Civil Liberties Union of Southern California and Imperial Valley Agricultural Labor Disturbances: 1930, 1934* (San Francisco: R and E Research Associates, 1977). Jerold S. Auerbach, *Labor and Liberty: The La Follette Committee and the New Deal* (Indianapolis and New York: The Bobbs–Merrill Co., Inc., 1966), investigated the role that the La Follette Committee played in bringing to light many abuses suffered by farm and other laborers in America.

3. Fuller, "Supply of Agricultural Labor," pp. 19782–83.

4. Ibid., pp. 19784–85 and 19789.

5. Ibid., p. 19798.

6. Ibid., p. 19807; and interview with Varden Fuller (Occidental, California, May 9, 1979).

7. Fuller interview.

8. Fuller, "Supply," p. 19878.

9. Ibid., p. 19822.

10. Fuller interview.

11. Fuller interview.

12. Fuller interview and "Supply," p. 19879.

13. Interview with Carey McWilliams (Berkeley, April 6, 1980).

14. McWilliams, *Factories in the Field*, pp. 11–22 and 29–39.

15. Ibid., p. 48.

16. Ibid., pp. 22–25.

17. Ibid., pp. 60–65.

18. Ibid.

19. California State Bureau of Labor Statistics, *Second Biennial Report of the Bureau of Labor Statistics of California for the Years 1885–1886* (Sacramento: State Printing Office, 1886), pp. 44–45.

20. Ibid., p. 46.

21. Ibid., p. 47.

22. B[ertram] Schrieke, *Alien Americans: A Study of Race Relations* (New York: The Viking Press, 1936), p. 23.

23. McWilliams, *Factories*, pp. 70–71.

24. Ibid., p. 71.

25. Ibid., p. 73, in which he quoted Eaves, *Labor Legislation*, p. 187. In her footnote, Eaves stated that "among places taking such action were Eureka, Truckee, Redding, Santa Cruz, Boulder Creek, Nicolaus, in California; [and] Tacoma in Washington," but she did not give any sources for this information.

26. McWilliams, *Factories*, pp. 82–92.

27. Ibid., 88.

28. Paul S. Taylor, "Foundations of California Rural Society," *California Historical Society Quarterly* 24 (1945): 193–228.

29. Ibid., pp. 196–97.

30. Ibid., p. 197.

31. Ibid., p. 194.

32. Ibid., p. 193.

33. Ibid., pp. 297–98, in which he quoted U.S. Congress, Joint Special Committee, *Report*, pp. 787, 768, and 778.

34. Ibid., p. 209.

35. Ibid., p. 215.

36. Ibid., p. 218.

37. Ibid., pp. 220–21.

38. Ibid., p. 221.

39. Fisher's study was an interpretive work, two chapters of which were based on research done by others. Chapter 2 relied heavily on an anonymous manuscript entitled "The Labor Contracting System in California Agriculture" and prepared under the auspices of the Federal Writers' Project, Oakland, California, in the 1930s. Both it and another study, "Oriental Labor Unions and Strikes in California Agriculture," contained information from various contemporary newspapers, periodicals such as the *Pacific Rural Press* and the *California Fruit Grower*, reports of the California Bureau of Labor Statistics, and several key federal documents, including the reports of the U.S. Immigration Commission published in 1911, and *Report of the [U.S.] Industrial Commission on Agriculture and Agricultural Labor* published in 1901.

40. Clark Kerr and John T. Dunlop, "Foreword," in L. Fisher, *Harvest Labor*

Market, p. viii.

41. Fisher, *Harvest Labor Market*, p. 2.
42. Ibid., p. 4.
43. ibid., pp. 7–9.
44. Ibid., p. 7.
45. Ibid., p. 10.
46. Ibid., p. 11.
47. Ibid., pp. 13–15.
48. Ibid., p. 20.
49. Chiu, *Chinese Labor*, p. 81.
50. Ibid.
51. Ibid., pp. 84–85.
52. Ibid., p. 86.
53. Ibid., p. 87.
54. "Land monopoly" and "industrial agriculture" or "agribusiness" were among the chief forces affecting the lives of farm laborers.
55. A distinction should be drawn among the terms "commercial agriculture," "industrial agriculture," and "agribusiness." The first refers to the production of crops for sale; the second refers to a system of agriculture based on the principles of factory production, including the employment of gangs of agricultural wage laborers; the third refers to what economists have called "vertical integration" in the cultivation, harvesting, processing, distribution, and the sales promotion of crops.
56. Daniel, *Bitter Harvest*, pp. 17–18.
57. Ibid., pp. 37–39.
58. Ibid., p. 19.
59. Ibid., pp. 21–23.
60. Ibid., p. 26.
61. Ibid., p. 27.
62. Ibid., p. 31.
63. Ibid., p. 34–35.
64. Ibid., p. 36.
65. Ibid., p. 38.
66. Ibid., p. 285.
67. Ibid., p. 27–28.
68. The authors based their theoretical exposition on the following works of Nicos Poulantzas: *Political Power and Social Classes* (London: New Left Books, 1973); *Classes in Contemporary Capitalism* (London: New Left Books, 1975); "The Problem of the Capitalist State," *New Left Review* 58 (1969): 67–78; and "The State and the Transition to Socialism," interview with Henri Weber, *Socialist Review* 38 (1978): 9–36.
69. Majka and Majka, *Farm Workers*, pp. 13–19.
70. Ibid., pp. 10 and 31–32.
71. These eight works were Seward, *Chinese Immigration*; Coolidge, *Chinese Immigration*; Cross, *Labor Movement*; Saxton, *Indispensable Enemy*; Eaves, *Labor Legislation*; Bancroft, *History*; McWilliams, *Factories*; and Herbert Hill, "Anti-Oriental Agitation and the Rise of Working Class Racism," *Transaction: Social*

Science and Modern Society 10 (1973): 43–54.

72. Majka and Majka, *Farm Workers*, pp. 20 and 31–32.

CHAPTER NINE

1. U.S. Bureau of the Census, *Ninth Census of the United States: Population, 1870* (Washington, D.C.: Government Printing Office, 1872), p. 799.

2. McGowan, *Sacramento Valley*, vol. 1, p. 244; and my tally from U.S. National Archives, Record Group 29, "Census of the United States Population" (manuscript), 1880, Colusa County, California.

3. R. L. Adams, *Seasonal Labor Requirements for California Crops*, pp. 22–23.

4. California State Bureau of Labor Statistics, *Second Biennial Report 1885–86*, pp. 44–64.

5. U.S. Congress, Joint Special Committee, *Report* (1877), p. 931.

6. For citations on farm tenancy, see note 2 of chapter 3.

7. Fuller interview.

8 The following account is based on statistical computations in Chan, "Bittersweet Harvest," pp. 330–54.

9. *Pacific Rural Press*, October 6, 1883.

10. Ibid.

11. Thomas Edward Malone, "The California Irrigation Crisis of 1886: Origins of the Wright Act, 1887"; and Henrik Teilmann, "The Role of Irrigation Districts in California's Water Development."

12. U.S. Bureau of the Census, *Thirteenth Census of the United States: Agriculture, 1910*, vol. 5, p. 845.

13. William E. Smythe, *The Conquest of Arid America* (1899; repr. ed., Seattle: University of Washington Press, 1969), p. 43. For an assessment of Smythe, see Lawrence B. Lee, "William Ellsworth Smythe and the Irrigation Movement: A Reconsideration," *Pacific Historical Review* 41 (1972): 289–311.

14. Paul S. Taylor has written extensively about the federal government's failure to impose acreage limitation on land irrigated by water from projects built with federal funds. Two of his more important articles on this topic are "Excess Land Law: Pressure versus Principle," *California Law Review* 47 (1959): 499–541; and "Excess Land Law: Calculated Circumvention," *California Law Review* 52 (1964): 978–1014.

15. For histories of California marketing cooperatives, see W. W. Cumberland, *Cooperative Marketing: Its Advantages as Exemplified in the California Fruit Growers Exchange* (Princeton: Princeton University Press, 1917); H. E. Erdman, *The California Fruit Growers Exchange: An Example of Cooperation in the Segregation of Conflicting Interests*; Albert J. Meyer, "History of the California Fruit Growers Exchange, 1893–1920"; Erich Kraemer and H. E. Erdman, *History of Cooperation in the Marketing of California Fresh Deciduous Fruits*, Bulletin no. 557 (Berkeley: University of California, College of Agriculture, Agricultural Experiment Station, 1933); Fred W. Powell, "Co-operative Marketing of California Fresh Fruit"; A. J. Schoendorf, *Beginnings of Cooperation in the Marketing of California Fresh Deciduous Fruits and History of the California Fruit Exchange* (Sacramento: Inland Press, 1946); H. E. Erdman, "The Development and Significance of

California Cooperatives, 1900–1915"; Herbert W. Free, "California Apricot Growers Union" (M.A. thesis, University of California, Berkeley, 1941); Philip J. Webster, "An Analysis of the Development of Co-operative Marketing Policies in the California Prune and Apricot Growers Association" (Ph.D. diss., University of California, Berkeley, 1930); "History of California Almond Growers Exchange," *California Cultivator* 40 (1913): 614; Carl A. Scholl, "An Economic Study of the California Almond Growers Exchange" (Ph.D. diss., University of California, Berkeley, 1927); and Fred K. Howard, *History of Raisin Marketing in California* (Fresno: Sun Maid Raisin Growers, n.d.).

16. California State Agricultural Society, *Transactions*, 1872, pp. 257–58, quoted in Chiu, *Chinese Labor*, p. 82.

17. California State Bureau of Labor Statistics, *First Biennial Report of the Bureau of Labor Statistics of California for the Years 1883–1884* (Sacramento: State Printing Office, 1884), p. 167.

18. Ibid.

19. Computed from U.S. Department of Agriculture, Division of Statistics, *Wages of Farm Labor in the United States, Results of Eleven Statistical Investigations, 1866–1899* (Washington, D.C.: Government Printing Office, 1901), p. 10.

20. *San Francisco Post*, October 12, 1877.

21. Jamieson, *Labor Unionism*, pp. 46–47.

22. Some of the more important guilds in Chinese American communities were the Tung Duck Tong for cigarmakers, the Fook Sing Tong for shoemakers, the Sai Fook Tong for laundrymen, and the Gam Yee Tong for garment workers.

23. *San Francisco Bulletin*, August 18, 1880.

24. Jamieson, *Labor Unionism*, p. 46, n. 8.

25. *Pacific Rural Press*, November 3, 1883.

26. *Pacific Rural Press*, October 11, 1884.

27. Jamieson, *Labor Unionism*, p. 46, n. 8.

28. California State Bureau of Labor Statistics, *Ninth Biennial Report of the Bureau of Labor Statistics of California for the Years 1899–1900* (Sacramento: State Printing Office, 1900), p. 15.

29. *Pacific Rural Press*, August 28, 1884.

30. *Wheatland Graphic*, March 27, 1886.

31. *Pacific Rural Press*, October 6, 1883.

32. Interview with Earl Ramey (Marysville, August 11, 1980).

33. van Löben Sels interview.

34. *Pacific Rural Press*, March 6, 1886, and March 13, 1886.

35. "Orchards and Orientals," *Oakland Tribune*, June 2, 1952.

36. Ibid.

37. California State Bureau of Labor Statistics, *Fourth Biennial Report of the Bureau of Labor Statistics of California for the Years 1889–1890* (Sacramento: State Printing Office, 1890), p. 93.

38. Ibid., pp. 90 and 92.

39. U.S. National Archives, Record Group 29, "Census of the United States Population" (manuscript), 1880, Alameda County, California.

40. Sutter County Historical Society, *Newsletter*, January 1966, pp. 2 and 4.

41. California State Bureau of Labor Statistics, *Fourth Biennial Report*, pp. 97–101.

42. U.S. Immigration commission, *Reports*, vol. 24, pp. 251–52.

43. Hugh S. Jewett, "Reminiscences," file 6, "Yellow Sam" (Berkeley: Bancroft Library, University of California, Berkeley).

44. Ibid.

CHAPTER TEN

1. Interview with Pardow Hooper (San Francisco, March 14, 1979).

2. Frank Lortie, "Historic Sketch of the Town of Locke"; van Löben Sels manuscript; and testimony of George D. Roberts in U.S. Congress, Joint Special Committee, *Report*, p. 433.

3. Hooper interview.

4. *Pacific Rural Press*, November 3, 1883.

5. Lortie, "Locke," p. 16.

6. Kung-chuan Hsiao, *Rural China: Imperial Control in the Nineteenth Century* (1960; paper ed., Seattle: University of Washington Press, 1967), pp. 26–29 and 43–83.

7. Robert A. Nash, "The 'China Gangs' in the Alaska Packers Association Canneries, 1892–1935," in *The Life, Influence and the Role of the Chinese in the United States, 1776–1960* (San Francisco: Chinese Historical Society of America, 1967), p. 266.

8. Yen-p'ing Hao, *The Comprador in Nineteenth-Century China: Bridge between East and West* (Cambridge: Harvard University Press, 1970), p. 2.

9. Ibid.

10. U.S. Immigration Commission, *Reports*, vol. 24, p. 188.

11. Ivan Light, *Ethnic Enterprise in America: Business and Welfare among Chinese, Japanese and Blacks*.

12. Satow interview.

13. U.S. Immigration Commission, *Reports*, vol. 24, pp. 421–23.

14. The biography of Lee Bing is based on Jean Rossi, "Lee Bing: Founder of California's Historical Town of Locke"; and on Frank Lortie, "Historic Sketch of the Town of Locke," pp. 8–9. Rossi's account was derived from interviews with Lee Bing's son, Ping Lee.

15. Lortie, "Sketch," dated the construction of Chan Tin San's store to 1912. However, the lease of the lot on which Tin San's store was built was dated 1914.

16. Sacramento County, "Leases," Book N, p. 356.

17. Lortie, "Sketch," p. 8.

18. Ibid., p. 6.

19. Hooper interview.

20. *Pacific Rural Press*, October 7, 1876.

21. Ibid.

22. *Pacific Rural Press*, August 26, 1882.

23. U.S. Immigration Commission, *Reports*, vol. 24, p. 213.

24. Monterey County, "Leases," Book I, p. 25.

25. Donald H. Kulp, *Country Life in South China* (New York: Columbia University Teachers College, Bureau of Publications, 1925), pp. 91–92.

26. The following biographical sketch of Yuen Yeck Bow is from William Turner Ellis, *Memoirs: My Seventy-two Years in the Romantic County of Yuba, California* (Eugene, Ore.: John Henry Nash for University of Oregon, 1939), pp. 12–13. Ellis was mayor of Marysville for two terms and served as head of the city's Levee Commission for many years.

27. The following account of Wong Gee is from Book, *Chinese in Butte County*, pp. 33–35.

28. *Stockton Independent*, October 23, 1876.

29. Ibid.

30. U.S. National Archives, Record Group 29, "Census of the United States Population" (manuscript), 1900, Fresno County, California.

CHAPTER ELEVEN

1. *San Francisco Bulletin*, July 14, 1876.

2. Ibid.

3. Ibid.

4. McGowan, *Sacramento Valley*, p. 324.

5. *Chico Enterprise*, March 2, 1877.

6. Ibid., March 9, 1877.

7. Ibid., March 9 and March 30, 1877.

8. Ibid., March 16, 1877.

9. Ibid., March 30, 1877.

10. *Yuba City and Marysville Independent Herald*, June 14, 1951.

11. *San Francisco Daily Morning Call*, August 20, 1877.

12. Ibid.

13. Ibid.

14. McGowan, *Sacramento Valley*, p. 326.

15. Ibid., p. 327.

16. *Marysville Daily Appeal*, February 26, 1886.

17. Ibid.

18. *Wheatland Graphic*, February 27, 1886.

19. Ibid.

20. Ibid.

21. *Wheatland Graphic*, March 6, 1886.

22. *Wheatland Graphic*, April 10, 1886.

23. *Pacific Rural Press*, April 24, 1886.

24. *Hayward Journal*, May 8, 1886, reprinted in *Pacific Rural Press*, May 15, 1886.

25. *Pacific Rural Press*, May 15, 1886.

26. *Pacific Rural Press*, April 24, 1886.

27. *Pacific Rural Press*, April 10, 1886.

28. McWilliams, *Factories in the Field*, pp. 74–77.

29. Ibid., p. 77.

30. For a list of counties whose lease and mortgage documents were used to

compile the total number of acres leased by Chinese that were recorded, see table 37.

31. *Daily Alta California*, September 10, 1877.

32. Carleton H. Parker, *The Casual Laborer and Other Essays* (New York: Harcourt, Brace and Howe, 1920), pp. 70–73.

33. Lucie Cheng Hirata, "Free, Indentured, Enslaved: Chinese Prostitutes in Nineteenth-Century America."

34. It is part of Chinese American folklore that only merchants could afford to have their wives in residence.

35. The following analysis is based on statistical computations in Chan, "Bittersweet Harvest," pp. 246–51.

APPENDIX

1. Margo Anderson Conk, *The United States Census and Labor Force Change: A History of Occupation Statistics, 1870–1940* (Ann Arbor, Michigan: University Microfilms International, 1978), p. 9.

2. Conk, *Census*, pp. 9–10.

3. Ibid., p. 12.

4. Owen C. Coy, *Guide to the County Archives of California* (Sacramento: California Historical Survey Commission, California State Printing Office, 1919), though published more than six decades ago, remains useful. Coy gives succinct summaries of how various county offices were organized, beginning with the pre-American period. Though many county offices have long since reorganized themselves, Coy's descriptions are still helpful in understanding each county's functions and, consequently, the kinds of documents which may be found within each. In general, I have found that nineteenth-century documents have been preserved better than twentieth-century ones. Perhaps this reflects the presumption that many county employees seem to have that handwritten documents have greater historical value. The destruction wrought by the 1906 San Francisco earthquake and fire has created a major gap in California's county archival holdings, since San Francisco's pre-1906 records are no longer available.

5. Personal communication, staff members in the Sacramento County Recorder's office (Sacramento, October 10, 1978).

6. Personal communication, James Johnstone, San Joaquin County Recorder (Stockton, November 10, 1978).

Selected Bibliography

County, State, and Federal Archival Materials

CALIFORNIA COUNTY ARCHIVAL MATERIALS

Counties differ in how their records are organized and kept. The date of establishment of each county, the natural resources each possessed, the size of the county, the relative tendency of its people to record transactions, the amount of space allotted the recorder's office, the predilections of the county recorder and his or her clerks to dispose of those old records which they are not legally required to keep, whether there had been any fires, and how many moves the office of the county recorder has had to make since the county's establishment, all affected the volume of archival documents still available. There was inconsistency in what kinds of documents were entered in what kinds of books, as well as in the titles of the books. The list below gives the accessible records that were consulted in this study.

Alameda Co.
  "Leases." Books A–Z, 27–35. 1857–1920.
  "Lessee-Lessor Index." Vols. 1–8. 1857–1920.
  Oakland: Alameda Co. Recorder's Office.

Amador Co.
  "Deeds." Books A–N, 1–33. 1854–1910.
  "Deeds Grantee-Grantor Index." Vols. 1–5. 1854–1910.
  "Miscellaneous Records." Books A–L. 1861–1935. (Contains leases.)
  "Miscellaneous Records Index." Vols. 1–2. 1861–1935.
  "Mining Claims." Book A. 1862–68.
  "Mining Claims Index." Vol. 2. 1862–68.
  Jackson: Amador Co. Recorder's Office.

Butte Co.
  "Deeds." Books A–Z, 27–53. 1850–1900.
  "Deeds Grantee-Grantor Index." Vols. 1–5. 1850–1900.
  "Leases." Books A–H. 1860–1927.

"Lessee-Lessor Index." 1860–1927.
Oroville: Butte Co. Recorder's Office.

Calaveras Co.
"Deeds." Books A–X. 1852–77.
"Deeds Grantee-Grantor Index." Vols. A1–A2. 1852–77.
"Leases and Bills of Sale." Books A–F. 1853–1922.
"Leases and Bills of Sale Index." Vols. A–B. 1853–1922.
"Mining Claims." Books A–H. 1852–92.
"Mining Claims Index." Vols. 1–5. 1852–92.
San Andreas: Calaveras Co. Recorder's Office.

Colusa Co.
"Leases." Books A–D, 1–4. 1855–1927.
"Lessee-Lessor Index." Vol. A. 1855–1927.
"Personal Property Mortgages." Books A–Z, 1–9. 1877–1924.
"Personal Property Mortgagee-Mortgagor Index." Vols. A–B. 1877–1924.
Colusa: Colusa Co. Recorder's Office.

Contra Costa Co.
"Chattel Mortgages." Books 1–34. 1852–1924.
"Leases." Books 1–12. 1862–1924.
"General Index." 1849–1917.
Martinez: Contra Costa Co. Recorder's Office.

El Dorado Co.
"Bills of Sale and Miscellaneous." Books A–B. 1896–1928.
"Deeds." Books A–Z, 27–50. 1850–99.
"Deeds Grantee-Grantor Index." 4 sets of unnumbered vols. 1850–99.
"Leases." Books A–D. 1856–1927.
(No lease index.)
"Mining Locations." Books A–M. 1852–1905.
"Mining Locations Index." Vols. 1–2. 1852–1905.
"Miscellaneous Records." Books A–C. 1896–1928.
"Miscellaneous Records Index." Vol. 1. 1896–1928.
Placerville: El Dorado Co. Recorder's Office.

Fresno Co.
"Deeds." Books A–Z, 27–728. 1856–1920.
"Deeds Grantee-Grantor Index." Vols. 1–31. 1856–1913. Unnumbered
vols., 1914–20.
"Leases." Books A–T. 1869–1920.
"Lessee-Lessor Index." Vols. 1–2. 1869–1920.
Fresno: Fresno Co. Recorder's Office.

Glenn Co.
"Chattel Mortgages." Books 17, 19, and 21. 1891–1927.
"Crop Mortgages." Books 5, 9, 14, and 30. 1891–1927.
"Leases." Books 1–4. 1891–1927.
"Lessee-Lessor Index." Vol. 1. 1891–1927.

"Personal Property Mortgages." Books 1–4, 6–8, 10–13, 15–16, 18, 20, and
  22–29. 1891–1927.
"Personal Property Mortgagee-Mortgagor Index." Vols. 1–2. 1891–1927.
Willows: Glenn Co. Recorder's Office.

Imperial Co.
  "Chattel Mortgages." Books 1–233. 1907–23.
  "Chattel Mortgagee-Mortgagor Index." Vols. 2, 3, and 5. 1909–12, 1913–
    16, and 1922–23. (Vols. 1 and 4 are missing.)
  "Leases." Books 1–12. 1907–23.
  "Lessee-Lessor Index." Vol. 1. 1907–23.
  "Miscellaneous Records." Books 1–10. 1896–1923.
  "Miscellaneous Records Index." Vols. 1–2. 1896–1923.
  El Centro: Imperial Co. Recorder's Office.

Kern Co.
  "Deeds." Books 1–9. 1867–79.
  "Leases." Books 1–4. 1867–99.
  "General Consolidated Index." 1867–99. (Computerized; no vol. no.)
  Bakersfield: Kern Co. Recorder's Office.

Kings Co.
  "Chattel Mortgages." Books 7–31. 1906–20.
  "Leases." Books 1–5. 1893–1925.
  (Lease index at the front of each book.)
  "Personal Property Mortgages." Books 1–6. 1893–1906.
  "Personal Property Mortgagee-Mortgagor Index." Vols. 1–3. 1893–1921.
  Hanford: Kings Co. Recorder's Office.

Los Angeles Co.
  "Deeds." Books 1–7474. 1850–1920.
  "Deeds Grantee-Grantor Index." Unnumbered vols. 1850–1920.
  "Leases." Books 1–143. 1850–1923.
  "Lessee-Lessor Index." Vols. 1–6. 1850–1923.
  Los Angeles: Los Angeles Co. Recorder's Office.

Madera Co.
  "Chattel Mortgages." Books 1–33. 1893–1922.
  "Chattel Mortgagee-Mortgagor Index." Vol. 1. 1893–1922.
  "Leases." Books 1–3. 1893–1922.
  "Lessee-Lessor Index." Vol. 1. 1893–1922.
  Madera: Madera Co. Recorder's Office.

Merced Co.
  "Deeds." Books A–Z, 27–153. 1857–1920.
  "Deeds Grantee-Grantor Index." Unnumbered vol. 1857–1920.
  "Leases." Books B–G. 1874–1921. (Book A is missing.)
  "Lessee-Lessor Index." Vol. 1. 1857–1921.
  "Mortgages" (Real Property). Books 33–56. 1901–13.
  Merced: Merced Co. Recorder's Office.

Monterey Co.
  "Chattel Mortgages." Books A–Z, 1–15. 1869–1916.
  "Chattel Mortgagee-Mortgagor Index." Vols. 1–4. 1869–1916.
  "Leases." Books A–H. 1866–1916.
  "Lessee-Lessor Index." Vol. 1. 1866–1916.
  Salinas: Monterey Co. Recorder's Office.

Napa Co.
  "Leases." Books A–G. 1850–1918.
  "Lessee-Lessor Index." Vol. 1. 1850–1918.
  "Personal Property Mortgages." Books B–V. 1877–1927.
  "Personal Property Mortgagee-Mortgagor Index." Vols. 1–2. 1877–1927.
  Napa: Napa Co. Recorder's Office.

Nevada Co.
  "Deeds." Books 1–140. 1856–1921.
  "Deeds Grantee-Grantor Index." Vols. 1–8. 1856–1921.
  "Leases." Books 1–4. 1856–1921.
  "Lessee-Lessor Index." Vol. 1. 1856–1921.
  "Mining Claims." Books 1–37. 1856–1979.
  "Mining Claims Index." Vols. 1–2. 1856–1979.
  Nevada City: Nevada Co. Recorder's Office.

Orange Co.
  "Leases." Books 1–63. 1891–1926.
  "Lessee-Lessor Index." Vols. 1–2. 1891–1926.
  Santa Ana: Orange Co. Recorder's Office.

Placer Co.
  "Deeds." Books A–Z, AA–ZZ. 1851–1920.
  "Deeds Grantee-Grantor Index." Unnumbered vols. 1851–1920.
  "Leases." Books B–M. 1870–1920. (Book A is missing.)
  "Lessee-Lessor Index." Vols. 1–2. 1870–1920.
  "Personal Mortgages." Books 1–27. 1857–1923.
  "Personal Mortgagee-Mortgagor Index." Vol. 1, 1–A, and 2. 1857–1923.
  Auburn: Placer Co. Recorder's Office.

Riverside Co.
  "Chattel Mortgages." Books 1–68. 1893–1920.
  "Chattel Mortgagee-Mortgagor Index." 1893–1920.
  "Leases." Books 1–14. 1893–1928.
  "Lessee-Lessor Index." Vols. 1–2. 1893–1928.
  Riverside: Riverside Co. Recorder's Office.

Sacramento Co.
  "Chattel Mortgages." Books K, 27, 37, 50, 56, 80, 110, 121, 140, 161, 179,
    191, 202, and 210. 1860–1925.
  "Chattel, Crop, Personal Property Mortgagee-Mortgagor Index." Unnumbered vols. 1860–1925.
  "Crop Mortgages." Books 66, 72, 84, 89, 96, 105, 109, and 172. 1890–1923.

"Deeds." Books 1–933. 1850–1941.

"Deeds Grantee-Grantor Index." Unnumbered index cards and volumes. 1850–1941.

"Leases." Books A–S. 1853–1941.

(No lease index.)

"Personal Property Mortgages." Books 67, 77, 86, 93, 102, 111, 124, 131, 135, 142, 147, 151, 155, 159, 164, 168, 170, 174, 176, 178, 180, 182, 184, 185, 188, 189, 192, 196, 198, 200, 204, 205, 207, 209, 213, 214, 216, 217, and 218. 1892–1925.

Sacramento: Sacramento Co. Recorder's Office, and Sacramento Museum and History Commission.

San Bernardino Co.

"Chattel Mortgages." Books 1–25. 1900–25.

"Chattel Mortgagee-Mortgagor Index." Vols. 1–2. 1900–25.

"Leases." Books B–I. 1880–1923.

"Lessee-Lessor Index." Vol. 1. 1880–1923.

San Bernardino: San Bernardino Co. Recorder's Office.

San Diego Co.

"Leases." Books 1–35. 1869–1931.

"Lessee-Lessor Index." Vols. 1–2. 1869–1931.

San Diego: San Diego Co. Recorder's Office.

San Joaquin Co.

"Book G of Miscellaneous." (Contains leases.) Books G1–G49. 1850–1921.

"Book G of Miscellaneous Index." Vols. 1–4. 1850–1921.

"Book I of Mortgages." Books I1–I73. 1850–1921.

"Book I of Mortgages Index." Vols. 1–6. 1850–1921.

"Reclamation Districts." Files, 1898–1920.

Stockton: San Joaquin Co. Recorder's Office, and

Lodi: San Joaquin Co. Museum.

San Luis Obispo Co.

"Leases." Books A–J. 1880–1920.

"Lessee-Lessor Index." Vol. 1. 1880–1920.

San Luis Obispo: San Luis Obispo Co. Recorder's Office.

San Mateo Co.

"Leases." Books 1–9. 1875–1923.

"Lessee-Lessor Index." Vol. 1. 1875–1923.

Redwood City: San Mateo Co. Recorder's Office.

Santa Barbara Co.

"Leases." Books A–L. 1854–1921.

"Lessee-Lessor Index." Vols. A–B. 1854–1921.

"Personal Property Mortgages." Books A–Z, 1–23. 1857–1923.

"Personal Property Mortgagee-Mortgagor Index." Vols. A–E. 1857–1923.

Santa Barbara: Santa Barbara Co. Recorder's Office.

Santa Clara Co.
    "Chattel Mortgages." Books 1–52. 1865–1922.
    "Leases." Books B–S. 1867–1922. (Book A is missing.)
    (No chattel mortgage or lease index.)
    "Miscellaneous Records." Books A–Z, 1. 1868–1895.
    "Miscellaneous Records Index." Vols. A–D, F–G. 1868–1910. (Index
      Vol. E is missing.)
    San Jose: Santa Clara Co. Recorder's Office.

Santa Cruz Co.
    "Leases." Books 1–9. 1870–1920.
    "Lessee-Lessor Index." Vol. 1. 1870–1920.
    Santa Cruz: Santa Cruz Co. Recorder's Office.

Shasta Co.
    "Agreements." Books 1–5. 1858–1920.
    "Agreements Index." Vols. 1–2. 1858–1920.
    "Deeds." Books 1–94. 1864–1906.
    "Deeds Grantee-Grantor Index." Unnumbered vols. 1864–1906.
    "Leases." Books 1–5. 1879–1920.
    "Lessor Index." Vol. 1. 1856–1913. (Lessee index is missing.)
    "Miscellaneous Records." Books A–Z, 27–45. 1854–1920.
    "Miscellaneous Records Index." Vols. 1–5. 1851–99.
    Redding: Shasta Co. Recorder's Office.

Sierra Co.
    "Bank and Water Claims." Books A–H. 1852–99.
    "Contracts and Agreements." Books A–F. 1854–1907.
    "Contracts and Agreements Index." Vol. 1. 1854–1932.
    "Deeds." Books A–Z, 1–30. 1852–1932.
    "Deeds Grantee-Grantor Index." Vols. 1–9. 1852–1932.
    "Leases." Book A. 1881–1910.
    "Lessee-Lessor Index." Vol. 2. 1860–1931.
    "Mortgages and Leases Index." Vol. 1. 1864–1934.
    "Placer Claims Index." Vols. 1–3. 1852–99.
    "Tax Deeds." Books 1–6. 1875–1919.
    Downieville: Sierra Co. Recorder's Office.

Solano Co.
    "Chattel Mortgages." Books 1–39. 1849–1920.
    "Chattel Mortgagee-Mortgagor Index." Vols. 1–3. 1849–1920.
    "Leases." Books 1–15. 1853–1919.
    "Lessee-Lessor Index." Vol. 1. 1853–1919.
    Fairfield: Solano Co. Recorder's Office.

Sonoma Co.
    "Chattel Mortgages." Books A–Z, 1–4. 1874–1921.
    "Chattel Mortgagee-Mortgagor Index." Vols. 1–3. 1874–1921.
    "Leases." Books A–O. 1854–1921.
    "Lessee-Lessor Index." Vols. 1–2. 1854–1921.
    Santa Rosa: Sonoma Co. Recorder's Office.

Stanislaus Co.
    "Leases." Books 1–8. 1869–1920.
    "Lessee-Lessor Index." Vol. 1. 1869–1920.
    "Miscellaneous Records." Books 1–10. 1862–1920.
    "Miscellaneous Records Index." Vols. 1–2. 1862–1920.
    Modesto: Stanislaus Co. Recorder's Office.

Sutter Co.
    "Crop and Chattel Mortgages." Books A–Z, 1–5. 1850–1927.
    "Crop and Chattel Mortgagee-Mortgagor Index." Vols. A–C. 1850–1927.
    "Leases." Books A–K. 1856–1927.
    "Lessee-Lessor Index." Vols. 1–2. 1856–1927.
    "Miscellaneous Records." Books A–H. 1856–1926.
    "Miscellaneous Records Index." Vol. 1. 1856–1926.
    Yuba City: Sutter Co. Recorder's Office.

Tehama Co.
    "Deeds." Books A–Z, 1–34. 1856–1903.
    "Deeds Grantee-Grantor Index." Vols. 1–7. 1856–1903.
    "Leases." Books A–D, 1878–1927.
    "Lessee-Lessor Index." Vol. 1. 1878–1927.
    Red Bluff: Tehama Co. Recorder's Office.

Tulare Co.
    "Deeds." Books A–Z, 27–206. 1852–1913.
    "Deeds Grantee-Grantor Index." Vols. 1–51. 1855–1923.
    "Leases." Books A–I. 1864–1921.
    "Lessee-Lessor Index." Vols. 1–2. 1864–1921.
    Visalia: Tulare Co. Recorder's Office.

Tuolumne Co.
    "Deeds." Books 1–21. 1850–95.
    "Deeds Grantee-Grantor Index." Vols. 1–5. 1850–95.
    "Leases." Books 1–3, 1873–1926.
    "Lessee-Lessor Index." Vol. 1. 1873–1926.
    Sonora: Tuolumne Co. Recorder's Office.

Yolo Co.
    "Leases." Books A–H. 1854–1922.
    "Lessee-Lessor Index." Vols. A–B. 1854–1922.
    Woodland: Yolo Co. Recorder's Office.

Yuba Co.
    "Bonds, Contracts and Agreements." Books 1–2. 1866–95.
    "Certificates of Sale." Books 2–3. 1920–27.
    "Chattel Mortgages." Books 1–15. 1857–1923.
    "Chattel Mortgagee-Mortgagor Index." Vols. 1–2. 1857–1923.
    "Deeds." Books 1–55. 1850–1906.
    "Deeds Grantee-Grantor Index." Unnumbered vols. 1850–1906.
    "Leases." Books 1–4. 1863–1924.
    "Lessee-Lessor Index." Vol. 1. 1863–1924.

"Mining Deeds." Book 1. 1865.
"Pre-emptions." Books 1–2. 1865–81.
"Tax Deeds." Books 1–3. 1876–1910.
Marysville: Yuba Co. Recorder's Office.

CALIFORNIA STATE ARCHIVAL MATERIALS

California State Lands Commission. Cardex for the disposal of public lands.

California Secretary of State. California State Archives. Incorporation papers for:
Chinese American Farms, inc. 1919.
Empire Navigation Co., inc. 1906.
English-Wallace Co., inc. 1903.
Holland Land and Water Co., inc. 1905.
Middle River Farming Co., inc. 1901.
Middle River Navigation and Canal Co., inc. 1902.
Rindge Land and Navigation Co., inc. 1905.
Roberts Island Improvement Co., inc. 1903.
Tide Land Reclamation Co., inc. 1869.
Venice Island Land Co., inc. 1906.
Victoria Island Co., inc. 1898.
Western Co., inc. 1908.

FEDERAL ARCHIVAL MATERIALS

U.S. Department of Commerce. Bureau of the Census. "Census of Productions of Agriculture" (manuscript), 1860, 1870, and 1880. (In the California State Archives, Sacramento; a microfilmed copy is in the Bancroft Library, University of California, Berkeley.)
U.S. Department of the Interior. Bureau of Land Reclamation. Records and plats for the disposal of federal public lands in California.
U.S. National Archives. Record Group 29. "Census of U.S. Population" (manuscript schedules for California), 1860, 1870, 1880, 1900, and 1910.

The Chinese in California Agriculture, 1860s to 1920s

California. Bureau of Labor Statistics. *Second Biennial Report of the Bureau of Labor Statistics of California for the Years 1885–1886*. Sacramento: State Printing Office, 1886.
———. *Fourth Biennial Report of the Bureau of Labor Statistics of California for the Years 1889–1890*. Sacramento: State Printing Office, 1890.
Castle, Allen. "Locke: A Chinese Chinatown." *The Pacific Historian* 24 (1980): 1–7.
Chen, Jack. *The Chinese of America*. San Francisco: Harper and Row, 1980.
Chiu, Ping. *Chinese Labor in California, 1850–1880: An Economic Study*. Madison:

The State Historical Society of Wisconsin, 1963.

Chu, George. "Chinatowns in the Delta: The Chinese in the Sacramento–San Joaquin Delta, 1870–1960." *California Historical Society Quarterly* 49 (1970): 21–37.

Daniel, Cletus E. *Bitter Harvest: A History of California Farmworkers, 1870–1941.* Ithaca: Cornell University Press, 1981.

Fisher, Donald J. "A Historical Study of the Migrant in California." M.A. thesis, University of Southern California, 1945.

Fisher, Lloyd H. *The Harvest Labor Market in California.* Cambridge: Harvard University Press, 1953.

Fuller, Varden. "The Supply of Agricultural Labor as a Factor in the Evolution of Farm Organization in California." In U.S. Congress. Senate. Committee on Education and Labor. *Hearings Pursuant to Senate Resolution 266, Exhibit 8762-A.* 76th Cong., 3rd sess., 1940, pp. 19777–898.

Hallagan, William S. "Labor Contracting in Turn-of-the-Century California Agriculture." *Journal of Economic History* 40 (1980): 757–76.

Heintz, William F. "The Role of Chinese Labor in Viticulture and Wine Making in Nineteenth-Century California." M. A. thesis, California State University, Sonoma, 1977.

Hewes, Laurence I. "Some Migratory Labor Problems in California's Industrialized Agriculture." Ph.D. diss., George Washington University, 1945.

"Hua-ch'iao tsai mei-kuo chia-shen nung-ts'un ti-wei chih chin-hsi" (The Past and Present Position of Overseas Chinese in the Villages of California in the United States). *Hua-ch'iao Jih-pao* (May 1935).

Jones, Lamar B. "Labor and Management in California Agriculture, 1864–1964." *Labor History* 2 (1970): 23–40.

Leung, Peter C. Y. *One Day, One Dollar: Locke, California, and the Chinese Farming Experience in the Sacramento Delta.* El Cerrito, Calif.: Chinese/Chinese American History Project, 1984.

Lortie, Frank. "Historic Sketch of the Town of Locke." Sacramento: State Department of Parks and Recreation, 1980.

Lydon, Sandy. *Chinese Gold: The Chinese in the Monterey Bay Region.* Santa Cruz: Capitola Book Company, 1985.

McGowan, Joseph A. *History of the Sacramento Valley.* 3 vols. New York and West Palm Beach: Lewis Historical Publishing Co., 1961.

McWilliams, Carey. *Factories in the Field: The Story of Migratory Farm Labor in California,* 1939. Repr. ed. New York: Archon Books, 1969.

Majka, Linda C., and Theo J. Majka. *Farm Workers, Agribusiness, and the State.* Philadelphia: Temple University Press, 1982.

Rossi, Jean. "Lee Bing: Founder of California's Historical Town of Locke." *The Pacific Historian* 20 (1976): 351–66.

Seward, George F. *Chinese Immigration, in Its Social and Economic Aspects.* 1881. Repr. ed. New York: Arno Press, 1970.

Speth, Frank Anthony. "A History of Agricultural Labor in Sonoma County, California." M.A. thesis, University of California, Berkeley, 1938.

Taylor, Paul S. "Foundations of California Rural Society." *California Historical Society Quarterly* 24 (1945): 193–228.

United States. Circuit Court of the United States for the Ninth Circuit and State of California. *Edward Woodruff v. The North Bloomfield Gravel Mining Company, et al.* 1883.

United States. Congress. Senate. Immigration Commission. *Report of the Immigration Commission: Immigrants in Industries.* Pt. 25. "Japanese and Other Immigrant Races in the Pacific Coast and Rocky Mountain States." Vol. 24. "Agriculture." 61st Cong., 2nd sess., 1911, Senate Doc. 633.

Wagers, Jerard D. "History of Agricultural Labor in California Prior to 1880." M.A. thesis, University of California, Berkeley, 1957.

Yip, Christopher L. "A Time for Bitter Strength: The Chinese in Locke, California." *Landscape* 22 (1978): 3–13.

Selected Bibliography on the History of the Chinese in California, 1850s to 1920s

The literature in Chinese American history is substantial. Only writings that deal with the Chinese in California in the pre-1920 period and some of the more important general studies that have no particular geographic focus are included here. Issue numbers for journal articles are given only when a journal does not use continuous pagination for all issues in a volume. For more inclusive listings, the following bibliographies are available.

Amerasia Journal 4 (1977): 156–58; 5 (1978): 157–61; 6 (1979): 110–13; 7 (1980): 161–65; 8 (1981): 176–80; 9 (1982): 153–58; and 10 (1983): 131–53.

Asian American Studies, University of California, Davis. *Asians in America: A Selected Annotated Bibliography, An Expansion and Revision.* Davis: University of California, Davis, Asian American Studies Division, 1983.

Chan, Sucheng. "Selected Bibliography on the Chinese in America, 1850–1920." *The Immigration History Newsletter* 16 (1984): 7–15.

Cowan, Robert Ernest, and Boutwell Dunlap. *Bibliography on the Chinese Question.* San Francisco: A. M. Robertson, 1909.

Fujimoto, Isao, Michiyo Yamaguchi Swift, and Rosalie Zucker. *Asians in America: A Selected Annotated Bibliography.* Davis: University of California, Davis. Asian American Studies Division, 1971.

Griffin, A. P. C. *Library of Congress Select List of References on Chinese Immigration.* Washington, D.C.: Government Printing Office, 1904.

Hansen, Gladys, and William F. Heintz. *The Chinese in California: A Bibliographic History.* Portland, Ore.: Richard Abel and Co., Inc., 1970.

Lum, William Wong. *Asians in America: A Bibliography of Master's Theses and Doctoral Dissertations.* Davis: University of California, Davis, Asian American Studies Division, 1970.

Ng, Pearl. *Writings on the Chinese in California.* San Francisco: R and E Research Associates, 1972.

Yoshitomi, Joan, Paul M. Ong, Karen Ko, and Joanne T. Fujita. *Asians in the Northwest: An Annotated Bibliography.* Seattle: Northwest Asian American Studies Research Group, 1978.

*   *   *

Anderson, Harold P. "Wells Fargo and Chinese Customers in Nineteenth-Century California." In *Eastern Banking: Essays in the History of the Hongkong and Shanghai Banking Corporation*, ed. Frank H. H. King, pp. 735–52. London: Athlone Press, 1983.

Armentrout-Ma, Eve. "The Big Business Ventures of Chinese in North America, 1850–1930." In *The Chinese American Experience*, ed. Genny Lim, pp. 101–12.

———. "Chinese in California's Fishing Industry, 1850–1941." *California History* 60 (1981): 142–57.

———. "Conflict and Contact between the Chinese and Indigenous Communities in San Francisco." In *The Life, Influence and the Role of the Chinese in the United States, 1776–1960*, pp. 55–70. San Francisco: Chinese Historical Society of America, 1976.

Asing, Norman. "To His Excellency Governor Bigler from Norman Asing." *Daily Alta California*, May 5, 1855.

Bancroft, Hubert Howe. "The Folly of Chinese Exclusion." *North American Review* 179 (1904): 263–68.

———. *History of California*. Vol. 7: 1860–1890, pp. 335–69. Vol. 24 of *The Works of Hubert Howe Bancroft*. San Francisco: The History Company, 1890.

Barth, Gunther. *Bitter Strength: A History of the Chinese in the United States, 1850–1870*. Cambridge: Harvard University Press, 1964.

Book, Susan W. *The Chinese in Butte County, California, 1860–1920*. San Francisco: R and E Research Associates, 1976.

Caldwell, Dan. "The Negroization of the Chinese Stereotype in California." *Southern California Quarterly* 53 (1971): 123–31.

California. State Legislature. "Report of Joint Select Committee Relative to the Chinese Population of the State of California." *Appendix to the Journals of the Senate and Assembly*, vol. 3. Sacramento: State Printing Office, 1862.

———. Senate. Special Committee on Chinese Immigration. *Chinese Immigration: Its Social, Moral, and Political Effect*. Sacramento: State Printing Office, 1878.

Carranco, Lynwood. "Chinese Expulsion from Humboldt County." *Pacific Historical Review* 30 (1961): 329–40.

———. "Chinese in Humboldt County, California: A Study in Prejudice." *Journal of the West* 12 (1973): 139–62.

Cather, Helen V. *The History of San Francisco Chinatown*. San Francisco: R and E Research Associates, 1974.

Chan, Sucheng. "The Chinese in California Agriculture, 1860–1900." In *The Chinese American Experience*, ed. Genny Lim, pp. 67–84.

———. "Chinese Livelihood in Rural California: The Impact of Economic Change, 1860–1880." *Pacific Historical Review* 53 (1984): 273–307.

———. "Public Policy, U.S.–China Relations, and the Chinese American Experience: An Interpretive Essay." In *Pluralism, Racism, and the Search for Equality*, ed. Edwin G. Clausen and Jack Bermingham, pp. 5–38. Boston: G. K. Hall, 1981.

———. "Using California Archives for Research in Chinese American History." In *Annals of the Chinese Historical Society of the Pacific Northwest*, ed. Douglas W. Lee, pp. 49–55. Seattle: Chinese Historical Society of the Pacific Northwest, 1983.

Chen, Jack. *The Chinese of America*. San Francisco: Harper and Row, 1980.

"The Chinese in California: Letter of the Chinamen to His Excellency Governor Bigler." *Living Age* 34 (1852): 32–35.

Chinn, Thomas W., Him Mark Lai, and Philip C. Choy, eds. *A History of the Chinese in California: A Syllabus*. San Francisco: Chinese Historical Society of America, 1969.

Chiu, Ping. *Chinese Labor in California, 1850–1880: An Economic Study*. Madison: The State Historical Society of Wisconsin, 1963.

Choy, Philip P. "San Francisco Chinatown's Historical Development." In *The Chinese American Experience*, ed. Genny Lim, pp. 126–29.

Clark, E. P. "Case of Yick Wo." *Nation* 60 (1895): 438–39.

Cole, Cheryl L. "Chinese Exclusion: The Capitalist Perspective of the Sacramento Union, 1850–1882." *California History* 57 (1978): 8–31.

Coolidge, Mary Roberts. *Chinese Immigration*. New York: Henry Holt and Co., 1909.

Courtney, William J. *San Francisco Anti-Chinese Ordinances, 1850–1900*. San Francisco: R and E Research Associates, 1974.

Cross, Ira B. *A History of the Labor Movement in California*. Berkeley and Los Angeles: University of California Press, 1935.

Daniels, Roger. "Westerners from the East: Oriental Immigrants Reappraised." *Pacific Historical Review* 35 (1966): 373–83.

De Falla, Paul M. "Lantern in the Western Sky." *Historical Society of Southern California Quarterly* 42 (1960): 57–88 and 161–85.

DuFault, David V. "The Chinese in the Mining Camps of California: 1818–1870." *Historical Society of Southern California Quarterly* 41 (1959): 155–70.

Eaves, Lucile. *A History of California Labor Legislation*. Repr. ed. New York: Johnson Reprint Corp., 1966.

Eberhard, Wolfram. "Economic Activities of a Chinese Temple in California." *Journal of the American Oriental Society* 82 (1962): 362–71.

Etter, Patricia A. "The West Coast Chinese and Opium Smoking." In *Archaeological Perspectives on Ethnicity in America: Afro-American and Asian American Culture History*, ed. Robert L. Schuyler, pp. 97–101. Farmingdale, N.Y.: Baywood Publishing Co., Inc., 1980.

Evans, William S., Jr. "Food and Fantasy: Material Culture of the Chinese in California and the West, circa 1850–1900." In *Archaeological Perspectives on Ethnicity in America: Afro-American and Asian American Culture History*, ed. Robert L. Schuyler, pp. 89–96. Farmingdale, N.Y.: Baywood Publishing Co., Inc., 1980.

Fong, Kum Ngon. "The Chinese Six Companies." *Overland Monthly* 23 (1894): 518–26.

Genzoli, Andrew M., and Wallace Martin. "Expulsion: The Story of the Chinese." In *Redwood Bonanza: A Frontier's Reward*. Eureka, Calif.: Schooner Features, 1967.

Gibson, Otis. *The Chinese in America*. Cincinnati: Hitchcock and Walden, 1877.

Gioia, John. "A Social, Political and Legal Study of Yick Wo v. Hopkins." In

The Chinese American Experience, ed. Genny Lim, pp. 211–20.

Greenwood, Roberta S. "The Chinese on Main Street." In *Archaeological Perspectives on Ethnicity in America: Afro-American and Asian American Culture History*, ed. Robert L. Schuyler, pp. 113–23. Farmingdale, N.Y.: Baywood Publishing Co., Inc., 1980.

———."The Overseas Chinese at Home: Life in a Nineteenth-Century Chinatown in California." *Archaeology* 31 (1978): 42–49.

Griego, Andrew, ed. "Rebuilding the California Southern Railroad: The Personal Account of a Chinese Labor Contractor, 1884." *Journal of San Diego History* 25, 4 (1979): 324–37.

Healy, Patrick J. *Reasons for Non-Exclusion with Comments on the Exclusion Convention.* San Francisco: printed for the author, 1902.

Hellwig, David J. "Black Reactions to Chinese Immigration and the Anti-Chinese Movement: 1850–1910." *Amerasia Journal* 6 (1979): 25–44.

Hill, Herbert. "Anti-Oriental Agitation and the Rise of Working-class Racism." *Transaction: Social Science and Modern Society* 10 (1973): 43–54.

Hirata, Lucie Cheng. "Chinese Immigrant Women in Nineteenth-Century California." In *Women of America: A History*, ed. Carol Ruth Berkin and Mary Beth Norton, pp. 223–44. Boston: Houghton Mifflin Co. and the Caxton Printers, Ltd., 1979.

———. "Free, Indentured, Enslaved: Chinese Prostitutes in Nineteenth-Century America." *Signs: Journal of Women in Culture and Society* 5 (1979): 3–29.

Ho Yow. "Chinese Exclusion, A Benefit or a Harm?" *North American Review* 173(1901): 314–30.

Hom, Gloria Sun, ed. *Chinese Argonauts: An Anthology of the Chinese Contributions to the Historical Development of Santa Clara County.* Los Altos, Calif.: Foothill Community College, 1971.

Hutchison, Percy Adams. "Hep Sin, Section Hand." *Collier's*, July 26, 1913.

Johnsen, Leigh Dana. "Equal Rights and the 'Heathen Chinee': Black Activism in San Francisco, 1865–1875." *Western Historical Quarterly* 11 (1980): 57–68.

Kalisch, Philip A. "The Black Death in Chinatown: Plague and Politics in San Francisco, 1900–1904." *Arizona and the West* 14 (1972): 113–36.

Kung, S. W. *Chinese in American Life: Some Aspects of Their History, Status, Problems, and Contributions.* Seattle: University of Washington Press, 1962.

Kwang, Chang Ling. "Why Should the Chinese Go?" *The Argonaut*, August 2, 1878.

Lai, Chun-chuen. *Remarks of the Chinese Merchants of San Francisco, upon Governor John Bigler's Message and Some Common Objections, with Some Explanations of the Character of the Chinese Companies, and the Laboring Class in California.* San Francisco: Office of The Oriental, 1855.

Lai, Him Mark. "China Politics and the U.S. Chinese Communities." In *Counterpoint: Perspectives on Asian America*, ed. Emma Gee, pp. 152–59. Los Angeles: University of California, Los Angeles, Asian American Studies Center, 1976.

———. "A Historical Survey of the Chinese Left in America." In *Counterpoint: Perspectives on Asian America*, ed. Emma Gee, pp. 63–80. Los Angeles: Uni-

versity of California, Los Angeles, Asian American Studies Center, 1976.
———. "Island of Immortals: Chinese Immigrants and the Angel Island Immigration Station." *California History* 57 (1978): 88–103.
Lai, Him Mark, Genny Lim, and Judy Yung. *Island: Poetry and History of Chinese Immigrants on Angel Island, 1910–1940.* San Francisco: Hoc Doi, 1980.
Lai, Him Mark, and Philip P. Choy. *Outlines History of the Chinese in America.* San Francisco: Chinese-American Studies Planning Group, 1973.
Lan, Dean. "The Chinatown Sweatshops: Oppression and an Alternative." *Amerasia Journal* 1 (1971): 40–57.
Langenwalter, Paul E. II. "The Archaeology of 19th Century Chinese Subsistence at the Lower China Store, Madera County, California." In *Archaeological Perspectives on Ethnicity in America: Afro-American and Asian American Culture History*, ed. Robert L. Schuyler, pp. 102–12. Farmingdale, N.Y.: Baywood Publishing Co., Inc., 1980.
Layres, Augustus. "Facts Upon the Other Side of the Chinese Question with a Memorial to the President of the U.S. from Representative Chinamen in America." N.p., 1876.
Lee, Ming How, et al. "To His Excellency U.S. Grant, President of the United States: A Memorial from Representative Chinamen in America." N.p., n.d.
Lee, Rose Hum. *The Chinese in the United States of America.* Hong Kong: Hong Kong University Press, 1960.
Light, Ivan. *Ethnic Enterprise in America: Business and Welfare among Chinese, Japanese, and Blacks.* Berkeley and Los Angeles: University of California Press, 1972.
———. "The Ethnic Vice Industry, 1880–1944." *American Sociological Review* 42 (1977): 464–79.
Lim, Genny, ed. *The Chinese American Experience.* San Francisco: Chinese Historical Society of America and the Chinese Cultural Center, 1984.
Locklear, William R. "The Celestials and the Angels: A Study of the Anti-Chinese Movement in Los Angeles to 1882." *Historical Society of Southern California Quarterly* 42 (1960): 239–56.
Lou, Raymond. "The Chinese American Community of Los Angeles: 1870–1900: A Case of Resistance, Organization, and Participation." Ph.D. diss., University of California, Irvine, 1982.
———."Community Resistance of Los Angeles Chinese Americans, 1870–1900: A Case Study of Gaming." In *The Chinese American Experience*, ed. Genny Lim, pp. 160–69.
Lyman, Stanford M. *The Asian in the West.* Reno and Las Vegas: University of Nevada System, Desert Research Institute, 1970.
———. *Chinese Americans.* New York: Random House, 1974.
———. "The Chinese Diaspora in America, 1850–1943." In *The Life, Influence and the Role of the Chinese in the United States, 1776–1960*, pp. 128–46. San Francisco: Chinese Historical Society of America, 1976.
———. "Conflict and the Web of Group Affiliation in San Francisco's Chinatown, 1850–1910." *Pacific Historical Review* 43 (1974): 473–99.
———."The Structure of Chinese Society in Nineteenth-Century America." Ph.D. diss., University of California, Berkeley, 1961.

McClellan, Robert. *The Heathen Chinee: A Study of American Attitudes toward China, 1890–1905*. Columbus: Ohio State University Press, 1971.

McEvoy, Arthur F. "In Places Men Reject: Chinese Fishermen at San Diego, 1870–1893." *Journal of San Diego History* 23, 4 (1977): 12–24.

McKee, Delber L. *Chinese Exclusion versus the Open Door Policy, 1900–1906: Clashes over China Policy in the Roosevelt Era*. Detroit: Wayne State University Press, 1977.

———. "'The Chinese Must Go!' Commissioner General Powderly and Chinese Immigration, 1897–1902." *Pennsylvania History* 44 (1977): 37–51.

McLeod, Alexander. *Pigtails and Gold Dust: A Panorama of Chinese Life in Early California*. Caldwell, Idaho: The Caxton Printers, Ltd., 1947.

MacPhail, Elizabeth C. "San Diego's Chinese Mission." *Journal of San Diego History* 23, 2 (1977): 9–21.

Mann, Ralph. *After the Gold Rush: Society in Grass Valley and Nevada City, California, 1849–1870*. Stanford: Stanford University Press, 1982.

———. "Community Change and Caucasian Attitudes toward the Chinese: The Case of Two California Mining Towns, 1850–1870." In *American Working Class Culture*, ed. Milton Cantor, pp. 397–422. Westport, Conn.: Greenwood Press, 1979.

Mark, Gregory Yee. "Racial, Economic and Political Factors in the Development of America's First Drug Laws." *Issues in Criminology* 10 (1975): 49–72.

Mason, William. "The Chinese in Los Angeles." *Museum Alliance Quarterly* 6 (1967): 15–20.

"Memorial: Six Chinese Companies. An Address to the Senate and House of Representatives of the United States." 1877. Repr. ed. San Francisco: R and E Research Associates, 1970.

Miller, Stuart Creighton. *The Unwelcome Immigrant: The American Image of the Chinese, 1785–1882*. Berkeley and Los Angeles: University of California Press, 1969.

Minke, Pauline. *Chinese in the Mother Lode*. San Francisco: R and E Research Associates, 1974.

Morley, Charles, trans. with foreword. "The Chinese in California As Reported by Henryk Sienkiewicz." *California Historical Society Quarterly* 34 (1955): 301–16.

Nash, Robert A. "The Chinese Shrimp Fishery in California." Ph.D. diss., University of California, Los Angeles, 1973.

Nee, Victor G., and Brett de Bary. *Longtime Californ': A Documentary Study of an American Chinatown*. New York: Pantheon Books, 1972.

Ng Poon Chew. *The Treatment of the Exempt Classes of Chinese in the United States: A Statement from the Chinese in America*. San Francisco: Chung Sai Yat Po, 1908.

Ochs, Patricia Mary. "A History of Chinese Labor in San Luis Obispo County and A Comparison of Chinese Relations in This County with the Anti-Chinese Movement in California, 1869–1894." M.A. thesis, California State Polytechnic College, San Luis Obispo, 1966.

Ong, Paul M. "Chinese Labor in Early San Francisco: Racial Segmentation and Industrial Expansion." *Amerasia Journal* 8 (1981): 69–92.

———. "An Ethnic Trade: The Chinese Laundries in Early California."

Journal of Ethnic Studies 8 (1981): 95–113.

Paul, Rodman W. "The Origin of the Chinese Issue in California." *The Mississippi Valley Historical Review* 25 (1938): 181–96.

Perkins, Dorothy J. "Coming to San Francisco by Steamship." In *The Chinese American Experience*, ed. Genny Lim, pp. 26–33.

Reynolds, C. N., "The Chinese Tongs." *American Journal of Sociology* 40 (1935): 612–23.

Riddle, Ronald. *Flying Dragons, Flowing Streams: Music in the Life of San Francisco Chinese*. Westport, Conn.: Greenwood Press, 1983.

Rodecape, Lois. "Celestial Drama in the Golden Hills: The Chinese Theater in California, 1849–1869." *California Historical Society Quarterly* 23 (1944): 97–116.

Rohe, Randall E. "After the Gold Rush: Chinese Mining in the Far West, 1850–1890." *Montana* 32 (1982): 2–19.

Sandmeyer, Elmer Clarence. *The Anti-Chinese Movement in California*. Urbana: University of Illinois Press, 1939.

———. "California Anti-Chinese Legislation and the Federal Courts: A Study in Federal Relations." *Pacific Historical Review* 5 (1936): 189–211.

Saxton, Alexander. "The Army of Canton in the High Sierra." *Pacific Historical Review* 35 (1966): 141–52.

———. *The Indispensable Enemy: Labor and the Anti-Chinese Movement in California*, Berkeley and Los Angeles: University of California Press, 1971.

Schwendinger, Robert J. "Investigating Chinese Immigrant Ships and Sailors." In *The Chinese American Experience*, ed. Genny Lim, pp. 16–25.

Seager, Robert II. "Some Denominational Reactions to Chinese Immigration to California, 1856–1892." *Pacific Historical Review* 28 (1959): 49–66.

Seward, George F. *Chinese Immigration, in Its Social and Economic Aspects*. 1881. Repr. ed. New York: Arno Press, 1970.

Shankman, Arnold. "Black on Yellow: Afro-Americans View Chinese Americans, 1850–1935." *Phylon* 39 (1978): 1–17.

Shimabukuro, Milton, "Chinese in California State Prisons, 1870–1890." In *The Chinese American Experience*, ed. Genny Lim, pp. 221–23.

"The Six Chinese Companies." *Overland Monthly* 1 (1868): 221–27.

Speer, William. *Plea in Behalf of the Immigrants from China*. San Francisco: Sterett and Co., 1856.

Spier, Robert F. "Food Habits of Nineteenth-Century California Chinese." *California Historical Society Quarterly* 37 (1958): 79–84 and 129–36.

Spoehr, Luther W. "Sambo and the Heathen Chinee: Californians' Racial Stereotypes in the Late 1870s." *Pacific Historical Review* 42 (1973): 185–203.

Sung, Betty Lee. *Mountain of Gold: The Story of the Chinese in America*. New York: Macmillan Co., 1967.

Sylva, Seville. *Foreigners in the California Gold Rush*. San Francisco: R and E Research Associates, 1972.

Tachibana, Judy M. "Outwitting the Whites: One Image of the Chinese in California Fiction and Poetry, 1849–1924." *Southern California Quarterly* 61 (1979): 379–89.

Tang, Vincente. "Chinese Women Immigrants and the Two-Edged Sword of

Habeas Corpus." In *The Chinese American Experience*, ed. Genny Lim, pp. 48–56.

Tow, J. S. *The Real Chinese in America*. 1923. Repr. ed. San Francisco: R and E Research Associates, 1970.

Trauner, Joan B. "The Chinese as Medical Scapegoats in San Francisco, 1870–1906." *California History* 57 (1978): 70–87.

Tsai, Shih-shan H. *China and the Overseas Chinese in the United States, 1868–1911.* Fayetteville, Ark.: University of Arkansas Press, 1983.

U.S. Congress. Joint Special Committee to Investigate Chinese Immigration. *Report*, 44th Cong., 2nd sess., 1877.

Weiss, Melford S. *Valley City: A Chinese Community in America*. Cambridge, Mass.: Schenkman Publishing Co., 1974.

Wells, Mariann Kaye. *Chinese Temples in California*. San Francisco: R and E Research Associates, 1971.

Williams, Stephen. "The Chinese in the California Mines, 1848–1860." M.A. thesis, Stanford University, 1930.

Williamson, Joann. "Chinese Studies in Federal Records." In *The Life, Influence and the Role of the Chinese in the United States, 1776–1960*, pp. 6–24. San Francisco: Chinese Historical Society of America, 1976.

Wilson, Carol Green. *Chinatown Quest: One Hundred Years of Donaldina Cameron House*. San Francisco: California Historical Society, 1974.

Wong, Charles Choy. "The Continuity of Chinese Grocers in Southern California." *Journal of Ethnic Studies* 8 (1980): 63–82.

Woo, Wesley S. "Protestant Work Among the Chinese in the San Francisco Bay Area, 1850–1920." Ph.D. diss., Graduate Theological Union, Berkeley, 1983.

Wood, Ellen R. "Californians and Chinese: The First Decade." San Francisco: R and E Research Associates, 1974.

Wu, Cheng-Tsu, ed., with an introduction. *"Chink!": A Documentary History of Anti-Chinese Prejudice in America*. New York: World Publishing Co., 1972.

Yen, Tzu-Kuei. "Chinese Workers and the First Transcontinental Railroad of the United States of America." Ph.D. diss., St. John's University, 1977.

Yu, Connie Young. "From Tents to Federal Housing Projects: Chinatown's Housing History." In *The Chinese American Experience*, ed. Genny Lim, pp. 130–38.

Selected Bibliography on the History of California Land Reclamation, Agricultural Development, and Farm Labor, 1840s to 1920s

The items in this section of the bibliography focus on those aspects of California agriculture which provided the socioeconomic framework within which Chinese agriculturalists operated. For a more complete bibliography on the history of California agriculture, consult Orsi, Richard J., comp., *A List of References for the History of Agriculture in California*. Davis: University of California, Agricultural History Center, 1974.

Adams, Edward F. "Early History of Agriculture in California." *San Francisco Chronicle*, Dec. 31, 1905, pp. 2–4.

———. "Our Great Reclamation Problem: The Swamp and Overflowed Lands of the Sacramento and Lower San Joaquin." *Transactions of the Commonwealth Club of California* 1 (1904): 3–21, and 4 (1909): 200–32.

Adams, Frank. "California Irrigation Development." *Transactions of the Commonwealth Club of California* 15 (1920): 331–36.

———. "The Historical Background of California Agriculture." In *California Agriculture*, ed. Claude B. Hutchison, pp. 1–50. Berkeley and Los Angeles: University of California Press, 1946.

Adams, R. L. "Farm Help for the Coming Harvest." Berkeley: University of California, Giannini Foundation Agricultural Economics Library, mss., n.d.

———. "Farm Labor." *Journal of Farm Economics* 19 (1937): 913–25.

———. "The Farm Labor Problem." *University of California Chronicle* 22 (1918).

———. "The Farm Labor Situation in California." Berkeley: University of California, College of Agriculture, Agricultural Experiment Station, mss., 1917.

———. *Seasonal Labor Requirements for California Crops*. Berkeley: University of California, College of Agriculture, Agricultural Experiment Station, Bulletin 623, 1938.

Adams, R. L., and T. R. Kelly. *A Study of Farm Labor in California*. Berkeley: University of California, College of Agriculture, Agricultural Experiment Station, Circular 193, 1918.

Adams, R. L., and W. H. Smith, Jr. *Farm Tenancy in California and Methods of Leasing*. Berkeley: University of California, College of Agriculture, Agricultural Experiment Station, Bulletin 655, 1941.

Allen, R. H. "The Spanish Land Grant System as an Influence in the Agricultural Development of California." *Agricultural History* 9 (1935): 127–42.

Alward, Dennis M., and Andrew F. Rolle. "The Surveyor-General Edward Fitzgerald Beale's Administration of California Lands." *Southern California Quarterly* 53 (1971): 113–22.

Atherton, George A. "Reclamation and Development in the Sacramento–San Joaquin Delta." *Agricultural Engineering* 12 (1931): 129–30.

Baker, Charles C. "Mexican Land Grants in California." *Historical Society of Southern California Annual Publications* 9 (1914): 235–43.

Baur, John E. "The Health Seekers and Early Southern California Agriculture." *Pacific Historical Review* 20 (1951): 347–63.

Beckman, H. J. "Century of Agricultural Progress." *California Cultivator* 94 (1947): 742–43.

Bell, George L. "The Wheatland Hop Field Riots." *The Outlook* 107 (1914): 118–23.

Bennett, M. K. "Climate and Agriculture in California." *Economic Geography* 15 (1939): 153–64.

Bidwell, John. "California's Productive Interests." *Transactions of the State Agricultural Society, 1881*. In California. State Legislature. *Appendix to the Journals of the Senate and Assembly*, 1881, vol. 2. Sacramento, 1881.

Bleyhl, Norris A. "A History of the Production and Marketing of Rice in California." Ph.D. diss., University of Minnesota, 1955.

Bowen, William A. "Evolution of a Cultural Landscape: The Valley Fruit District of Solano County, California." M.A. thesis, University of California, Berkeley, 1966.

Brauton, Ernest, "Seventy-five Years of Agriculture." *California Cultivator* 65 (1925): 187, 194–95.

Browne, J. Ross. "Agricultural Capacity of California." *The Overland Monthly* 10 (1873): 297–314.

————. *Reclamation of Marsh and Swamp Lands, and Projected Canals for Irrigation in California: With Notes on the Canal System of China and Other Countries, Addressed to the Legislature of California....* San Francisco: Alta California Printing House, 1872.

Butterfield, Harry M. "Builders of California Horticulture—Past and Present." *California Horticultural Society Journal* (1961): 2–7.

California State Legislature. Assembly. Committee on Problems of School Land Warrants. "Report." In *Assembly Journal*, 1859, pp. 531–32. Sacramento, 1859.

California State Legislature. Commission on Reclamation of Swamp and Overflowed Lands in Sacramento Valley. "Report." In *Appendix to the Journals of the Senate and Assembly*, 1869–70, vol. 2. Sacramento, 1870.

California State Legislature. Joint Committee on the Condition of Public and State Lands Lying Within the Limits of the State. "Report." In *Appendix to the Journals of the Senate and Assembly*, 1871–72, vol. 2. Sacramento, 1872.

California State Legislature. Senate. Committee on Sale of Marsh and Tide Lands. "Report." In *Appendix to the Journals of the Senate and Assembly*, 1865–66, vol. 3. Sacramento: 1866.

California State Legislature. Senate. Committee on State Land Sales. "Report." In *Appendix to the Senate Journal*, 1856. Sacramento, 1856.

California. State Board of Agriculture. *Report of the State Board of Agriculture for the Year 1911*. Sacramento: Superintendent of State Printing, 1912.

Cance, Alexander E. "Immigrants and American Agriculture." *Journal of Farm Economics* 7 (1925): 106–8.

Carosso, Vincent P. *The California Wine Industry, 1830–1895: A Study of the Formative Years*. Berkeley and Los Angeles: University of California Press, 1951.

Carr, Ezra S. *The Patrons of Husbandry on the Pacific Coast*. San Francisco: A. L. Bancroft and Co., 1875.

Chipman, Norton P. "Annual Address Delivered Before the State Agricultural Society at Sacramento, California, September 16, 1886." In California State Legislature. *Appendix to the Journals of the Senate and Assembly*, 1881, vol. 3. Sacramento, 1881.

————. "The Fruit Industry of California: Its Growth and Development." *Transactions of the California State Agricultural Society during the Year 1889*. Sacramento: Office of State Printing, 1890.

Ciriacy-Wanthrup, S. V. "Major Economic Forces Affecting Agriculture with Particular Reference to California." *Hilgardia* 18 (1947): 1–76.

Clark, Robert. "The Labor History of Fresno, 1886–1910." M.A. thesis,

California State University, Fresno, 1976.

Cleland, Robert G. *The Cattle on a Thousand Hills: Southern California, 1850–1880.* San Marino: The Huntington Library, 1951.

Cleland, Robert G., and Osgood Hardy. *March of Industry.* Los Angeles: Powell Publishing Co., 1929.

Colby, Charles C. "The California Raisin Industry—A Study in Geographic Interpretation." *Annals of the Association of American Geographers* 14 (1924): 49–108.

Cole, Chester F. "Rural Occupance Patterns in the Great Valley Portion of Fresno County, California." Ph.D. diss., University of Nebraska, 1951.

Conner, Forrest. "The Opposition to Hydraulic Mining in California." M.A. thesis, University of California, Berkeley, 1950.

Cooper, Erwin. *Aqueduct Empire: A Guide to Water in California.* Glendale: Arthur H. Clark Co., 1968.

Cooper, Margaret Aseman. "Land, Water, and Settlement Patterns in Kern County, California, 1850–1900." M.A. thesis, University of California, Berkeley, 1954.

Corbett, Francis J. "The Public Domain and Mexican Land Grants in California." M.A. thesis, University of California, Berkeley, 1959.

Cordray, William W. "Claus Spreckels of California." Ph.D. diss., University of Southern California, 1955.

Cosby, Stanley W. *Soil Survey of the Sacramento–San Joaquin Delta Area, California.* In United States Department of Agriculture, Bureau of Plant Industry. Berkeley: University of California Agricultural Experiment Station, Series 1935, 21, 1941.

Cox, Edwin E. "Farm Tenancy in California." In *Land Settlement in California.* Transactions of the Commonwealth Club of California 11 (1916): 444–56.

Crawford, L. A., and Edgar B. Hurd. *Types of Farming in California Analyzed by Enterprise.* Berkeley: University of California, College of Agriculture, Agricultural Experiment Station, Bulletin 654, 1941.

Cronise, Titus Fey. *The Agricultural and Other Resources of California.* San Francisco: A. Roman and Co., 1870.

———. *The Natural Wealth of California, Comprising Early History ... Agriculture and Commercial Products ... Agricultural Advantages....* San Francisco: H. H. Bancroft and Co., 1868.

"Crop Statistics for 1883." *Transactions of the State Agricultural Society, 1883.* In California State Legislature. *Appendix to the Journals of the Senate and Assembly,* 1885, vol. 2. Sacramento, 1885.

"Crop Statistics for 1884." *Transactions of the State Agricultural Society, 1884.* In California State Legislature. *Appendix to the Journals of the Senate and Assembly,* 1885, vol. 2. Sacramento, 1885.

Dana, Julian. *The Sacramento: River of Gold.* New York: Farrar and Rinehart, 1939.

Daniel, Cletus E. *Bitter Harvest: A History of California Farmworkers, 1870–1941.* Ithaca: Cornell University Press, 1981.

DeLay, Peter J. *History of Yuba and Sutter Counties.* Los Angeles: The Historical Record Co., 1924.

Dennis, Margaret. "The History of the Beet Sugar Industry in California." M.A. thesis, University of Southern California, 1937.

Eddy, Dale. "The History of the Upper Salinas Valley." M.A. thesis, University of Southern California, 1938.

Eigenheer, Richard A. "Early Perceptions of Agricultural Resources in the Central Valley of California." Ph.D. diss., University of California, Davis, 1976.

Erdman, H. E. *The California Fruit Growers Exchange: An Example of Cooperation in the Segregation of Conflicting Interests.* New York: Institute of Pacific Relations, 1933.

————. "The Development and Significance of California Cooperatives, 1900–1915." *Agricultural History* 32 (1958): 179–86.

Fabian, Bentham. *The Agricultural Lands of California: A Guide to the Immigrant as to the Productions, Climate and Soil of Every County in the State.* San Francisco: H. H. Bancroft and Co., 1869.

Fearis, Donald F. "The California Farm Worker, 1930–1942." Ph.D. diss., University of California, Davis, 1971.

Federal Writers' Project, Oakland, California. "The Contract Labor System in California Agriculture." Oakland, ca. 1938. (Typescript in the Bancroft Library, University of California, Berkeley.)

————. "The Industrial Workers of the World in California Agriculture." Oakland, ca. 1939. (Typescript in the Bancroft Library, University of California, Berkeley.)

————. "The Migratory Agricultural Worker and the American Federation of Labor to 1938 Inclusive." Oakland, ca. 1939. (Typescript in the Bancroft Library, University of California, Berkeley.)

————. "Oriental Labor Unions and Strikes, California Agriculture." Oakland, ca. 1939. (Typescript in the Bancroft Library, University of California, Berkeley.)

————. "Unionization of Filipinos in California Agriculture." Oakland, ca. 1939. (Typescript in the Bancroft Library, University of California, Berkeley.)

————. "Unionization of Migratory Labor, 1903–1930." Oakland, ca. 1939. (Typescript in the Bancroft Library, University of California, Berkeley.)

————. "Wage Chart by Crops, State of California, 1865–1938." Oakland, ca. 1939. (Typescript in the Bancroft Library, University of California, Berkeley.)

Field, Alston G. "Attorney-General Black and the California Land Claims." *Pacific Historical Review* 4 (1935): 234–45.

Fisher, Anne B. *The Salinas: Upside-Down River.* New York: Farrar and Rinehart, 1945.

Fisher, J. Donald. "A Historical Study of the Migrant in California." M.A. thesis, University of Southern California, 1945.

Fisher, Lloyd H. *The Harvest Labor Market in California.* Cambridge: Harvard University Press, 1953.

————. "The Harvest Labor Market in California." *Quarterly Journal of Economics* 65 (1951): 392–415.

Foote, A. D. "The Redemption of the Great Valley of California." *Transactions of the American Society of Civil Engineers* 66 (1910): 229–45.

Fuller, Varden. "The Supply of Agricultural Labor as a Factor in the Evolution of Farm Organization in California." U.S. Congress. Senate. Committee on Education and Labor. *Hearings Pursuant to Senate Resolution 266, Exhibit 8762-A*. 76th Cong., 3rd sess., 1940, pp. 19777–898.

Gardner, Barbara S. "The Economic Development of Kern County, California." Ph.D. diss., University of California, Los Angeles, 1969.

Gates, Paul W. "Adjudication of Spanish-Mexican Land Claims in California." *Huntington Library Quarterly* 21 (1958): 213–36.

———. "The California Land Act of 1851." *California Historical Quarterly* 50 (1971): 395–430.

———. "California's Agricultural College Lands." *Pacific Historical Review* 30 (1961): 103–22.

———. "California's Embattled Settlers." *California Historical Society Quarterly* 41 (1962): 99–130.

———. "Carpetbaggers Join the Rush for California Land." *California Historical Quarterly* 56 (1977): 99–127.

———. "The Fremont-Jones Scramble for California Land Claims." *Southern California Quarterly* 56 (1974): 13–44.

———. *History of Public Land Law*. United States Public Land Law Review Commission. Washington, D.C.: Government Printing Office, 1968.

———. "The Homestead Law in an Incongruous Land System." *American Historical Review* 41 (1936): 652–81.

———. *Land Policies in Kern County*. Bakersfield: Kern County Historical Society, 1978.

———. "Pre–Henry George Land Warfare in California." *California Historical Society Quarterly* 46 (1967): 121–48.

———. "Public Land Disposal in California." *Agricultural History* 49 (1975): 158–78.

———. "The Suscol Principle, Pre-emption, and California Latifundia." *Pacific Historical Review* 39 (1970): 453–72.

Gehre, J. Stephen. "The Geography of Selected Patterns of Agriculture in San Mateo County, California." M.A. thesis, California State University, San Francisco, 1968.

George, Henry. *Our Land and Land Policy, National and State*. San Francisco: White and Bauer, 1871.

Gilbert, Frank T. *History of San Joaquin County*. 1879. Repr. ed., Berkeley: Howell-North Books, 1968.

Gilliam, Harold. "San Joaquin County—the Conquest of the Delta." *San Francisco Chronicle This World Magazine*, Dec. 7, 1952, pp. 2–3, and 7.

Graham, D. J. "The Settlement of Merced County, California." M.A. thesis, University of California, Los Angeles, 1957.

Gregg, E., "History of the Famous Stanford Ranch at Vina." *Overland Monthly* 52 (1908): 334–38.

Gregor, Howard F. "Farm Structure in Regional Comparison: California and New Jersey Vegetable Farms." *Economic Geography* 45 (1969): 209–25.

————. "The Industrial Farm as a Western Institution." *Journal of the West* 9 (1970): 78–92.

————. "The Large Industrialized American Crop Farm: A Mid-Latitude Plantation Variant." *The Geographical Review* 60 (1970): 151–75.

————. "The Local-Supply Agriculture of California." *Annals of the Association of American Geographers* 47 (1957): 267–76.

————. "Regional Hierarchies in California Agricultural Production, 1939–1954." *Annals of the Association of American Geographers* 53 (1963): 27–37.

Griffen, Paul F., and C. Langdon White. "Lettuce Industry of the Salinas Valley." *Scientific Monthly* 81 (1955): 77–84.

Guinn, J. M. "The Passing of the Cattle Barons of California." *Historical Society of Southern California Annual Publications* 8 (1909–11): 51–60.

————. "The Passing of the Rancho." *Historical Society of Southern California Annual Publications* 10 (1915–16): 46–53.

"Half Century's Statistics of California Farming." *Pacific Rural Press* 101 (1921): 7 and 18.

Hardy, Osgood. "Agricultural Changes in California, 1860–1900." American Historical Association, Pacific Coast Branch. *Proceedings*, 1929, 216–30.

Hecke, G. H. "The Pacific Coast Labor Question from the Standpoint of a Horticulturalist." *Proceedings of the Thirty-third Fruit Growers' Convention* 33 (1907): 67–72.

Hewes, Laurence Ilsley, Jr. "Some Migratory Labor Problems in California's Industrialized Agriculture." Ph.D. diss., George Washington University, 1945.

Hinkel, Edgar J., and William E. McCann, eds. *History of Rural Alameda County* 2 vols. Oakland: Works Progress Administration, 1937.

History of Fresno County. San Francisco: W. W. Elliott, 1882.

Holmes, Edwin S., Jr. *Statistics on the Fruit Industry of California.* United States Department of Agriculture, Division of Statistics. Miscellaneous Series, Bulletin 23. Washington, D.C.: Government Printing Office, 1901.

"Hop Cultivation: Interesting History of Hop Growing in California." *San Francisco Post*, Sept. 20, 1879.

Hutchison, Claude B., ed. *California Agriculture.* Berkeley and Los Angeles: University of California Press, 1946.

An Illustrated History of San Joaquin County, California. Chicago: The Lewis Publishing Co., 1890.

"Impressions of the Character and Progress of California Agriculture." *San Francisco Chronicle*, Dec. 20, 1906.

"Irrigation in California." *Scientific American* 94 (1906): 457–58.

Irwin, William H. *Augusta Bixler Farms: A California Delta Farm from Reclamation to the Fourth Generation of Owners.* N.p.: William Hyde Irwin and Augusta Bixler Farms, 1973.

Jackson, Harold E. "A Geography of the Sacramento Valley Peach Belt." M.A. thesis, California State University, Chico, 1968.

Jamieson, Stuart. *Labor Unionism in American Agriculture.* U.S. Department of Labor, Bureau of Labor Statistics, Bulletin 836. Washington, D.C.: Government Printing Office, 1945.

Jelinek, Lawrence J. *Harvest Empire: A History of California Agriculture*. San Francisco: Boyd and Fraser Publishing Co., 1979.

Jenny, Hans, et al. "Exploring the Soils of California." In *California Agriculture*, ed. Claude B. Hutchison, pp. 317–93.

Jewell, Marion. "Agricultural Development in Tulare County, 1870–1900." M.A. thesis, University of Southern California, 1950.

Jones, Jenkin W. *How to Grow Rice in the Sacramento Valley*. United States Department of Agriculture, Farmers' Bulletin 1240. Washington, D.C.: 1924, rev. 1931.

Jones, Lamar B. "Labor and Management in California Agriculture, 1864–1964." *Labor History* 11 (1970): 23–40.

Jordan, Frank C. "Ten Years in California Agriculture." *California Cultivator* 57 (1921): 216.

Judge, Anna F. "Reclamation of Lands in California." M.A. thesis, University of California, Berkeley, 1924.

Kelley, Robert L. *Gold versus Grain: The Hydraulic Mining Controversy in California's Sacramento Valley; A Chapter in the Decline of the Concept of Laissez Faire*. Glendale: Arthur H. Clark Co., 1959.

———. "The Mining Debris Controversy in the Sacramento Valley." *Pacific Historical Review* 25 (1956): 331–46.

Ketteringham, W. J. "The Settlement Geography of the Napa Valley." M.A. thesis, Stanford University, 1961.

Key, Leon Goodwin. "The History of the Policies in Disposing of the Public Lands in California, 1769–1900." M.A. thesis, University of California, Berkeley, 1938.

Lapham, Macy H., and W. W. Mackie. *Soil Survey of the Stockton Area, California*. United States Department of Agriculture, Bureau of Soils. Washington, D.C.: Government Printing Office, 1906.

Lapham, Macy H., Aldert S. Root, and W. W. Mackie. *Soil Survey of the Sacramento Area, California*. United States Department of Agriculture, Bureau of Soils. Washington, D.C.: Government Printing Office, 1905.

Leonard, Charles Berdan. "History of the San Joaquin Valley." M.A. thesis, University of California, Berkeley, 1922.

Lowenstein, Norman. "Strikes and Strike Tactics in California Agriculture: A History." M.A. thesis, University of California, Berkeley, 1940.

Lund, John. "Martin J. Lund: Father of the Delta." *The Pacific Historian* 17 (1973): 36–42.

McClain, Marcia H. "The Distribution of Asparagus Production in the Sacramento-San Joaquin Delta." M.A. thesis, University of California, Berkeley, 1954.

McGowan, Joseph A. *History of the Sacramento Valley*. 3 vols. New York: Lewis Historical Publishing Co., 1961.

McWilliams, Carey. *Factories in the Field: The Story of Migratory Farm Labor in California*. 1939. Repr. ed., New York: Archon Books, 1969.

———. *Ill Fares the Land: Migrants and Migratory Labor in the United States*. Boston: Little, Brown, and Co., 1942.

Magnuson, Torstein A. "History of the Beet Sugar Industry in California."

Historical Society of Southern California Annual Publications 11 (1917–18): 68–79.

Majka, Linda C., and Theo J. Majka. *Farm Workers, Agribusiness, and the State.* Philadelphia: Temple University Press, 1982.

Malone, Thomas E. "The California Irrigation Crisis of 1886: Origins of the Wright Act." Ph.D. diss., Stanford University, 1965.

Manson, Marsden. "The Swamp and Marsh Lands of California." Technical Society of the Pacific Coast. *Transactions* 5 (1888): 83–99.

Margo, Joan. "The Food Supply Problem of the California Gold Mines, 1848–1858." M.A. thesis, University of California, Berkeley, 1947.

Meinig, Donald W. "The Growth of Agricultural Regions in the Far West, 1850–1910." *Journal of Geography* 54 (1955): 221–32.

A Memorial and Biographical History of the Counties of Fresno, Tulare, and Kern, California. Chicago: Lewis Publishing Co., 1892.

Meyer, Albert J., Jr. "History of the California Fruit Growers Exchange, 1893–1920." Ph.D. diss., The Johns Hopkins University, 1952.

Meyer, Edith C. "The Development of the Raisin Industry in Fresno County, California." M.A. thesis, University of California, Berkeley, 1932.

Moulton, Larry E. "The Vina District, Tehama County, California: Evolution of Land Utilization in a Small Segment of the Middle Sacramento Valley." M.A. thesis, California State University, Chico, 1969.

Nash, Gerald D. "The California State Land Office, 1858–1898." *Huntington Library Quarterly* 27 (1964): 347–56.

———. "Problems and Projects in the History of Nineteenth Century California Land Policy." *Arizona and the West* 2 (1960): 327–40.

———. "Stages of California's Economic Growth, 1870–1970." *California Historical Quarterly* 51 (1972): 215–30.

National Advisory Committee on Farm Labor. *Farm Labor Organizing, 1905–1967, A Brief History.* New York: National Advisory Committee on Farm Labor, 1967.

Newman, Ralph. "California's Hundred and Eighty Crops." *Pacific Rural Press* 110 (1925): 103.

Orsi, Richard J. "Selling the Golden State: A Study of Boosterism in Nineteenth-Century California." Ph.D. diss., University of Wisconsin, Madison, 1973.

Orton, W. A. "Potato Diseases in San Joaquin County, California." U.S. Department of Agriculture, Circular 23, n.d.

Osborne, Thomas J. "Claus Spreckels and the Oxnard Brothers: Pioneer Developers of California's Beet Sugar Industry, 1890–1900." *Southern California Quarterly* 54 (1972): 117–25.

Parker, Zelma. "History of the Destitute Migrant in California, 1840–1939." M.A. thesis, University of California, Berkeley, 1940.

Parsons, James J. "Hops in Early California Agriculture." *Agricultural History* 14 (1940): 110–16.

"Past, Present and Future of Fresno County, California." *Overland Monthly* (new series) 52 (1908): 575–78.

Paul, Rodman W. "The Beginnings of Agriculture in California: Innovation versus Continuity." *California Historical Quarterly* 52 (1973): 16–27.

————. "The Great California Grain War: The Grangers Challenge the Wheat King." *Pacific Historical Review* 27 (1958): 331–49.

————. "The Wheat Trade between California and the United Kingdom." *Mississippi Valley Historical Review* 45 (1958): 391–412.

Pearl, Milton A. "Historical View of Public Land Disposal and the American Land Use Pattern, California." *Western Law Review* 4 (1969): 65–75.

Peters, Hugh R. "California's Public Land and the Issue of Land Monopoly, 1868–79." M.A. thesis, California State University, Sacramento, 1966.

Pitt, Leonard. *The Decline of the Californios: A Social History of the Spanish-Speaking Californians, 1846–1890.* Berkeley and Los Angeles: University of California Press, 1966.

Polos, Nicholas C. "School Lands of California." *Southern California Quarterly* 51 (1969): 63–70.

Powell, Fred W. "Co-operative Marketing of California Fresh Fruit." *Quarterly Journal of Economics* 24 (1910): 392–418.

Prior, Robert. "A Historical Study of the Labor Contractor System in California Agriculture, 1868–1954." M.A. thesis, University of Southern California, 1954.

Raup, Hallock F. "Transformation of Southern California to a Cultivated Land." *Annals of the Association of American Geographers* 49 *Supplement* (1959): 58–78.

"Reclamation of Marsh Lands in California." *Pacific Rural Press* 30 (1885): 510.

"Reclamation of Swamp and Overflowed Lands in California." *Report of the Commissioner of Agriculture for the Year 1872.* Washington, D.C.: Government Printing Office, 1874.

Reed, Howard S. "Major Trends in California Agriculture." *Agricultural History* 20 (1946): 252–55.

Richards, William S. "Geographical Aspects of Rice Cultivation in California." M.A. thesis, University of California, Berkeley, 1969.

Roberts, Doyle. "A History of the Reclamation of the Delta Lands of California." M.A. thesis, University of the Pacific, 1951.

Robinson, W. W. *Land in California: The Story of Mission Lands, Ranchos, Squatters, Mining Claims, Railroad Grants, Land Scrip, [and] Homesteads.* Berkeley and Los Angeles: University of California Press, 1948.

Rogers, C. H. "Pajaro Valley, Its Great Apple Industry." *California Cultivator* 25 (1905): 27 and 42.

Ross, Ivy. "The Confirmation of Spanish and Mexican Land Grants in California." M.A. thesis, University of California, Berkeley, 1929.

Royce, G. C. *John Bidwell: Pioneer, Statesman, Philanthropist.* Chico: published by the author, 1906.

Saloutos, Theodore. "The Immigrant in Pacific Coast Agriculture, 1890–1940." *Agricultural History* 49 (1975): 182–201.

Sanderson, Lawson. "Land Ownership and Distribution in California, 1850–1950." M.A. thesis, University of Southern California, 1958.

Sawyer, Eugene T. *History of Santa Clara County, California, with Biographical Sketches of the Leading Men and Women of the County Who Have Been Identified with Its Growth and Development from the Early Days to the Present.* Los Angeles:

Historic Record Co., 1922.

Schwartz, Harry. *Seasonal Farm Labor in the United States—with Special Reference to Hired Workers in Fruit and Vegetable and Sugar-Beet Production.* Columbia University Studies in the History of American Agriculture, 11. New York: Columbia University Press, 1945.

Seiverston, Bruce. "Pajaro Valley, California: The Sequent Occupance of a Coastal Agricultural Basin." M.A. thesis, California State University, Chico, 1969.

Shaw, John A. "Commercialization in an Agricultural Economy: Fresno County, California, 1856–1900." Ph.D. diss., Purdue University, 1969.

———. "Railroads, Irrigation, and Economic Growth: The San Joaquin Valley of California." *Explorations in Economic History* 10 (1973): 211–27.

Smith, Arthur L. "Delta Experiments." *Bulletin of the American Geographical Society* 41 (1909): 729–42.

Smith, Wallace. *Garden of the Sun: A History of the San Joaquin Valley from 1772–1939.* Los Angeles: Powell Publishing Co., 1939.

Speth, Frank A. "A History of Farm Labor in Sonoma County, California." M.A. thesis, University of California, Berkeley, 1938.

Taylor, Paul S. "California Farm Labor: A Review." *Agricultural History* 42 (1968): 49–54.

———. "Foundations of California Rural Society." *California Historical Society Quarterly* 24 (1945): 193–228.

———. "Water, Land, and People in the Great Valley." *The American West* 5 (1968): 24–28 and 68–72.

Taylor, Paul S., and Tom Vasey. "Contemporary Background of California Farm Labor." *Rural Sociology* 1 (1936): 401–19.

———. "Historical Background of California Farm Labor." *Rural Sociology* 1 (1936): 281–95.

Teilmann, Hendrik. "The Role of Irrigation Districts in California's Water Development." *American Journal of Economics and Sociology* 22 (1963): 409–15.

Thickens, Virginia E. "Pioneer Agricultural Colonies in Fresno County." *California Historical Society Quarterly* 25 (1946): 17–38.

Thompson, Alvin H. "Aspects of the Social History of California Agriculture, 1885–1902." M.A. thesis, University of California, Berkeley, 1953.

Thompson, John. "Reclamation Sequence in the Sacramento-San Joaquin Delta." *California Geographer* 6 (1965): 29–35.

———. "The Settlement Geography of the Sacramento-San Joaquin Delta, California." Ph.D. diss., Stanford University, 1958.

Thompson, Kenneth. "Historical Flooding in the Sacramento Valley." *Pacific Historical Review* 29 (1960): 349–60.

Thompson, Thomas H., and Albert A. West. *History of Sacramento County.* 1880. Repr. ed., Berkeley: Howell-North Books, 1960.

———. *History of Sutter County.* Oakland: Thompson and West, 1879.

Tinkham, George H. *History of San Joaquin County, California.* Los Angeles: Historic Record Co., 1923.

Torbert, Edward N. "The Specialized Commercial Agriculture of the Northern Santa Clara Valley." *Geographical Review* 26 (1936): 247–63.

U.S. Department of Commerce, Bureau of the Census. *Ninth Census of the United States: Statistics of Agriculture, 1870.* Washington, D.C.: Government Printing Office, 1872.

―――. *Ninth Census of the United States: Wealth and Industry, 1870.* Washington, D.C.: Government Printing Office, 1872.

―――. *Tenth Census of the United States: Production of Agriculture, 1880.* Washington, D.C.: Government Printing Office, 1883.

―――. *Eleventh Census of the United States: Statistics of Agriculture, 1890.* Washington, D.C.: Government Printing Office, 1895.

―――. *Twelfth Census of the United States: Agriculture, 1900.* Washington, D.C.: Government Printing Office, 1902.

―――. *Thirteenth Census of the United States: Agriculture, 1910.* Washington, D.C.: Government Printing Office, 1913.

―――. *Fourteenth Census of the United States: Agriculture, 1920.* Washington, D.C.: Government Printing Office, 1922.

U.S. Industrial Commission. *Report of the Industrial Commission on Agriculture and Agricultural Labor,* vol. 10. Washington, D.C.: Government Printing Office, 1901.

Vandor, Paul. *History of Fresno County, California, with Biographical Sketches of the Leading Men and Women of the County who Have Been Identified with Its Growth and Development from the Early Days to the Present.* 2 vols. Los Angeles: Historic Record Co., 1919.

Wagers, Jerard D. "History of Agricultural Labor in California Prior to 1880." M.A. thesis, University of California, Berkeley, 1957.

Wallick, Phillip K. "An Historical Geography of the Salinas Valley." M.A. thesis, California State University, San Francisco, 1969.

Weintraub, Hyman. "The I.W.W. in California, 1905–1931." M.A. thesis, University of California, Los Angeles, 1947.

Weir, Walter W. "The Effects of a Rapidly Changing Environment on Crop History." *Agricultural Engineering* 2 (1930): 277–79.

―――. "Peat Lands of the Delta." *California Agriculture* 3 (1949): 6 and 15.

Wells, A. J. "Tilling the 'Tules' of California." *Review of Reviews* 30 (1904): 312–17.

Wentz, Ralph A. "The California Canned and Dried Fruit Industries, with Special Reference to their Dependence upon Exporting." M.A. thesis, University of California, Berkeley, 1925.

Whitten, Woodrow C. "The Wheatland Episode." *Pacific Historical Review* 17 (1948): 37–42.

Wildman, Ester T. "The Settlement and Resources of the Sacramento Valley." M.S. thesis, University of California, Berkeley, 1921.

Wills, Harry W. "Large Scale Farm Operations in the Upper San Joaquin Valley, California." M.A. thesis, University of California, Los Angeles, 1953.

Wilson, Edwin E., and Marion Clawson. *Agricultural Land Ownership and Operation in the Southern San Joaquin Valley.* Berkeley: U.S. Department of Agriculture, Bureau of Agricultural Economics, 1945.

Wilson, Iris. "William Wolfskill and the Development of Southern California." M.A. thesis, University of Southern California, 1957.

Winchell, Lilibourne A. *A History of Fresno County and the San Joaquin Valley, Narrative and Biographical*. Fresno: A. H. Cawston, 1933.

Historical Maps

Many of the lease and chattel mortgage documents stated the exact location and size of the land that Chinese leased in terms of U.S. survey coordinates. However, when such township and range coordinates were not given, the historical maps listed below, most of which gave the names of owners of tracts at the time the maps were made, proved helpful.

Allardt, G. F. "Official Map of Alameda Co." 1874.

Bannister, Alfred. "Map of Tulare Co." 1883.

Beasley, Thomas D. "Official Map of the South-west Portion of San Bernardino Co., California." 1891.

Beck, Warren A., and Ynez D. Haase. *Historical Atlas of California*. Norman: University of Oklahoma Press. 1974.

Bowers, A. B. "Map of Sonoma Co." 1863.

Brown, J. "Official Map of Amador Co., California." 1881.

Buckman, O. "Official Map of the Co. of Napa, California." 1895.

Budd, Henry B. "Map of the San Joaquin Delta Drawn Especially for the Byron Times." 1912.

California. State Mining Bureau. "Map of Tuolumne Co." 1903.

Cosby, Stanley W. "Soil Map: Sacramento-San Joaquin Delta Area, California." 1935.

Cox, St. John. "Map of the Co. of Monterey, California." 1877.

Crawford, L. A., and E. B. Hurd. "Type of Farming Areas, Lower San Joaquin River Valley." 1930.

———. "Type of Farming Areas, Sacramento River Valley." 1930.

———. "Type of Farming Areas, Upper San Joaquin River Valley." 1930.

Crossman and Cochrane. "Map of Sierra Co., California." 1867.

De Jarnatt and John Burress. "Official Map of Colusa Co., California." 1874.

De Jarnatt and Crane. "Official Map of Colusa Co., California." 1885.

De Pue and Co. "Map of Yolo Co." 1879.

Doherty, John. "Map of the City of Sacramento, Showing Swamp and Overflowed Lands." 1859.

Doyle, J. M. "Official Map of Yuba Co." 1887.

Durrenberger, Robert W. *Patterns on the Land: Geographical, Historical and*

Political Maps of California. Woodland Hills: Aegeus Publishing Co. 1965.

Edmunds, William. "Oroville." 1872.

Hall, William Hammond. "California State Engineering Department Detail Irrigation Map: Fresno Sheet." 1885.

Hare, Lou G. "Official Map of Monterey Co., California." 1898.

Harris, R. "Map of the Co. of San Luis Obispo." 1874.

Hartwell, J. G. "Map of Nevada Co." 1880.

Healey, Charles T. "Official Map of Santa Clara Co." 1866.

Henning, J. S. "Map of Solano Co." 1872.

Hermann Brothers. "Official Map of Santa Clara Co." 1890.

Lapham, Macy H. and W. W. Mackie. "Soil Map: California Stockton Sheet." 1905.

Leicht, Ferdinand von. "Official Map of Kern Co., California." 1888.

Luning, W. F. "Official Map of Tehama Co." 1920.

McCloud, J. J. "Official Map of the Co. of San Mateo, California." 1877.

McMahon, T. A. "Official Map of Contra Costa Co." 1908.

"Map of Sacramento and San Joaquin Rivers, Showing All Landings to Sacramento and Stockton and Roads Leading to Them." 1913.

Martin, Charles D. "Official Map of the Co. of Merced." 1888.

Nubaumer, George L. "Official Map of Alameda Co." 1889.

Pennington, J. "Official Map of Sutter Co., California." 1873.

Perley, George. "Map of the Co. of Stanislaus, California." 1895.

Phinney, C. M. "Official Map of the Co. of Sacramento, California." 1911.

Punnett Brothers. "Map of the Co. of El Dorado, California." 1895.

———. "Map of the Co. of Glenn, California." 1894.

———. "Official Map of Sutter Co., California." 1895.

Quail, F. E. "Official Map of San Joaquin Co., California." 1905.

———. "Official Map of San Joaquin Co., California." 1912.

Riecker, Paul. "Map of the Co. of Santa Barbara." 1889.

Ryan, H. L. "Map of San Diego Co., California." 1889.

"Sacramento Co. Map Books." 1870, 1881, 1892, 1900, 1906, 1912–13, 1917, 1921.

San Joaquin Valley Counties Association. "San Joaquin: The Gateway Co. of California." 1916.

Santa Barbara Abstract and Guaranty Co. "Official Map of Santa Barbara." 1909.

Shackelford, H. B. "Official Map of the Co. of Tehama." 1878.

Shepherd, Fred A. "Map of Sacramento Co." 1885.

Stevenson, H. J. "Map of Los Angeles Co." 1884.

Thomas Brothers. "Map of Sacramento Co. and Delta Region." n. d.

———. "Map of San Joaquin Co." n. d.

Thompson, R. A. "Map of Sonoma Co." 1884.

Thompson, Thomas H. "Map of Fresno Co." 1891.

Uren, Charles E. "Official Map of Placer Co., California." 1887.

Walker, Frank. "Walker's Official Map of Kings Co., California." 1893.

Walkup, W. B. "Official Map of Calaveras Co., California." 1895.

Weir, Walter W., and R. Earl Storie. "Soil Classification: Sacramento Valley."
    n. d.
————. "Soil Classification: Upper San Joaquin Valley." n. d.
Wescoatt, N. and W. S. Watson. "The Official Map of the City of Marysville."
    1856.

Interviews

Chin Gway (San Francisco, July 29, 1979, telephone interview conducted by Judy Yung on behalf of the author).

The fourth son of Chin Lung, Chin Gway was born in Macao in 1904 and immigrated to the United States in 1924. He worked in several of his father's businesses, served in the U.S. Merchant Marines, and operated a restaurant before he retired.

Chin Shou and John Chin (Milpitas, October 12, 1979).

Chin Shou, the fifth son of Chin Lung, and John Chin, son of Chin Bow and grandnephew of Chin Lung, were both born in Macao in 1906. They came to the United States in 1924. Chin Shou was a chrysanthemum grower in Milpitas, California, at the time of the interview; John Chin was retired from the civil service.

Varden Fuller (Occidental, May 9, 1979).

Varden Fuller, Professor Emeritus of Agricultural Economics, University of California, Berkeley and Davis, the author of many studies on California farm labor, was living in retirement in Occidental at the time he was interviewed.

R. Pardow Hooper (San Francisco, March 14, 1979).

Pardow Hooper, a professional farm-manager, worked in many places in the Sacramento-San Joaquin Delta and all over the Sacramento Valley. He had extensive contacts with Chinese tenants as well as white landlords, and knew Lee Bing well. Age eighty-two at the time of the interview, he was still working full time as a farm-investment consultant.

Joe Lung Kim (Marysville, August 9, 1980).

Joe Lung Kim's real family name was Chou, but since white Americans called his father "Mr. Kim" after the name of his laundry, Kim Wing, in Marysville, the family eventually adopted Kim as its name. Born in 1900 in Marysville, Joe Kim is the grandson of an immigrant who came in the 1860s. A former truck gardener, farmer, and merchant, Joe Kim was official keeper of the Marysville Bok Kai Temple, a California State Historical landmark, at the time he was interviewed.

Dennis Leary (Walnut Grove, June 6, 1979).

Born on his father's farm in the Sacramento Delta in 1898, Dennis Leary and his father have leased land to dozens of Asian tenants for a century. He was still actively farming when he was interviewed at age eighty-one.

Carey McWilliams (Berkeley, April 6, 1980).

The author of more than half a dozen popular histories of California, Carey McWilliams served as editor of the *Nation* for many years. While in Berkeley to give a lecture, he discussed with me how he came to write *Factories in the Field*. He died unexpectedly in June 1980.

Charan Singh Sandhu (Calipatria, May 22 and June 3, 1981).

Born in 1889 in Lahore District, Punjab, India, Charan Singh Sandhu went to China in 1907, where he worked as a night watchman. After saving enough money for his passage, he came to the United States in 1909. He worked as a farm laborer, tenant farmer, and owner-operator for seven decades in almost every major agricultural area in California. At the time he was interviewed, he was living in retirement in the Imperial Valley and had many stories to tell about the Chinese, Japanese, Korean, Armenian, Black, Chicano, and white farm laborers and farmers he has known.

Riichi Satow [*sic*] (Sacramento, May 30, 1979).

Born in 1895 in Chiba Province, Japan, Riichi Satow came to the United States as a "yobiyose" (one who was summoned to come to the United States by a relative) in 1912. He grew strawberries as a tenant farmer and owner-operator in the Florin area, just south of the city of Sacramento, for six decades.

Rindge Shima (Oakland, March 17, 1979).

The third and youngest child of George Shima, the Japanese "Potato King," Rindge Shima was born in 1908 and was named after the Rindge Tract in the Sacramento-San Joaquin Delta where his father farmed in 1908. An aerospace engineer, he is the patriarch of what remains of the Shima clan.

E. Cared van Löben Sels (Polder Amistad Ranch, Courtland, May 26, 1979).

The grandson of Pieter Justus ("P. J.") van Löben Sels, an employer of many Chinese and Japanese, Cared van Löben Sels was a third-generation owner-operator in the Sacramento Delta.

Index

Designer: Randall Goodall
Compositor: Asco Trade Typesetting Ltd.
Text: 11/13 Baskerville
Display: Baskerville
Printer: Braun-Brumfield, Inc.
Binder: Braun-Brumfield, Inc.